Sergey Bushuyev · Ronggui Ding ·
Mladen Radujkovic
Editors

Project Management in the Digital Transformation Era

The Proceedings of the 32nd World Congress of the International Project Management Association (IPMA)

Springer

Editors
Sergey Bushuyev
Kyiv National University of Construction
Kyiv, Ukraine

Mladen Radujkovic
Evropski Center
Alma Mater Europaea
Maribor, Slovenia

Ronggui Ding
School of Management
Shandong University
Jinan, Shandong, China

ISSN 2367-3370 ISSN 2367-3389 (electronic)
Lecture Notes in Networks and Systems
ISBN 978-3-031-34628-6 ISBN 978-3-031-34629-3 (eBook)
https://doi.org/10.1007/978-3-031-34629-3

This Springer imprint is published by the registered company Springer Nature Switzerland AG
The registered company address is: Gewerbestrasse 11, 6330 Cham, Switzerland

Lecture Notes in Networks and Systems 704

The series “Lecture Notes in Networks and Systems” publishes the latest developments in Networks and Systems—quickly, informally and with high quality. Original research reported in proceedings and post-proceedings represents the core of LNNS.

Volumes published in LNNS embrace all aspects and subfields of, as well as new challenges in, Networks and Systems.

The series contains proceedings and edited volumes in systems and networks, spanning the areas of Cyber-Physical Systems, Autonomous Systems, Sensor Networks, Control Systems, Energy Systems, Automotive Systems, Biological Systems, Vehicular Networking and Connected Vehicles, Aerospace Systems, Automation, Manufacturing, Smart Grids, Nonlinear Systems, Power Systems, Robotics, Social Systems, Economic Systems and other. Of particular value to both the contributors and the readership are the short publication timeframe and the world-wide distribution and exposure which enable both a wide and rapid dissemination of research output.

The series covers the theory, applications, and perspectives on the state of the art and future developments relevant to systems and networks, decision making, control, complex processes and related areas, as embedded in the fields of interdisciplinary and applied sciences, engineering, computer science, physics, economics, social, and life sciences, as well as the paradigms and methodologies behind them.

Indexed by SCOPUS, INSPEC, WTI Frankfurt eG, zbMATH, SCImago.

All books published in the series are submitted for consideration in Web of Science.

For proposals from Asia please contact Aninda Bose (aninda.bose@springer.com).

Contents

Project Governance and External Stakeholders: The Role of Social Technologies – A Literature Review

Mahdieh Dehghan Nayeri(✉) and Mohammad Hosayn Sobhiyah

Tarbiat Modares University, Tehran, Iran
{m.nayeri,sobhiyah}@modares.ac.ir, mnayeri@ut.ac.ir

Abstract. While the project governance literature has focused on internal stakeholders, with the emergence of social technologies, the historical rules of the literature are changing. Regarding these technologies, some intangible strategies are available, including the possibility to collect and share information, communicating data and communities, organizing issues to transfer them to organizations, accessing and organizing a large number of people simultaneously, establishing a social discourse, and distribution of power and value between organizations and external stakeholders; Everything that facilitates good governance. This paper reviews the literature in three project management journals, which contain a significant part of the project management literature, addresses the importance of developing a stakeholder-oriented approach in this new context, and shows the studied and neglected parts. The review showed that stakeholder engagement becomes more complex through facilitating, strengthening, and diversifying partnerships by online technologies. Moreover, mediating factors such as the relationship between online and offline stakeholders add to this complexity. The advantages of having potent stakeholders are more than its disadvantages. However, it is necessary to examine more deeply the interaction of managers with influential stakeholders.

Keywords: Social technology · External stakeholders · Governance

1 Introduction

According to various definitions of project governance, governance appears to be a multi-level phenomenon about aligning project goals with the organization's strategies to produce benefits for different stakeholders. Thus, governance focuses on aligning the interests of all key stakeholders, and all governance mechanisms seek to guide and balance the objectives and determine the rights and responsibilities of stakeholders [1]. These definitions indicate the strong relationship between project governance and stakeholders. On the other hand, according to the project governance literature, this literature concentrates on the rights, demands, and concerns of internal stakeholders; it ignores the non-financial aspects of stakeholders and those who impress or are impressed by the project or external stakeholders. Project governance research is also limited to internal stakeholders' management and most of its tools and potential, and the dominant

S. Bushuyev et al. (Eds.): IPMA 2021, LNNS 704, pp. 1–12, 2023.
https://doi.org/10.1007/978-3-031-34629-3_1

theories of governance, including Agency theory, Transaction cost economics, Stewardship theory, Resource dependence theory, and Stakeholders' theory, have mainly used in explaining management and engagement mechanisms related to this group of stakeholders [2]. Ignoring the non-financial relationships of stakeholders and paying no attention to the public people in project governance, regardless of the context of the projects, is observed even in democratic societies [3]. However, the significance of external stakeholders in the success of projects, particularly in large and public projects, is evident; because there are numerous demands and a wide range of requests in such projects, from gaining national credit to minimal manipulation of the environment. When project governance pays more attention to the relationship between itself and external stakeholders, project investors and teams receive more support from this group of stakeholders, and the short-term and long-term success of the project and the organization is not far-fetched.

However, it seems this inequality has decreased with the introduction of the Internet and social media. In the context of these media, the strategies that early look subtle are available, such as the possibility to collect and share information, communicating data and communities, organizing and framing the issues to transfer them to organizations and all of the relevant practitioners [4], accessing and engaging a large number of people simultaneously, establishing a social discourse [5], and distribution of power and value between the organization and external stakeholders. All the things that facilitate good governance. The purpose of this study is to encourage governors to think about their leading role in respecting and paying careful attention to social issues and developing a stakeholder-oriented approach in the new context. In this regard, we address the changing roles of external stakeholders and project community beyond management and engagement, and the significance of social technologies in the emergence of new developments, and also review the efforts made to use online social technologies in project management to search for the main themes in this field. In addition, we discuss the conducted studies and identify the neglected parts, which will be a guide for future studies.

2 Features and Roles of Project Community in the Stakeholder-Oriented Approach

The number of large projects is increasing all over the world to meet the growing needs of communities. The complex nature of the relationships between stakeholders [6] and numerous project participants, whose management is effortful, particularly in large projects, require more diverse governance mechanisms to include all external stakeholders, such as landowners, public services, and the project community. The new form of these stakeholders as small groups of two or more individuals with understanding and assessment based on shared values and objectives tend to unite with larger groups to develop strategies for direct involvement. It has led to transforming smaller groups into larger ones with a single identity. Moreover, various relationships of individuals in these groups with the project organization [7], lack of integrity in societies, their ever-changing nature [8] and their interest in self-governance [9], and lack of specific, limited, and demarcated contractual relationships with these stakeholders all have turned them into groups with endless and unpredictable demands.

In line with the differences between stakeholder management and management of stakeholders' approaches [10], external stakeholder groups, particularly the project community, have changed the prevailing view of project governance in their management and engagement. Such that, in public projects, society and people are also involved in decision-making processes. So, in this new approach organizations are expected addition to being accountable and more actively engaging with stakeholders, involve the stakeholders in governance structures and consequently decision-making -the framework of which is provided by governance-. Project governance considers a special significance for decision-making [11]. Decision-makers have the authority to start a project and each phase of it, make changes in the project, allocate resources, determine risk management strategies, manage contracts, etc. While some project information will be obtained contingently and difficult to exchange with stakeholders, clearly and without leading to different interpretations among stakeholders, the significance of transparency in dealing with external stakeholders -who have manifested many features of a social movement- throughout the project life-cycle is undeniable. Therefore, understanding the collective strategies of these stakeholders, which are dynamic, fluid, and changeable throughout the project life-cycle, analyzing stakeholders' networks [12] and providing the required space, and managing their partnership in decision-making by governors and project managers in a systematic manner is necessary to predict their behaviors.

3 Features and Roles of Social Technologies in the Stakeholder-Oriented Approach

The Internet was introduced in 1960 and began with a project aimed to connect the computers used by the military, defense contractors, and US university sites. As a result, in 1969, four computers succeeded in exchanging data with each other. After developing this network and the production of supercomputers in 1987, it became the core of a set of networks now we call the Internet [13]. The Internet itself became the context for the emergence, growth, and pervasiveness of social technologies in the mid-years of the first decade of this century [4]. Research on these technologies is a new topic. Researchers have defined social technologies as a group of applications based on the Internet that makes it possible to produce and exchange the content generated by users. These technologies include social networks such as Facebook and LinkedIn, and blogs such as Blogger, microblogs such as Twitter, wikis such as Wikipedia, and RSS, Tags, Mashups, Voice over IP such as Skype, cloud saving and sharing such as Dropbox, and cloud edit tools such as Google Drive [14]. All of them are a part of the social software revolution. The number of users of this technology, who refer to it for entertainment, getting information and news, and socializing is increasing rapidly. The attractive, pleasant, and miscellaneous environment provided by the tools of this technology [14] has a significant impact on this lucky. Social media applications and websites such as YouTube, Facebook, Instagram, Pinterest, Snapchat, Linked In, Twitter, WhatsApp, and WeChat, are constantly increasing through social technologies. In addition to completely transforming communication methods and social interactions in most societies and providing the possibility of sharing information among people in a one-on-one, one-on-many, and fewer on more numbers [12], these platforms have provided an always available social environment with the help of smartphone technology -social media on mobile- where people actively communicate [15] at anytime and anywhere.

Before the Internet era, great movements such as environmental movements required extensive coordination and the help of journalists to legitimize [4]; but since the beginning of the social technology revolution, this trend has completely changed. People partnership in social media allows them to receive information, investigate the viewpoints of others, and add their point of view to the created circle; In other words, they exchange information with each other and develop virtual and temporary organizations, with a collective and joint opinion, while the costs of this exchange are at a minimum. Those, who communicate through networks, branches, and small groups with tools such as Facebook and Twitter, develop micro-networks and connect them [7]; This is another function of these media that play a significant role in forming great movements. Investigating the impact of new technologies on approaches related to stakeholders and the project community is crucial. Now the management of stakeholders has been changed and is accompanied by some opportunities and challenges. With these technologies, external stakeholders can access all organization data such as interviews, employers' evaluations, opinions about the organization, and other pure information, in just a few minutes. In networks such as Facebook and Twitter, friends and colleagues' circles can add evidence and document, and debate and transfer information to other blogs and websites. As soon as debates reach other networks and circles, a set of considerable evidence is collected. Lack of information symmetry between organizations and external stakeholders [4], which was considered normal, can be resolved using social technologies. Social technologies help smaller stakeholders communicate with theorists and intellectuals, and through these influencers, they somehow affect the lines of thought of stakeholders. Continuous monitoring of prevailing trends on the Internet and social media gives managers the opportunity of having knowledge of and close contact with changes of social demands of the organization in relationship with stakeholders. It enables managers to plan and act for changes before the external stakeholders change the organization. In addition, the technology, through optimizing the mechanisms and applying new tools, can reduce the risk of misunderstanding or distorting the demands of people -by personal desires of decision-makers in organizations-. Moreover, it provides the possibility to analyze and compare the frequency of different views of stakeholders on a subject.

Therefore, with the new information flow in projects, data can be collected and analyzed not only on pre-determined and fixed points but also continuously and repeatedly throughout the project life cycle. Social media content analysis or the study of online communities, which by the inspiration of the term "ethnography" has named "Netnography" [16], has a particular position in this approach. Considering the public access to data, data collecting, interpreting, and analyzing tools have grown, increasingly; such the software called "web-crawler" [12], or social network analysis techniques [17] that some researchers have addressed.

4 Review Method

In this paper, we investigate the role of social technologies in project management. The research focuses on reviewing peer-reviewed articles published in scholarly journals, which are reliable sources for reviewing literature [18]. We conducted the study by selecting and reviewing three reputable project management journals, which contain a large part of project management literature [19]. Table 1 gives the specifications of these three journals.

Table 1. Specifications of three project management journals

Journal title and online address	Publisher
International Journal of Managing Projects in Business (IJMPiB) https://www.emeraldgrouppublishing.com/ijmpb.htm	Emerald
International Journal of Project Management (IJPM) https://www.sciencedirect.com/journal/international-journal-of-project-management	Elsevier
Project Management Journal (PMJ) https://journals.sagepub.com/loi/pmx	Sage

All articles of these three journals available in the database of the journal's publisher website were searched from the beginning of online publication until February 2020 to identify papers that in their title, keywords, or abstract, the words Facebook, Twitter, on line, online, mobile, social media, media, Internet, WhatsApp, and WeChat exist. At first, the keywords were restricted to the Internet, online, and social media, then based on the found articles, some new words were also searched. As a result of this search, we identified two articles in IJMPiB, seven in IJPM, and six articles in PMJ. Then, using Citavi software, we reviewed the content of the papers. By searching the main themes concerning the topic of study -the role of social technologies in stakeholders' governance- we developed the following three major themes, the details of which are in Appendix A.

- Online/social technology for the project team
- Online/social technology for project community
- Large/public projects/teams/groups

This review showed that more than half of the identified articles have the inductive approach. The list of articles, research approaches, and the type of stakeholders every article has focused on (internal or external [20]) are in Appendix B. In appendices, the list of the articles is according to the journal and alphabetical order of the authors' names.

5 Findings and Discussion

The review showed that research on online technologies and their impact and responsiveness on knowledge and practices of project management and the resulting opportunities and challenges in project governance mechanisms are very few, and these few numbers focus more on internal stakeholders and project teams. The articles mainly address the significance of improving the communications of project teams, discovering and analyzing the intra-team relationships, increasing the desirability of work environments, and enhancing the speed and accuracy of knowledge management by technology to achieve project success. Only three articles addressed the interactions of emerging social technologies with management, engagement, and decision-making processes of external stakeholders in large projects. Large and public projects attract lots of attention from the people, have profound impacts on the environment, economy, and society, form a part of the identity and character of the local community of these projects [9], and due to the multiplicity and size of teams and groups [7], impose a lot of complex challenges on organizations associated with stakeholders. Therefore, these projects are usually the starting point of encountering or applying the latest changes and technologies in traditional practices, and they must consider and develop self-governance technologies [9] as a necessity.

It seems that project management studies have ignored the changes imposed by social technologies on all aspects of individual and group life of people [21]. However, scientific studies relevant to social media in political sciences have been started for many years and entered into the business science around 2005 to 2007 [5]. The results showed a lack of evidence and knowledge on applying social technologies in all areas of the project. The article "The fourth dimension of project management (1995) [22]" presented in the first British colloquium on project management, with the theme of a leap to the next generation, states that almost there is no revolutionary and significant change in project management practices in the last 25 years of the 20th century. As a result, despite optimizing project management methods, they have not led to the success of many projects in meeting the demands. The article has addressed the significance of Internet access in evolving project management practices in the next 25 years (2020). Over these 25 years (1995 to 2020), with the quick entry of new social technologies into human life -while the technologies facilitate constructive and positive dialogue between the various groups- these technologies have complicated the project environment. So, transforming many of the structures that had governed this environment over the last 25 years to equip project management knowledge and bridging it to the next 25 years is crucial.

Social technologies motivate and influence civil and political engagement and public governance, and consequently, they control project governance. What is called stakeholder engagement, in literature, becomes more complex through facilitating, strengthening, and diversifying partnerships by online technologies that provide pure information for potential stakeholders and turns them into official controllers of the organization. Moreover, mediating factors, such as the relationship between online and offline stakeholders [5] and the two-way path of social technologies in developing more active participation and management by organizations and internal stakeholders [4], add to this complexity. Social technologies have not been studied and analyzed as a contributing factor and tool on stakeholders' engagement -including employers and employees, local community, public organizations, or social groups- and their participation in governance decision-making. This issue reveals the need for conducting qualitative and quantitative researches. Case study researches in different contexts can discover these issues both in terms of the organization and stakeholders. Such studies can address a wide range of stakeholders' participation, from management and engagement to decision-making.

Mainly younger age groups, who have the knowledge and desire for online partnership, use social technologies such as social media platforms. The features of these platforms are different too. For instance, in comparing two popular platforms of Facebook and Twitter, although users' posts are very similar in these two platforms, Facebook allows more active participation of users, versus Twitter due to its limitations on the number of characters used for each post, has led sharing more concise, accurate, and valuable posts. Research on the features of different social media platforms and social media content analysis or the study of online communities (Netnography [16]), with attention to the context in which these platforms operate, has also been neglected in the literature. Table 2 presents the available knowledge and future knowledge needs.

Table 2. Role of social technologies in project governance and stakeholders: The available knowledge and guidelines for future studies

Theme	Available	Neglected
Online/social technology for the project team	• The use of the Internet and web-based platforms to strengthen knowledge management, facilitate communication, and help active participation of project team members in mega-projects with a large number of stakeholders	• The impact of social technologies on changing communication methods, exchanging, and participation of project teams • The impact of social media in motivating, creating belonging, and changing organizational and project behavior in project teams • The way of interaction of governors and project managers with potent external stakeholders through the new era of technology

(continued)

Table 2. (*continued*)

Theme	Available	Neglected
Online/social technology for project community	• The use of the Internet and web-based platforms to know about the project community demands and concerns (Thera are narrow studies)	• Features of social technologies and their impact and responsiveness on project management knowledge and practices, its resulting opportunities and challenges affecting the project governance mechanisms focusing on external stakeholders • Studying real cases in different contexts both in terms of the organization and stakeholders • The complexity of social technologies era environments from the stakeholder-oriented perspective and the factors adding to this complexity, such as the relationship between online and offline stakeholders
Large/public projects/teams/groups	• Discovering and analyzing the relationships within multiple and large teams and their impact on work environments in mega-projects	• Interactions between public project teams and external stakeholders and project community and developing a technology-based two-way relationship

6 Conclusion

Given that governance aims to align the interests of all key stakeholders, the strong relationship between project governance and stakeholders is undeniable. Since the project governance literature, like project management literature, is more focused on internal stakeholders, it seems necessary to review the relevant mechanisms. However, with the emergence of social technologies, the historical rules of the literature are changing. The purpose of this article is to encourage decision-makers at different levels of the organization to rethink their principal role in respecting and paying impressive attention to social issues in a new context. New features of external stakeholders and project communities define their new roles beyond just management and engagement, and the significance of the Internet and social technologies in the recent developments are undeniable. It is necessary to study the impacts of social technologies on effective strategies of external stakeholders in projects and the interaction of managers with the increasing power of stakeholders. The advantages of having powerful stakeholders, especially if governance mechanisms would attract their interest and commitment, overcome its disadvantages; and in this context, governors will have more authority in developing and implementing the strategies and would balance the interests of all stakeholders.

Reviewing the project management literature in three scholarly journals results in a lack of research on social technologies and their role and impact on project governance mechanisms. Future researchers should discover the relationship between organizations and external stakeholders in this new context. The stakeholder-oriented approach, which starts at the organizational level, is applied at the portfolio level and used in daily two-way and multi-way interactions at the project level, with new technologies, will change, and its mechanisms will be different. Large or public projects' organizations should recognize the significance and impact of the new era of technology on the governance of external stakeholders and their participation in decision-making processes in a less costly and effective way than the past movements, and they should become aware of the added value provided through that.

This research has faced some limitations due to the strategies it selected. The review has focused on the peer-reviewed articles published in scholarly journals. Although this focus restricts validity, it can add to the accuracy and rigor of the research. Authors' interpretations during the review of articles are another limitation of this study. Moreover, we reviewed three journals; Searching other databases and types of sources may add to the depth and richness of this study.

Appendices

Appendix A. Reviewed articles and their main themes

Reference			Technology for Project Team		Technology for Project Community		Large/Public		Platform
Ref. No.	Article (Autor, year)	Journal	Online	Social	Online	Social	Projects	Teams/Groups	
[6]	(Harley 2011)	IJMPiB	✔						
[14]	(Rosa et al. 2016)	IJMPiB		✔					
[22]	(Holbrough 1995)	IJPM	✔				✔		
[13]	(Tam 1999)	IJPM	✔				✔		
[23]	(Xue et al. 2007)	IJPM	✔					✔	
[24]	(Ruuska and Teigland 2009)	IJPM	✔		✔			✔	Website
[25]	(Ojiako et al. 2017)	IJPM	✔				✔		
[15]	(Zhang et al. 2018)	IJPM		✔			✔		WhatsApp WeChat
[9]	(Ninan et al. 2019)	IJPM		✔		✔	✔		Facebook Twitter
[26]	(Scanlin 1998)	PMJ	✔						Website
[8]	(Mead 2001)	PMJ	✔				✔		
[21]	(Giffin 2002)	PMJ	✔					✔	
[7]	(Williams et al. 2015)	PMJ				✔	✔		Twitter
[27]	(Petter and Carter 2017)	PMJ	✔					✔	
[12]	(Lobo and Abid 2019)	PMJ		✔		✔	✔		Facebook Twitter Website

Appendix B. Reviewed articles, their search approaches, and the type of stakeholders every paper has focused on (internal or external)

Ref. No.	Article (Autor, year)	Journal	Research Approach	Research Strategy	Stakeholder View (Internal/External)
[6]	(Harley 2011)	IJMPiB	Inductive	Multiple case study	Internal
[14]	(Rosa et al. 2016)	IJMPiB	Inductive	Interview Literature	Internal
[22]	(Helbrough 1995)	IJPM	-	Interview Literature	Internal
[13]	(Tam 1999)	IJPM	Deductive	Experiment	Internal
[23]	(Xue et al. 2007)	IJPM	-	Literature	Internal
[24]	(Ruuska and Teigland 2009)	IJPM	Inductive	Single case study	Internal/External
[25]	(Ojiako et al. 2017)	IJPM	Deductive	Survey	Internal
[15]	(Zhang et al. 2018)	IJPM	Inductive Deductive	Multiple case study Survey	Internal
[9]	(Ninan et al. 2019)	IJPM	Inductive	Single case study Grounded theory	Internal/External
[26]	(Scanlin 1998)	PMJ	Deductive	Survey	Internal
[8]	(Mead 2001)	PMJ	Deductive	Survey	Internal
[21]	(Giffin 2002)	PMJ	-	Literature	Internal
[7]	(Williams et al. 2015)	PMJ	Inductive	Single case study	External
[27]	(Petter and Carter 2017)	PMJ	Inductive	Action Research	Internal
[12]	(Lobo and Abid 2019)	PMJ	Inductive	Single case study	Internal/External

References

1. Turner, R., Müller, R.: The governance of organizational project management. In: Sankaran, S., Muller, R., Drouin, N. (eds.) Cambridge Handbook of Organizational Project Management, pp. 75–91. Cambridge University Press, Cambridge (2017)
2. Derakhshan, R., Turner, R., Mancini, M.: Project governance and stakeholders: a literature review. Int. J. Project Manage. **37**(1), 98–116 (2019). https://doi.org/10.1016/j.ijproman.2018.10.007

3. Di Maddaloni, F., Davis, K.: The influence of local community stakeholders in megaprojects: rethinking their inclusiveness to improve project performance. Int. J. Project Manage. **35**(8), 1537–1556 (2017). https://doi.org/10.1016/j.ijproman.2017.08.011
4. Jurgens, M., Berthon, P., Edelman, L., Pitt, L.: Social media revolutions: the influence of secondary stakeholders. Bus. Horiz. **59**(2), 129–136 (2016). https://doi.org/10.1016/j.bushor.2015.11.010
5. Hoffmann, C.P., Lutz, C.: The impact of online media on stakeholder engagement and the governance of corporations. J. Public Aff. **15**(2), 163–174 (2015). https://doi.org/10.1002/pa.1535
6. Harley, J.: Collaboration and the use of online collaborative toolsets in the project management environment. Int. J. Manag. Proj. Bus. **4**(2), 345–354 (2011). https://doi.org/10.1108/17538371111120289
7. Williams, N.L., Ferdinand, N., Pasian, B.: Online Stakeholder Interactions in the early stage of a megaproject. Proj. Manag. J. **46**(6), 92–110 (2015). https://doi.org/10.1002/pmj.21548
8. Mead, S.P.: Using social network analysis to visualize project teams. Proj. Manag. J. **32**(4), 32–38 (2001). https://doi.org/10.1177/875697280103200405
9. Ninan, J., Clegg, S., Mahalingam, A.: Branding and governmentality for infrastructure megaprojects: the role of social media. Int. J. Project Manage. **37**(1), 59–72 (2019). https://doi.org/10.1016/j.ijproman.2018.10.005
10. Freeman, R.E., Harrison, J.S., Wicks, A.C.: Managing for stakeholders. Survival, reputation, and success. The Business Roundtable Institute for Corporate Ethics series in ethics and leadership. Yale University Press, New Haven (2007)
11. Xie, L.-L., Xia, B., Hu, Y., Shan, M., Le, Y., Chan, A.P.C.: Public participation performance in public construction projects of South China: a case study of the Guangzhou Games venues construction. Int. J. Project Manage. **35**(7), 1391–1401 (2017). https://doi.org/10.1016/j.ijproman.2017.04.003
12. Lobo, S., Abid, A.F.: The role of social media in intrastakeholder strategies to influence decision making in a UK infrastructure megaproject: crossrail 2. Proj. Manag. J. **7**(1), 875697281986445 (2019). https://doi.org/10.1177/8756972819864456
13. Tam, C.M.: Use of the internet to enhance construction communication: total information transfer system. Int. J. Project Manage. **17**(2), 107–111 (1999). https://doi.org/10.1016/S0263-7863(97)00077-X
14. Rosa, D.V., Chaves, M.S., Oliveira, M., Pedron, C.: Target: a collaborative model based on social media to support the management of lessons learned in projects. Int. J. Manag. Proj. Bus. **9**(3), 654–681 (2016). https://doi.org/10.1108/IJMPB-12-2015-0120
15. Zhang, Y., Sun, J., Yang, Z., Wang, Y.: Mobile social media in inter-organizational projects: aligning tool, task and team for virtual collaboration effectiveness. Int. J. Project Manage. **36**(8), 1096–1108 (2018). https://doi.org/10.1016/j.ijproman.2018.09.003
16. Kozinets, R.: Netnography: The Essential Guide to Qualitative Social Media Research, 3rd edn. SAGE Publications Ltd., Thousand Oaks (2019)
17. Zheng, X., Le, Y., Chan, A.P.C., Hu, Y., Li, Y.: Review of the application of social network analysis (SNA) in construction project management research. Int. J. Project Manage. **34**(7), 1214–1225 (2016). https://doi.org/10.1016/j.ijproman.2016.06.005
18. Rowley, J., Slack, F.: Conducting a literature review. Manag. Res. News **27**(6), 31–39 (2004). https://doi.org/10.1108/01409170410784185
19. Martinsuo, M., Hoverfält, P.: Change program management: toward a capability for managing value-oriented, integrated multi-project change in its context. Int. J. Project Manage. **36**(1), 134–146 (2017). https://doi.org/10.1016/j.ijproman.2017.04.018
20. Parmar, B.L., Freeman, R.E., Harrison, J.S., Wicks, A.C., Purnell, L., de Colle, S.: Stakeholder theory: the state of the art. Acad. Manag. Ann. **4**(1), 403–445 (2010). https://doi.org/10.1080/19416520.2010.495581

21. Giffin, S.D.: A taxonomy of internet applications for project management communication. Proj. Manag. J. **33**(4), 39–47 (2002). https://doi.org/10.1177/875697280203300405
22. Helbrough, B.: Computer assisted collaboration - the fourth dimension of project management? Int. J. Project Manage. **13**(5), 329–333 (1995). https://doi.org/10.1016/0263-7863(95)00041-N
23. Xue, X., Wang, Y., Shen, Q., Yu, X.: Coordination mechanisms for construction supply chain management in the Internet environment. Int. J. Project Manage. **25**(2), 150–157 (2007). https://doi.org/10.1016/j.ijproman.2006.09.006
24. Ruuska, I., Teigland, R.: Ensuring project success through collective competence and creative conflict in public–private partnerships – a case study of Bygga Villa, a Swedish triple helix e-government initiative. Int. J. Project Manage. **27**(4), 323–334 (2009). https://doi.org/10.1016/j.ijproman.2008.02.007
25. Ojiako, U., Chipulu, M., Marshall, A., Williams, T.: An examination of the 'rule of law' and 'justice' implications in online dispute resolution in construction projects. Int. J. Project Manage. **36**(2), 301–316 (2017). https://doi.org/10.1016/j.ijproman.2017.10.002
26. Scanlin, J.: The internet as an enabler of the bell atlantic project office. Proj. Manag. J. **29**(2), 6–7 (1998). https://doi.org/10.1177/875697289802900202
27. Petter, S., Carter, M.: In a league of their own: exploring the impacts of shared work history for distributed online project teams. Proj. Manag. J. **48**(1), 65–80 (2017). https://doi.org/10.1177/875697281704800105

High Uncertainty Projects: Making 'go-no-go' Decision

Svetlana Mitish, Elena Sharova, Julia Shekhter, and Grigory Tsipes(✉)

IBS, 9B Dmitrovskoe shosse, Moscow 127434, Russia
gtsipes@ibs.ru

Abstract. Uncertainty in project management has always been regarded as one of the most important factors affecting the success of the project. However, today the uncertainty in projects is often so large that traditional project management approaches become ineffective. Such projects are well known and have different names - "complicated" and "chaotic" projects in the Cynefin model, "extreme" projects in DeCarlo model. A significant part of digital transformation initiatives fall into this category of projects. Indeed, in such projects the understanding of the value of the product is formed as a rule in not at the start but in the course of the project. On the other hand, high speed of technological changes creates additional turbulence for digital transformation projects. The specificity of such projects requires a very responsible approach when making a "go-no-go" decision. For the project contractor, there is a great risk of both financial and reputational losses. For the customer of the project, the main threat is the failure to obtain the expected benefits for the business.

The paper presents a practical approach to making a "go-no-go" decision to participate in a digital transformation project, based on several well-known models, such as the Stacy matrix and others. The proposed process, criteria and assessment model were formed based on the experience of the projects for the development, implementation and maintenance of digital products for the public and commercial customers.

Keywords: Uncertainty · Digital Transformation Project · Decision Making · Assessment Criteria

1 Introduction

The approach presented in the paper was developed and applied by IBS, one of the largest Russian IT companies. IBS has been developing internal project management standards for more than twenty years [1]. They are regularly updated taking into account the ongoing changes in the company and in the external environment. Nevertheless, there has been a growth in the number of troubled projects in the last 2–3 years.

The first idea about the reason for what is happening was the weakening of management discipline and the incorrect implementation of internal management regulations. However, the audit of these projects has shown that in some cases the use of standard and well-proven project management tools does not guarantee the success of the project.

S. Bushuyev et al. (Eds.): IPMA 2021, LNNS 704, pp. 13–24, 2023.
https://doi.org/10.1007/978-3-031-34629-3_2

There are not so many such cases, but these were, as a rule, critically important projects for the company. The analysis revealed two main reasons for this situation - a high degree of uncertainty in customer requirements and the need for non-obvious technical solutions.

These reasons are well known to practitioners. For example, the Stacy matrix [2] determines the level of project complexity exactly based on these parameters. However, the problem is that projects that initially looked quite regular, in fact, turned out to be not just complicated or confusing. It was quite possible to apply such definitions to them as turbulent or chaotic. And the negative consequences in these projects were not long in coming.

The authors of the approach considered in the paper did not set themselves the task of developing a general methodology and relied on well-known models - Stacy's matrix, Cynefin framework [3], DE Carlo's approach to managing extreme projects [4]. The task of developing the approach was to create a simple and reliable tool that allows to accurately determining the degree of uncertainty of the project before its start. And the second part of the approach is devoted to identifying tools that would reduce the uncertainty in the project and make it manageable.

2 Decision Making Process

The decision-making process to participate in a high uncertainty project is based on an assessment of the various characteristics of the project. Taken together, these assessments make it possible to understand and comprehend the situation in the future project and make a decision based on a large set of facts and opinions.

The process includes four steps. The first three steps are an assessment of the partial characteristics of the project, grouped into three categories, with two groups in each category. The first step is an assessment of the level of uncertainty of requirements and technologies. The second step is a self-assessment, which includes an assessment of the contractor's readiness for the project and the attractiveness of the project for him. Finally, the third step is the assessment of the customer, including the level of his interest and readiness to accept the "rules of the game" in projects with a high level of uncertainty.

All assessments are carried out using special questionnaires. The questions are formulated in the form of positive judgments. Respondents express their opinions on these judgments in a Likert scale ranging from "strongly disagree" to "strongly agree". The questions included in the questionnaire represent the typical situations that the company encounters in similar projects and which cause significant and sometimes dramatic changes in projects.

For each group of characteristics, partial risk indexes are calculated. The algorithms for calculating the indexes are specific for each category and group of characteristics and allow taking into account their significance for final score.

At the fourth step of the process, an integrated analysis of all six indexes is carried out and a final decision is made on the feasibility of participating in the project, as well as decisions on the use of special management tools.

3 Assessment Tools

3.1 Uncertainty Project Assessment

The assessment of the project's uncertainty is the starting point for the analysis. Our experience has shown that errors in assessing the project as a whole, "by feeling" occur quite often, and discrepancies with subsequent reality are very essential. At the same time, both underestimation and overestimation of the level of uncertainty take place. The use of formal criteria can reduce the level of subjectivity, and although assessments are still based on opinions, these opinions have to be verified by facts now.

To assess the uncertainty of the project, proposed questions are combined into two groups - the uncertainty of requirements and the uncertainty of technologies.

When assessing the level of uncertainty of requirements, not only the fuzziness or incompleteness of the formulations is taken into account, but also the potential difficulties in the process of their clarification and harmonization. Examples of questions in this part of the questionnaire:

- preliminary requirements for the product are not formulated or allow for arbitrary interpretations in a very wide range;
- due to the novelty of the product, the requirements cannot be formed on the basis of any known template, there are no "good practices";
- participation of the customer's specialists in the discussion of specifying and detailing requirements is not expected or will be extremely limited.

The uncertainty of technologies is also due to many factors - low level of technology readiness, lack of standards or established practices for their use, the presence of factors that indicate potential technological problems in the area of performance or safety. Examples of questions in this part of the questionnaire:

- the project assumes the use of technologies with a low level of readiness or not covering a significant number of specific project tasks;
- the project assumes the use of new technologies for which there is no standard way of integration into a joint system;
- the characteristics of the data and users of the system indicate the high possibility of a significant increase in requirements for the level of security.

To calculate the final risk index, a special matrix is used. In the matrix partial uncertainty indexes are used as coordinates, reduced to a 10-point scale (see Fig. 1). The asymmetry of the matrix is due to the different level of threat to the project of the corresponding measurements. In this case, the uncertainty of the requirements poses a greater danger to the project than the uncertainty of technologies.

The matrix shows four zones corresponding to different levels of threat. Highly risky projects include, first of all, projects falling into the "red" zone, which is described by inequation (1)

$$\frac{x^2}{R^2} + \frac{y^2}{r^2} \geq 1 \tag{1}$$

where x – technology requirement index; y – requirement uncertainty index; R = 10, r = 7 – ellipse semi-axes based on expert assessments.

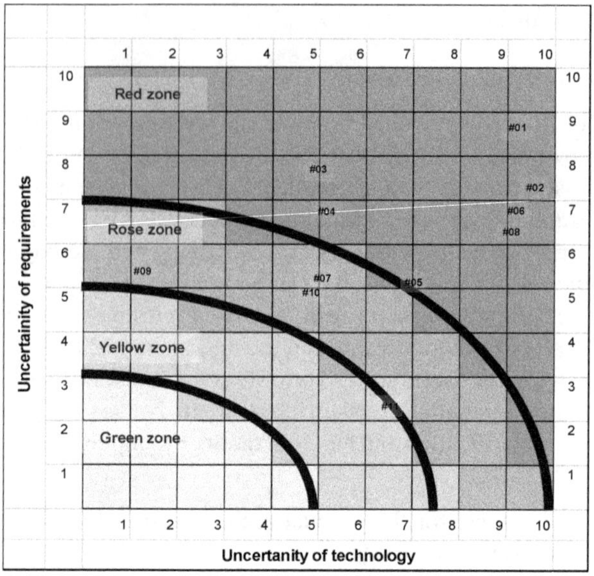

Fig. 1. Uncertainty project index

The boundaries of the "pink", "yellow" and "green" zones are described by similar formulas. Not only the "red", but also the "pink" zone requires increased attention. Projects that fell into these two zones as a result of the uncertainty assessment must go through the following assessment steps.

3.2 Contractor Self-assessment

In a large company that implements projects in a matrix organizational structure, situations inevitably arise when different business units participating in the project do not have the same ideas about the benefits of participating in a proposed project. Likewise, these units may have different levels of readiness to perform the tasks assigned to them in this project. The contractor can make an objective self-assessment only taking into account the views of all internal stakeholders of the project.

For self-assessment, as well as for assessing the uncertainty of the project, two groups of questions are proposed - the readiness of the contractor for the project and the attractiveness of the project for the contractor.

When assessing a company's level of readiness for a project, the focus is on the availability of qualified resources, as well as their ability to work in an environment of constant change and associated stress. The subject of evaluation is also the contractor's level of interest in the project and its willingness to accept the "rules of the game" in projects with a high degree of uncertainty. Examples of questions in this part of the questionnaire:

- assignments to key roles in the project (project manager, functional team leader, business architect, technical architect) are correspond to the role profiles;

- for each important expert position in the team, an experienced specialist is assigned with the priority of this project over other projects;
- the number of new employees in the team is limited.

Assessment of the level of attractiveness of a project is based on an analysis of the balance of planned benefits (not only direct profit, but also, for example, product development, partnerships, and so on) and, on the other hand, threats arising in the adverse events in the project. Examples of questions in this part of the questionnaire:

- the project allows to build long-term partnerships with a key customer;
- the project allows to develop a new market (new product, new industry);
- possible losses in the project in the event of negative scenarios are not significant.

To calculate the final self-assessment index, an asymmetric matrix is also used (see Fig. 2). In this case, we believe that the low attractiveness of the project poses a greater threat than the low level of readiness. The explanation for this is quite simple - the level of readiness can be increased quite quickly, and in most cases, we understand how this can be done.

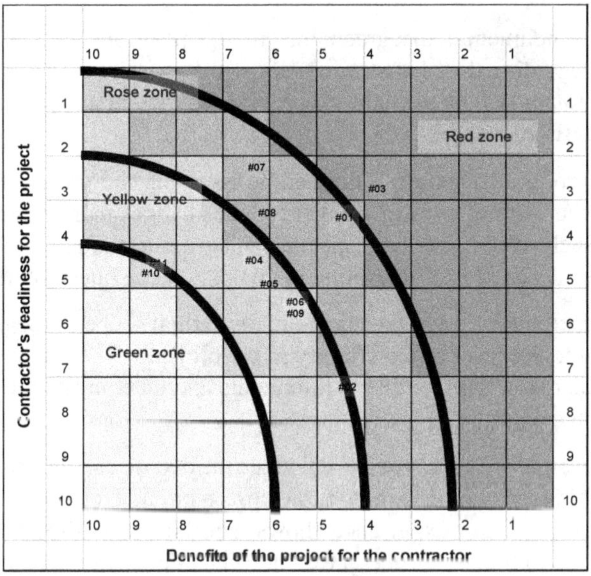

Fig. 2. Contractor self-assessment index

Note that the questions are formulated in such a way that a lower index value corresponds to a higher risk level. Therefore, for clarity, this matrix uses an inverted coordinate system. The "red" zone in this matrix is described by the inequation (2)

$$\frac{(10-x)^2}{r^2} + \frac{(10-y)^2}{R^2} \geq 1 \tag{2}$$

where x – contractor's readiness index; y – benefits of the project for contractor index;

R = 10, r = 8 – ellipse semi-axes based on expert assessments.

The boundaries of the "pink", "yellow" and "green" zones are described by similar formulas. If the self-assessment index falls into the "red" or "pink" zone, the overall risk of the project increases, and vice versa, falling into the "yellow" or "green" zone reduces it.

3.3 Customer Assessment Index

Success in the implementation of high uncertainty projects is largely determined by the desire and ability of the customer to comply with certain "rules of the game". There can be both subjective and objective obstacles to achieving a balance between, for example, flexibility of requirements and the desire for a fixed price, between the desire to do as much as possible and the desire to complete the project on time.

To assess the customer, two groups of questions are also used - interest and involvement of the customer and his agility in making decisions.

The key points for the first group of the questions are fast and competent feedback, short cycles of agreement and decision-making. Examples of questions in this part of the questionnaire:

- there are no fundamental disagreements in the positions of the customer's key stakeholders regarding the idea and concept of the project;
- experts of customer will be available to provide competent and prompt feedback and answers to questions.

The second group of questions is aimed at assessing the presence and rigidity of regulatory and other formal constraints that reduce changes management effectiveness and efficiency. Such restrictions are especially inherent in the public sector and state-owned companies. Examples of questions in this part of the questionnaire:

- regulatory constraints allow for a change in the formal scope of the project to a scale corresponding to the level of uncertainty in the project;
- the customer is ready to prioritize requirements and take into account cost and time constraints when determining deliverables.

An asymmetric matrix is also used to calculate the overall customer assessment index (see Fig. 3). In this case, we believe that low customer engagement is more dangerous than formal constraints. Our experience shows that the real desire of the stakeholders to get results allows to find the necessary compromises even in the conditions of highly bureaucratic structures.

In this matrix, as well as in the previous case, the inverted coordinate system is used. The "red" zone in this matrix is described by the inequation (3)

$$\frac{(10-x)^2}{R^2} + \frac{(10-y)^2}{r^2} \geq 1 \tag{3}$$

where x – customer interest and involvement index; y – customer agility index; R = 10, r = 8 – ellipse semi-axes based on expert assessments.

The boundaries of the "pink", "yellow" and "green" zones are described by similar formulas. As well as for the self-assessment index, getting the customer assessment index

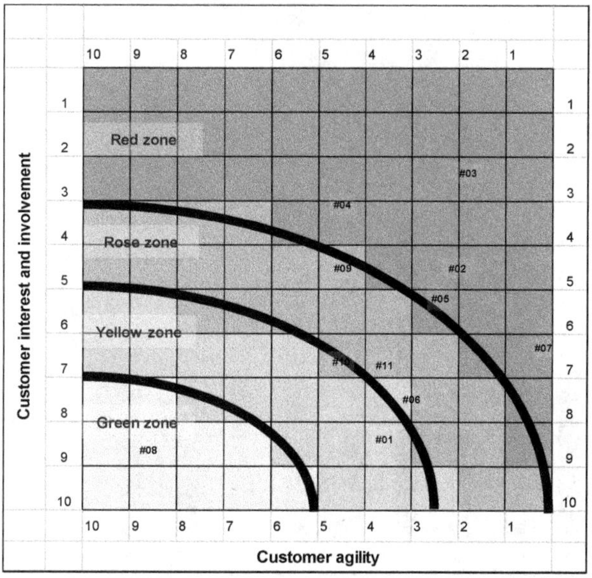

Fig. 3. Customer assessment index

in the "red" or "pink" zone increases the overall project risk, and vice versa, getting into the "yellow" or "green" zone reduces it.

4 Analysis and Decision Making

A high level of uncertainty is an important signal about the need for a special approach to management of the project. Important, but not the only one. As noted above, the internal and external context of a project can significantly affect the overall assessment of a project.

For example, in project # 09, the level of uncertainty does not fall into the "red" zone and seems to be quite acceptable (see Fig. 4). However, the results of self-assessment and customer assessment show that the overall project risk is comparable to the risks of other projects with a much higher level of uncertainty.

Another illustration of this situation are shown by comparing projects # 08 and # 04. With approximately the same level of uncertainty, the cumulative level of risk of these projects differs by almost two times. Therefore, the management tools for these projects will differ significantly.

In order to understand what these tools should be, we use partial indexes, discussed earlier and shown in the diagrams below for the example of project # 03. Figure 5 shows the overall indexes of uncertainty, self-assessment of the contractor and the assessment of the customer; Fig. 6 shows - partial indexes that allow to identify high-risk areas and select adequate management tools.

It is important to note that we assess not only negative risks, but also positive ones. Therefore, the developed management toolkit are aimed not only at mitigating risks, but

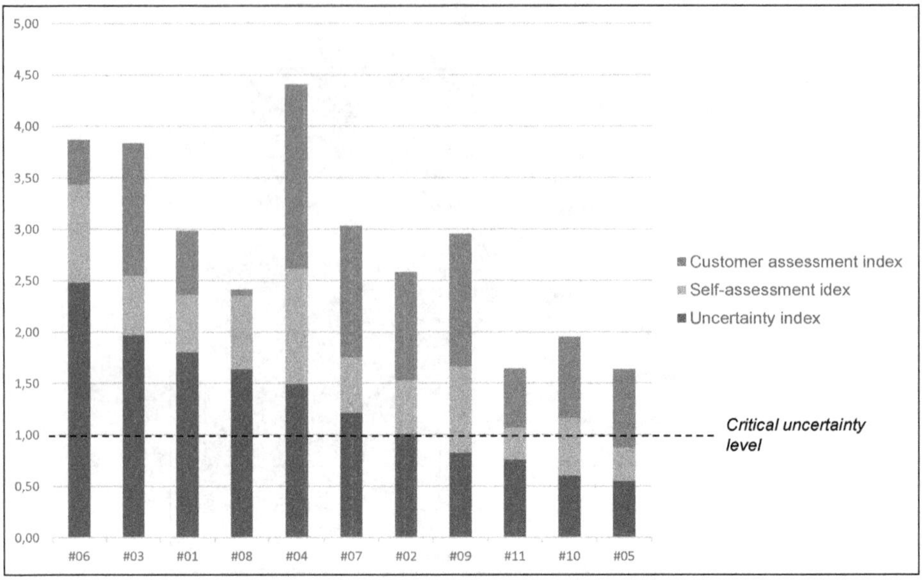

Fig. 4. Comparative analysis of the level of risk of projects

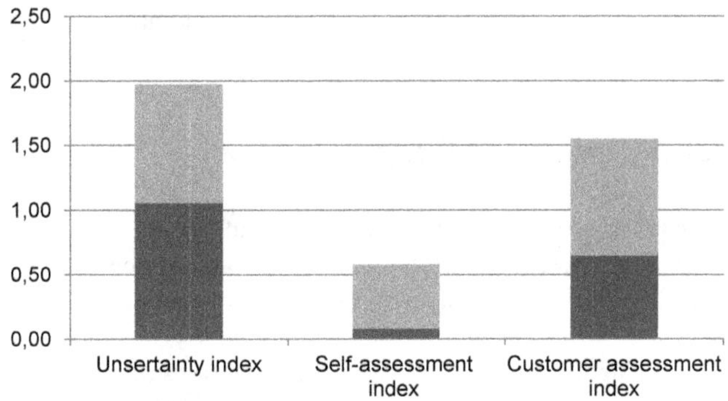

Fig. 5. Overall project risk indexes

also at strengthening opportunities. For example, if the attractiveness of the project for contractor is in the ‘red’ zone (noted in Fig. 5), the project sponsor is required to provide a plan for realizing the expected benefits over the medium term.

Overall, the management toolkit for such projects is described in the form of a checklist containing recommendations for each of the six blocks of project grades. We will consider two elements of this checklist in more detail.

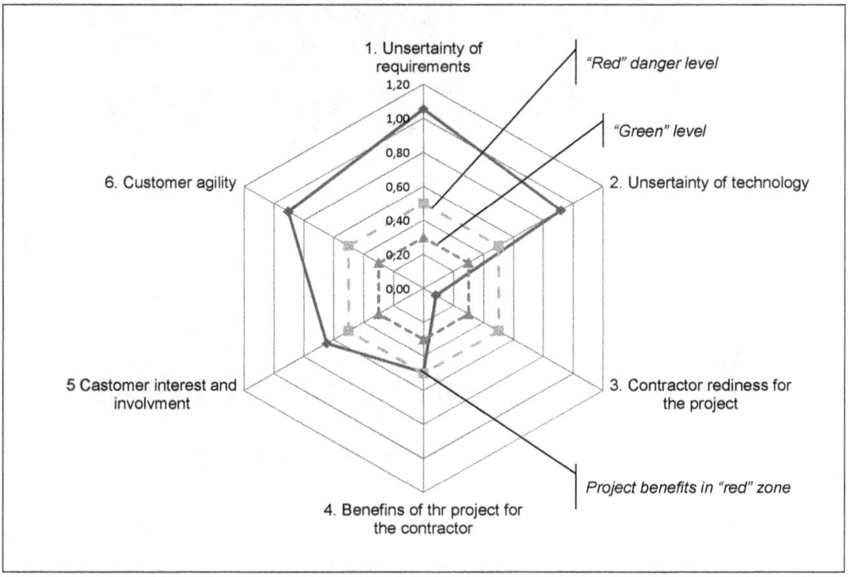

Fig. 6. Partial project risk indexes

5 Management Tools

5.1 Project Manager Profile for High Uncertainty Projects

People are a critical success factor for high uncertainty projects. Following [4], we can definitely say that the leader of an extreme project is not a profession, but a vocation. Not every person is able to work effectively, and just feel at least relatively comfortable in the face of constant changes in the project. Such projects occur quite often in the practice of our company, so the question of which of the employees can handle such projects is purely practical for us.

Thus, the first actual task is to form a general idea of what qualities the manager of such a project should have. This work was done in two stages. First, experts in the field of personnel management formed a detailed list of values, motivators, positive attitudes and competencies that correspond to the characteristics of high uncertainty projects. In Fig. 7, these elements are presented in aggregates, but it's easy to see that they describe, rather, a personality type than a professional qualities.

Selecting the characteristics, the experts proceeded from the fact that the manager of such a project should not just be a serious professional. This goes without saying, and a lot of attention is paid to the professional training of managers in the company [5]. We are also talking about a special personality of the managers of such projects. Here are some examples to illustrate this: "has a variety of personal and business interests and goals," "willing to take risks," "values freedom, the ability to make decisions and influence independently," "enjoys working in a changing environment."

There were 34 such characteristics in total, and at the second step, with the support of a focus group of business leaders, these characteristics were ranked based on the

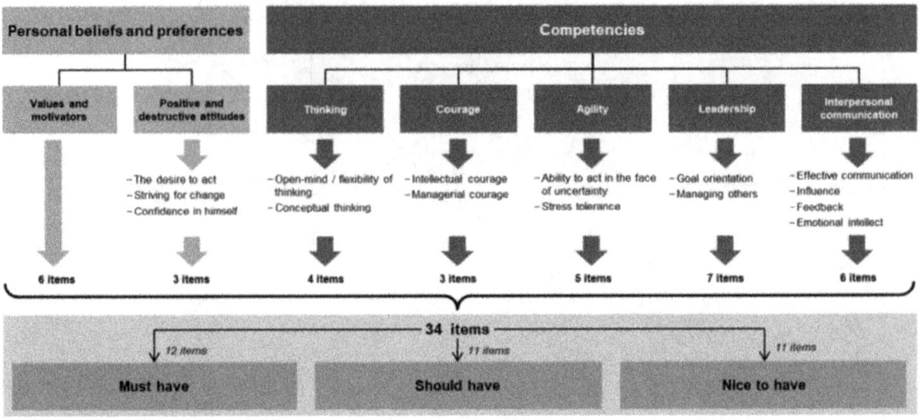

Fig. 7. Project manager profile for high uncertainty projects

peculiarities of the company's projects and the conditions in which these projects are being implemented. As a result, the project manager profile for high uncertainty projects includes twelve obligatory characteristics.

Among these characteristics, there are several difficult to develop or not at all developed qualities, and this requires serious changes in the processes of recruiting and professional development. In order to identify the most suitable candidates for this role and create a talent pool, the HR service uses various tools for analyzing vacancies [6] - an essay, a structured interview, and a value-motivational questionnaire. Project managers included in the talent pool are provided with individual development programs.

5.2 Ecosystem of High Uncertainty Projects

In high uncertainty projects not only the project manager, but also the entire team gets into difficult conditions. Firstly, it is a hard mode of work for a long time - regular overtime, often work seven days a week, the inability to allocate time for personal needs not related to work. Secondly, a high level of stress due to constant changes and customer dissatisfaction, often expressed in a rather harsh form.

These factors creates a high risk of burnout and a decrease in motivation for the project team, leads to a loss of interest in working in the project and even in the company. Can be added that such projects initially have a bad reputation, employees perceive them as a threat to their career, personal life and even health. Therefore, the formation and retention of the team is an important issue in such projects.

There are two main tasks to be accomplished - creating a work environment conducive to effective work, and maintaining the team's morale and positive emotional attitude. The tools used to solve these problems are combined into a set of services "Project HR". It supports forming and maintaining an optimal project environment, based on the specific conditions and circumstances of the project. Of particular note is the "psychologist" service aimed at controlling team burnout and implementing corrective actions.

Of course, such projects demands a special system of material and non-material motivation. The system should ensure constant attention to the project and recognition of the merits of the team and individual employees, not only after finishing of the project, but also directly during its implementation.

6 Conclusion

Almost twenty projects have been assessed using this approach, and some conclusions can be done.

1. Organization of assessment procedures

The projects were assessed in a 360° format. The focus groups included all the key project participants - account manager, sales manager, project sponsor, project manager and key technical specialists. In the course of the assessment, it was found that the opinions of the participants in the discussion differed greatly on many points, and sometimes radically. In some cases it was due to the heterogeneous structure of the project - for example, indeed, in some areas of the project, the uncertainty was higher, in some – lower. However, quite often these discrepancies allowed to reveal areas of incomplete information or even dangerous delusions.

The main lesson of approbation of the approach was the rejection of arithmetic mean estimates, pressure and forced equalization of estimates. Each case of serious disagreement should be discussed, the reasons should be identified and analyzed, and all participants in the discussion should agree with the final assessment.

2. Project management team

As noted earlier, the project manager profile for high uncertainty projects includes twelve mandatory characteristics. It seems that this is not much, but in fact, it is difficult to expect that such people will be found in the company at once and in the right amount.

Therefore, in the short term, the company focused on the formation of balanced project management teams, in which gaps in the project manager's profile are compensated for at other levels of management – project sponsor or functional groups leaders. And for the medium term, programs for the development of the competence of the project manager, focused on this profile, have been formed.

Acknowledgments. The authors thank the team of executives, managers and consultants of IBS, who made a great contribution to the development and implementation of the methodology: Grigory Kocharov, Dmitry Galagan, Yuri Yevtushik, Ekaterina Prokhorova, Alexandr Sokolov, Marina Torgovkina, Valentina Markova, Andrey Kiselev, Elena Sayapina, Maxim Shvedov, Nikita Shumov, and Dmitry Vasiliev.

References

1. Tovb, A., Tsipes, G.: Corporate standard for project management practices. In: Ottman, R., Grau, N., Schelle, H. (eds.) 16th World Congress on Project Management, IPMA 2002. VisionWorks Congress GmbH, Germany (2002)
2. Stacey, R., Mowles, Ch.: Strategic management and organisational dynamics: the challenge of complexity to Ways of Thinking about Organizations, 7th edn. Pearson, London (2015)

3. Snowden, D.: Cynefin - Weaving Sense-Making into the Fabric of Our World. Cognitive Edge, Singapore (2020)
4. DeCarlo, D.: eXtreme Project Management: Using Leadership and Tools to Deliver Value in the Face of Volatility. Jossey-Bass, USA (2004)
5. Tsipes, G., Echkalova, N., Sharova, E., Tovb, A.: Corporate university as a driver of project culture and competence development. In: Serpel, A., Ferrada, X. (eds.) 29th World Congress International Project Management Association (IPMA) 2015. Elsevier, The Netherlands (2016)
6. Prein, E.P., Goodstein, L.D., Goodstein, J., Gamble, L.G.: APractical Guide to Job Analysis. Wiley, Hoboken (2009)

Diagnostics of the Very Responsible Projects

Arkadiy I. Maron[1](✉) and Maxim A. Maron[2]

[1] National Research University Higher School of Economics, Moscow, Russia
amaron@hse.ru

[2] «FFIN Bank» LLC, Moscow, Russia

Abstract. When performing any work of the project there can be a mistake, which is also a deviation from the established requirements. The rational choice of control points is an effective method of decrease in time of restoring a regularity of implementation of the project when mistakes in works appeared. In this paper several methods of arrangement of control points in the very responsible projects – the projects directed to obtaining especially responsible result - are offered. They are based on the principle of maximization of the information. The diagnostic model of the project and the mathematical model of emergence of mistakes in works of the project are developed for using these methods. The criteria for comparison of options of arrangement of control points is developed as well as the algorithms and the software for the experimental check of effectiveness of the offered methods and their practical application to arrangement of control points in real projects. The experiment directed to assessment of effectiveness of each of the offered methods is made. Results of an experiment showed expediency of use of all methods as a part of a complex and its high performance.

Keywords: Informational Approach · Control Points · Diagnostics of Projects

1 Introduction

Let's consider the project as a result of which the unique product is created. This product has to meet the established requirements that is a quality indicator of it. The product will have the required level of quality in only one case and it when all works of the project are performed correctly. However, when performing any work of the project there can be a mistake, which is also a deviation from the established requirements. Verification of compliance of the final product to the established requirements is an obligatory stage of implementation of any project. If mistakes in works are revealed only at this stage, then a lot of time for their localization will be required. Localization here will be detection of the works performed incorrectly. In practice the similar situation almost inevitably leads to failure to meet time constraints of end of the project and the corresponding material losses. Therefore, it is necessary to carry out checks of correctness of performance of work during implementation of the project. If to carry out check of compliance of result to all established requirements for each work and to correct errors, then compliance of the final product of the project to all to the established requirements will be guaranteed,

S. Bushuyev et al. (Eds.): IPMA 2021, LNNS 704, pp. 25–34, 2023.
https://doi.org/10.1007/978-3-031-34629-3_3

but at the same time terms of the project and its cost can increase to unacceptable values. Performance of full check only when the final product is received, and full check of all works in the course of implementation of the project are two extreme options. The first of them leaves risk of failure to meet time constraints of the project on condition of appearing of mistakes in works, and the second considerably increases project duration.

At the same time there is more rational strategy of detection of mistakes in works of the project based that works of the project are connected not only by the connections defining an order of their realization in time but also by the usage of their results. Let's explain it on the simplest example. Let's say that two works of the project – A and B – are connected by dependence "finish to start". The result of work A is used in work B which result can't be the right if it is based on the wrong result of work of A. It is enough to carry out the complete check of realization of work of B that in case of the negative result of check to draw a conclusion that at least in one of works A or B mistake takes place. The strategy of timely detection of mistakes in works of the project at which each of works is checked for compliance to the major parameters (partial check), and the complete checks are carried out after some works chosen at the stage of planning the project is reasonable. We will further call a set of the works at which the complete checks will be carried out as set of the control points of the project.

For each large project there is a huge number of possible sets of control points. The rational choice of control points is one of the most important problems of diagnostics of projects.

One of the most important special cases of projects are the very responsible projects (VRP) – the projects directed to obtaining especially responsible result. They differ in a type of dependence of losses on a delay time of implementation of the project. Unlike other projects where in most cases there is a linear this relation [11], in VRP most often the nonlinear dependence is present.

Appearing of mistakes when performing the works of the project is a risk. Localization of mistakes by control points is one of methods of decrease in influence of risk of appearing of mistakes in works on results of the project. The refusal of the intermediate verification is adoption of risk. There are questions of how many has to be control points, and how to choose them in order that decrease in consequences of realization of risk of late detection of mistakes in works of the project was more favorable than its adoption [2].

2 Degree of Results Availability

The primal meaning of diagnostics of projects in the existing literature on project, is the definition of the indexes which allow to establish that there was a deviation from the basic plan demanding rescheduling or making of other management decisions [3, 16]. The significant role in the solution of these questions is played by works of Anshin V. M., Bogdanova V. V., Jaafari A., Ilyin V. V., Tovb A. C., Tsarkov I. N., Tsipes G.L., De Marco A., Thakurta R. Calculation of standard indicators of a deviation for cost and time is realized in MS Project and other project management applications which can be considered as the CASE tools corresponding to PMBOK [1, 6].

Articles devoted to stability of projects are closer to such works. In them the indexes, methods and algorithms are given which allow to define the moment when in the project deviations happened of enough force to be impossible to that project executed with the given time limits and costs [9].

Other direction of researches which should be analyzed on a possibility of application of the methods offered in them for the solution of the considered problem are the works devoted to quality assurance in projects [5, 12, 15]. To issues of decrease in probabilities of risks in general and mistakes at implementation of projects, in particular [8], it is devoted much more works, than to a problem of decrease in consequences of their realization. It is bound to the fact that first of all organizational measures are applied to decrease in probabilities of risks, and questions of management are dominating in works on project management [4, 14].

At the same time there are accurate recommendations about decrease in probability of emergence of mistakes in works for VRP. It is functional inspection. The methodology of object-oriented programming which significantly simplifies localization of mistakes by means of control points is developed for projects of development of software products. Here it should be noted especially Avdoshin S. M., Kruka E. A. and Lipayev V. V. works.

Fullestly questions of validation of functioning of technical and program systems and also issues of localization of malfunctions are resolved in the field of the science called technical diagnostics. Here works of Parkhomenko P. P. and his pupils are fundamental.

Despite existence of analogy between logical model of the continuous technical system and the network of the project, projects and technical systems are in essence various subjects of the diagnosis. The analysis of a possibility of application to projects of methods of optimization of troubleshooting process showed the following. Most methods of optimization which are developed in technical diagnostics directed on a problem of creation of the conditional algorithms of troubleshooting when an optimality criterion is mean time of searching. Johnson P, Caribbean V. V., Parkhomenko P. P., Sogomonyan E. S., Halcheva V. F. works are devoted to this question.

Johnson R. offered information approach to creation of the conditional algorithm of troubleshooting. He offered in each state which can arise in the course of searching, choose that check for which the ratio of an entropy of result to duration is minimum [7]. The noticed lack of this heuristic method which received the name "Johnson's criterion" was the instability. Small deviations on duration significantly influence the choice of checks, and the entropy which is slightly changing at the change of probabilities practically doesn't influence the choice. Further the methods guaranteeing achievement of an optimum at creation of the conditional algorithms of troubleshooting were offered. They are based on a method of a dynamic programming and a method of branches and borders. Because of computing complexity these methods don't allow to find for acceptable time an optimum algorithm of troubleshooting for a technical system in which the number of possible malfunctions (n) is more than several tens. So, the solution of this task by method of a dynamic programming assumes stay and storage of the conditional optimum managements for 2n states which can arise when troubleshooting. The effectiveness of a method of branches and borders depends on a probability distribution of refusals of elements and duration of checks. In some cases, it can be more difficult than the brute force.

Real projects contain hundreds, and quite often thousands of works, and target function is nonadditive. In such projects it is necessary to develop heuristic methods for the choice of control points. These methods have to be rather simple in the computing plan.

3 Description of the Research Methodology

For statement and the solution of a problem of the choice of control points it is necessary to establish connection between possible errors and results of possible checks, that is to offer diagnostic model of the project. This model has to be based on the data, available to the project manager when scheduling.

The following diagnostic model of the project based on its mathematical model in the form of the focused graph (the network of a project) is offered.

4 Diagnostic Model of the Project

The Diagnostic Model of the Project (DMP) is a form of representation of connection between mistakes in works of the project and results of the complete checks of works at which the project is submitted as the graph of network model where nodes represent works, and edges – not only logical communications between works, but also places of realization of possible complete checks. At the same time $\prod^j$ complete check of work j has a positive result – π_1^j then only when mistakes are absent in this work and all works preceding it. Otherwise this check has the negative result – π_0^j.

The offered DMP allows, using graph theory methods, to construct the table of possible mistakes (TPM) in which for each complete check its result at each possible mistake will be specified.

Also, for statement and the solution of a problem of the choice of control points it is necessary to determine probabilities of emergence of mistakes in works of the project. These probabilities can't be received purely statistically as the project is a unique set of works by its definition. The mathematical model of emergence of mistakes in works of the project is necessary. It is offered to consider that duration of works, used when scheduling the project, reflect on their complexity. So, if work duration is high, then probability of emergence of a mistake is high as well. This assumption will allow to calculate probabilities of mistakes, without resorting to the detailed analysis of each of works.

5 Distribution of the Conditional Probabilities of Mistakes in Works of VRP in the Presence of One Mistake

VRP are characterized by nonlinear dependence of losses from a delay time of the project and use of drastic measures of monitoring up to single operation that strongly reduces the probability of emergence of mistakes [10]. Proceeding from it, we will consider that in projects of this kind if the mistake takes place, then it only one. Owing to the assumption that in the project there is only one mistake the model given earlier can't be applied. Let's consider that duration of work reflects its complexity. Then it is logical to accept as the probability of existence of a mistake in work of i provided that in one of works of the project there is a mistake the value equal to the ratio of its duration t_i to total duration τ all works of the project (1)

$$p_i = \frac{t_i}{\tau} \tag{1}$$

6 Optimality Criterion of Set of Control Points for VRP

Let's consider that at nonlinear dependence of losses from restoring time set of m control points S^*, which breaks a set of all works of the project into subsets in such a way that the maximum restoring time of a regularity of implementation of the project corresponding to it is minimum – less, than at any other choice of m control points (CP) in this project is optimum (2).

$$S^* \rightarrow \min_{S} \max_{G \in \{G\}_S} T_G = \min_{S} \max_{G \in \{G\}_S} \sum_{i \in G} t_i \tag{2}$$

where S^* – the optimum CP, S – a set of the possible CP, G the subgraph consisting of the nodes corresponding to works to which the mistake is defined, $\{G\}_S$ – the set of subgraphs in G corresponding to set S, T_G – restoring time of a regularity of implementation of the project when determining a mistake to within G, i – number of node of a subgraph of G corresponding to work of the project with same number, t_i – duration of work i.

7 Mathematical Problem Definition of the Choice of Control Points in VRP

There is VRP with the works which are strongly connected by results. The network of the project is set. The graph corresponding to it consists of n + 1 node in which one node (with number n + 1) ending. For validation of performance of work, it is possible to establish m ≪ n of control points for realization of the intermediate checks. Final check after realization of the latest work of the project is carried out surely. Duration of work with number i is equal to t_i (i = 1, 2, …, n 1).

It is required to define a set of control points, being guided by a minimax optimality criterion.

7.1 Method of the Serial Conditional Choice of Control Points (SCC)

The following method of the serial choice of control points in VRP is offered in [Maron A., Maron M., 2012].

On each step of $k = 1, 2, \ldots, k, \ldots, m$ we choose check Π^{r_k}, to which there corresponds the maximum of an entropy of result (3)

$$\Pi^{r_k} \rightarrow \max_{j \in G\left(\pi_1^{r_1}; \pi_1^{r_2}; \ldots; \pi_1^{r_{k-1}}\right)} H\left(\pi^j / \pi_1^{r_1}; \pi_1^{r_2}; \ldots; \pi_1^{r_{k-1}}\right), \tag{3}$$

where $H\left(\pi^j/\pi_1^{r_1};\pi_1^{r_2};\ldots;\pi_1^{r_{k-1}}\right)$ - the entropy of result π^j of check $\prod^j$ considering that the final check signals that there is a mistake in the project and predecessor checks have positive results: $\pi_1^{r_1};\pi_1^{r_2};\ldots;\pi_1^{r_{k-1}}$; $G(\pi_1^{r_1};\pi_1^{r_2};\ldots;\pi_1^{r_{k-1}})$ – a set of works in which, if have mistake, the checks above will have positive results.

7.2 The Modified Method of the Serial Conditional Choice of CP (ModSCC)

The method explained above is based only on information approach to the choice of checks. Respectively, a set of checks will be predetermined by topology of the graph of works of the project and a probability distribution. As a result the first check will carry out half splitting the graph of works on probabilities. Following will carry out similar splitting for a subgraph of works, not being logical predecessors of work to which there corresponds the first check, and etc. As a result before the first check it can be already executed too large volume of works of the project. If after check it turns out that some work is performed irregularly, then total time of searching and elimination of a mistake can be larger. Proceeding from it, the modified method of the serial conditional choice of control points was offered.

The following method of solution is offered.

On each step of k = 1, 2, ..., k, ..., m we choose check Π^{r_k} (4), to which there corresponds the maximum of value W_k^j (5)

$$\Pi^{r_k} \to \max_{j\in G\left(\pi_1^{r_1};\pi_1^{r_2};\ldots;\pi_1^{r_{k-1}}\right)} W_k^j. \tag{4}$$

Here

$$W_k^j = \frac{H\left(\pi^j/\pi_1^{r_1};\pi_1^{r_2};\ldots;\pi_1^{r_{k-1}}\right)}{1+log_2\left(1+T\left(\pi_0^j/\pi_1^{r_1};\pi_1^{r_2};\ldots;\pi_1^{r_{k-1}}\right)\right)}, \tag{5}$$

where (6)

$$T\left(G\left(\pi_0^j/\pi_1^{r_1};\pi_1^{r_2};\ldots;\pi_1^{r_{k-1}}\right)\right) = \sum_{i\in G\left(\pi_0^j/\pi_1^{r_1};\pi_1^{r_2};\ldots;\pi_1^{r_{k-1}}\right)} t_i. \tag{6}$$

7.3 Method of the Serial Unconditional Choice of CP (SUC)

In the methods of the serial conditional choice of control points in VRP explained above each following control point was chosen so that it gave maximum information believing that checks in the previous points have positive results. The unconditional method of the serial choice of control points was offered. According to it on each step that control point is chosen at which the set consisting of it and earlier chosen control points, gives a maximum of information on what work of the project has the mistake. This method is more difficult from the computing point of view, than earlier offered conditional methods.

On k = 1 step, 2, ..., k, ..., m we choose check of that work in which if to establish CP, check, taking into account earlier established CP, gives a maximum of information on the place of emergence of a mistake (7)

$$\Pi^{r_k} \rightarrow I(N, s_k) = \max_{j \notin S_{k-1}} I(N, s_{k-1} \cup \pi^j), \tag{7}$$

where $I(N, s_{k-1} \cup \pi^j)$ – an amount of information about number N – work of the project with the mistake, which gives set of results $s_k = \{\pi_1^{r_1}; \pi_1^{r_2}; \ldots; \pi_1^{r_{k-1}}; \pi^j\}$ checks in control points (8).

$$S_k = S_{k-1} \cup \Pi^j = \{\Pi^{r_1}; \Pi^{r_2}; \ldots, \Pi^{r_{k-1};j}\}. \tag{8}$$

As a result of realization of this study it is proved that information can be calculated on the simplified formula (9)

$$I(N, s_k) = -\sum_{i \in G_u(s_k)} p_i \log_2 p_i - \sum_{G_f \in G_a(s_k)} \sum_{i \in G_f} p_i \log_2 \sum_{i \in G_f} p_i, \tag{9}$$

where p_i – a probability of a mistake in work i, $G_u(s_k)$ – a subset of works of a project which are defined unambiguously, $G(s_k)$ – a subset of works of a project which can't be defined unambiguously, G_f – a subset of works of a project which are indiscernible.

8 Assessment of Effectiveness of the Offered Methods of Arrangement of CP

Assessment of effectiveness of methods of CP arrangement in VRP was carried out in two steps. At the first stage the expediency of complex application of all three methods developed in work was checked. It was for this purpose necessary to check lack of dominance of one of methods over two others (existence obviously of the best method) and lack obviously of the inferior method in comparison with two others. The Monte-Carlo method was applied to check. Projects with different number of works, a random number of logical communications between works and random distribution of durations of works were generated. The quantity of control points m was accepted equal to 10–15% of number of works n. Practice shows that in this range there is a maximum number of works of the project in which it is expedient to carry out the complete check.

For each received project arrangement of control points was carried out by each of three offered methods: by the method of the serial conditional choice by modified method of the serial conditional choice and by method of the serial unconditional choice. The maximum restoring time of a regularity of implementation of the project at each of three received options of arrangement of control points was calculated. The method giving option of arrangement of control points to which there corresponded the least value of this maximum time admitted to the best in this realization of an experiment.

High computing performance of the offered methods allowed to make a representative experiment with the projects having rather large number of works. It was as a result established that there are no dominating and worst methods. Each of the offered methods of arrangement can yield the best result for the specific project (Fig. 1). It is expedient to apply all three methods in a complex and to choose the best arrangement for the real project according to the offered minimax criterion.

At the second stage the effectiveness of a complex from three offered heuristic methods was estimated. As well as at the first stage, the Monte-Carlo method was applied. In each realization of an experiment the deviation of value of target function expressed as a percentage corresponding to option of arrangement of control points received as a result of application of a complex calculated from its minimum possible value corresponding to optimum arrangement of control points for this project. Parameters of projects were generated the same as it is described earlier. Optimum arrangement of control points for each generated project was by method of brute force of all possible options. Need of use of the brute force was imposed by a tight restriction for number of works in the generated projects. The maximum size of projects in an experiment was 40 works.

Results allow to speak about high performance of the offered methods. So, the experiment with a large number of realization for projects with thirty works showed that the value of target function corresponding to arrangement of control points by means of a complex doesn't deviate its minimum value corresponding to strictly optimum arrangement, more than for 65% in 90% of cases (Fig. 2). Let's note that the greatest possible deviation from an optimum for projects with such number of works may be up to 1200%. Moreover, the deviation from an optimum when using a complex practically doesn't change with increasing of number of works in projects while with increasing of number of works of the project the greatest possible deviation from an optimum quickly increases.

On the basis of these results author found it possible to use for arrangement of control points in real projects. The heuristic methods are confirmed by adoption deeds. The authors developed the program realizing the offered methods for projects, control of which is exercised with application of MS – Project.

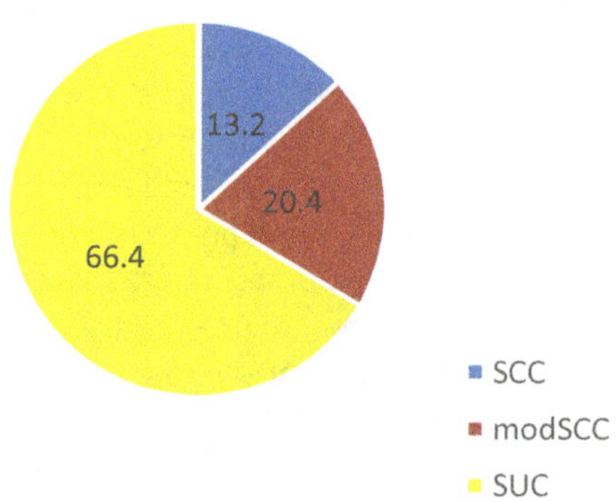

Fig. 1. Present of best results.

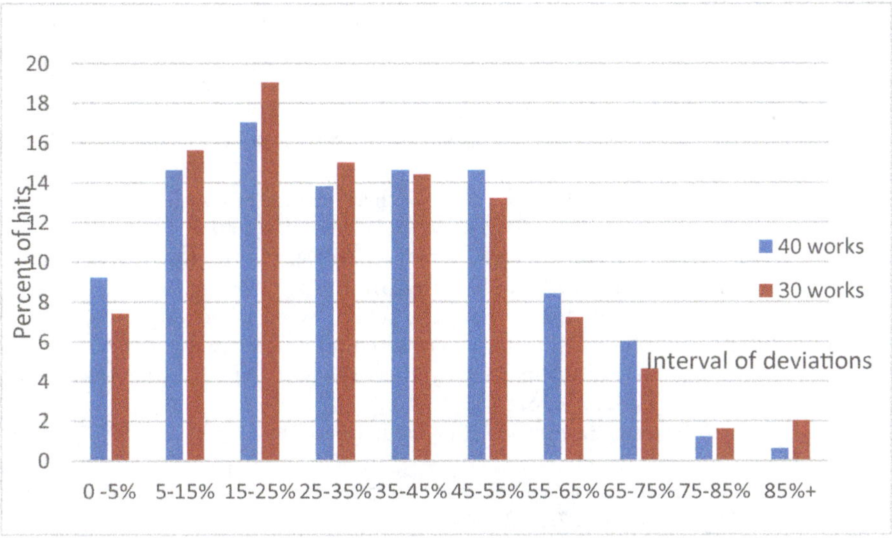

Fig. 2. Percent of hits in the group of deviation from optimal.

9 Conclusion

The foregoing allows us to draw the following conclusions.

1. Rational choice of the control points can significantly reduce the losses from the errors in the VRP work.
2. The effectiveness of the methods for selecting control points in the VRP, created by the authors, can be considered experimentally confirmed. These approaches are heuristic. This heuristic is based on K. Shannon's theory of information [13]. Verification is an experiment. It is necessary to carry out the experiment that gives the maximum information about the object under study. This provision is the basis of the proposed methods.
3. Software that implements these methods is developed.

References

1. Aleskerov, F., Chistyakov, V., Kalyagin, V.: The threshold aggregation. Econ. Lett. **107**(2), 261–262 (2010)
2. Cioffi, D.F.: Completing projects according to plans: an earned-value improvement index. J. Oper. Res. Soc. **57**(3), 290–295 (2006)
3. De Marco, A., Briccarello, D., Rafele, C.: Cost and schedule monitoring of industrial building projects: case study. J. Constr. Eng. Manag. **135**(9), 853–862 (2009)
4. Futrell, R., Shafer, L.: Quality Software Project Management, 1st edn. Prentice Hall, Hoboken (2002)
5. Herroelen, W., Leus, R.: Robust and reactive project scheduling: a review and classification of procedures. Int. J. Prod. Res. **42**(8), 1599–1620 (2004)

6. Jahangirian, M., et al.: Simulation in manufacturing and business: a review. Eur. J. Oper. Res. **203**(1), 1–13 (2010)
7. Jonson, R.: An information theory approach to diagnostic. In: Proceedings of 6th National Symposium on Reliability and Quality Control, pp. 102–109 (1960)
8. Leus, R., Herroelen, W.: Stability and resource allocation in project planning. IIEE Trans. **36**(7), 667–682 (2004)
9. Lyneis, J.M., Cooper, K.G., Els, S.A.: Strategic management of complex projects: a case study using system dynamics. Syst. Dyn. Rev. **17**(3), 237–260 (2001)
10. Maron, M.A.: The choice of control points of projects taking into account possible change of structure of works. Bus. Inform. **36**(2), 57–61 (2016)
11. Maron, M.A.: Diagnostics of projects. Eur. Res. Stud. J. **21**(1), 18–30 (2018)
12. Sakka, O., Barki, H., Côté, L.: Relationship between the interactive use of control systems and the project performance: the moderating effect of uncertainty and equivocality. Int. J. Proj. Manag. **34**(3), 508–522 (2016)
13. Shannon, C.E.: A mathematical theory of communication. Bell Syst. Tech. J. **27**, 379–423 (1948)
14. Taylor, H.: Risk management and problem resolution strategies for IT projects: prescription and practice. Proj. Manag. Q. **37**(5), 29 (2006)
15. Thakurta, R.: Impact of scope creep on software project quality. Vilakshan XIMB J. Manag. **10**(1) (2013)
16. Williams, T., et al.: The effects of design changes and delays on project costs. J. Oper. Res. Soc. **46**(7) (1995)
17. Tsipes, G., Echkalova, N., Sharova, E., Tovb, A.: Corporate university as a driver of project culture and competence development. In: 29th World Congress International Project Management Association (IPMA) 2015. Elsevier, The Netherlands (2016)
18. Prein, E.P., Goodstein, L.D., Goodstein, J., Gamble, L.G.: A Practical Guide to Job Analysis. Wiley, Hoboken (2009)

Ethical Risks in Digitally Managed Project Teams

Olga Ilina and Lev Tsipes(✉)

Higher School of Economics, Shabolovka Ulitsa, 26, Moscow 119049, Russia
levgtsipes@edu.hse.ru

Abstract. A new global scenario and "a new digital world" is coming, and we cannot ignore it. Project managers leading projects online face an added complexity, which causes additional ethics issues to cope with. How to manage employees' work/life balance? How to adjust compensation and benefits to the new working conditions? How to communicate with remotely working employees and monitor them? While tackling these and many other new issues that affected corporate ethics, some project managers may rely on standards and legislation, while others lead by example. A combination of the authors' personal experience combined with an academic literature review and a survey of more than 50 participants of 17 projects was done to support the conclusions and recommendations presented in this paper.

Keywords: Ethics · Project · Digital Era · Project Management

1 Introduction

The perspectivity of remote, distributed, or distant (in this paper these adjectives are used as synonyms) work was indicated quite long time ago, even when such progressive instruments like Zoom or MS Teams were not introduced. For instance, back in 1994, Kostner [10] indicated "dramatic paradigm shift in groupwork". The researcher stated that for new (for that time) business realities (e.g., fast changing environment or employing outsourcing) distributed teams are an essentiality that project managers should implement. It is claimed that with the help of technology many positive impacts can be deduced (like shorter meetings or speeding decision-making process) so that distant teams are no lesser effective than traditional ones.

Basic issues associated with distant work were also indicated in the previous millennium. For instance, Adams & Adams claim that creating trust and group identity, spreading, and understanding information, making understandable structures, and preventing undesirable informal groups are the main struggles project managers have to address. The authors also refer to technology as a main problem-solving tool for these cases [1].

There are some more contemporary works on distribute project management as well. For example, Michael Huber claims that working remotely may bring feeling of isolation to workers, which can be neglected by reducing unnecessary e-presence, offline bounding

S. Bushuyev et al. (Eds.): IPMA 2021, LNNS 704, pp. 35–49, 2023.
https://doi.org/10.1007/978-3-031-34629-3_4

with project team members and attending trainings. [8] Vergini [20] also highlights the importance of addressing isolation issues, as well as providing distribute team with all necessary up-to-date technology.

Milhauser's research [12] has indicated that there is a high risk for distributed teams to face difficulties because of cultural and general diversity. Because of these difficulties decision-making and facilitating also becomes more complex than in traditional teams. Das [7] lists challenges for remote teams more specifically, like difficulty in assessing candidate's skills distantly, unwillingness of some to report bad news when they are not forced, religion-specific communicational struggles etc.

However, despite long interest in distributed teams and related intercultural challenged, there are not so many studies of codes of conduct in such project teams. This topic is of big relevance due to pandemics of Covid-19, which forced project teams all over the world unexpectedly shift to distant workstyle. Since many project managers have never had experience of managing so many people remotely, they have faced a variety of issues related to this mode.

In this research the problem of adapting project teams to distant work in ethical dimension is considered. The aim of the paper is to outline especially sensitive ethical questions that arise during the remote teamwork to be included in project team codes of conduct. The object of this study are project teams. The subject of the study are ethical relations between project team members.

The main methods for this research are analyzing the literature and surveying project team members. In the survey, employees will be suggested to answer how much has one or another ethical issue affected them when switching to the remote work and compare them to when they worked traditionally. As a tool Google Forms is be employed and to survey ~90 Russian project team members.

It is believed that the issue addressed in this study is of utmost significance from both scientific and practical viewpoints. Enterprises, when considering hybrid or fully distant workstyle for their project teams, can use the results of this research to draw drafts for project teams code of conduct or include recommendations from this study into their corporate code of conduct if applicable. For scholars investigating project management ethics, this study can serve as an overview of the most common ethical issues remote project teams face, which can be used for further research.

2 Ethicality Management in Project Management Teams

Project management is a specific field of general management with its own peculiarities, including ethical ones. Project management is tied with specific goals as well as with limited resources and time, which makes it stand out from daily operations. Also, in project teams, members usually come from various departments and organizations, which also contributes to project ethics specificity. Hence, project management ethics need to have their own research, yet with consideration of general managerial ethics [11].

However, project management is still a fresh field and, consequently, is studied scholarly less than business. Therefore, ethical part of project management may also lack attention. Moreover, project management is also often considered to be a tool for practitioners, shifting focus onto efficiency and productivity, overshadowing the ethical part of the field [11].

One of the main guides in project management is IPMA Competence Baseline [9] that "provides the official definition of the competences expected from project management personnel by the IPMA for certification using the universal IPMA four-level-certification system". In the Ethics part they highlight that one of the key features a code of conduct can provide is releasing employees from moral conflicts with the help of guidelines. A code of conduct can have a legal basis or be incorporated into employment contracts. The important thing is that applicants should be informed that by accepting an offer of a given company they also accept its code of conduct. What may be concluded is that it may be useful to introduce corporate ethics to applicants during the hiring process to avoid interpersonal conflicts within employees.

Anyways, what should be embraced in the Code of conduct? In the IPMA Competence Baseline they recommend evaluating from ethical perspective:

1. Respect (towards ethical issues and ethical values);
2. Integrity (consistently follows ethical standards);
3. Information usage (treats information in an ethical way);
4. Transparence, fairness (transparent when drawing ethical norms);
5. Solidarity (is eager to act for the sake of their team);
6. Empathy (is happy when their colleagues are).

Code of Ethics & Professional Conduct of PMI [4] includes "aspirational" and mandatory standards. Mandatory standards set limitations, requirements and prohibitions, noncompliance with which will lead to disciplinary punishments. "Aspirational" standards are technically mandatory as well, however, they are harder to monitor and measure, so they are more of a lighthouse.

Besides history of the code, glossary and other parts, PMI's code covers the following ethical blocks:

1. Responsibility;
2. Respect;
3. Fairness;
4. Honesty.

Each block contains three sections: definition, aspirational standards, and mandatory standards.

There are four aspects of project management ethicality highlighted in Code of Ethics for The Project Management Profession [5]. Not to cite all the paragraphs, the key points are:

1. *Professionalism.* Project managers (PMs) must be highly professional and keep the honor of the profession clean by being responsible, obeying the law, participating in professional trainings regularly and promoting pro-ethical behavior;
2. *Employee relationship.* In work project managers (PMs) should raise productivity with minimal costs, promote professional development within employees, treat them equally, provide safe working conditions and be open for feedback;

3. *Information.* PMs must keep employers' and clients' information confidential, prevent any conflict of interest, not accept any form of bribes and be honest with clients about potential project costs and quality;
4. *Community contribution.* PMs should raise public awareness about the profession and avoid doing anything that could threaten the reputation of the community.

All these and optionally other ethical features might be fixed in the project code of ethics, which aims to set a common understanding of the ethics for project team and stakeholders. When designing one, project manager should [22]:

1. Outline desired and undesired ethical behavior;
2. Find out how the organization, as a whole, supports these;
3. Encourage all of the project team to contribute to the project code of conduct;
4. Dedicate some time during meetings to ethical problems regularly;
5. Ensure communicational transparency;
6. Outline rewards and punishments;
7. Include the code into the project plan.

Point (2) mentioned above is compulsory: project ethics are a part of corporate ethics. Hence, a good organizational base is a must, which PM must ensure. To do that, The Integrity Questionnaire can be used. It includes the following ten questions that give an overview on organizational ethical health (all 10 questions are cited [22]):

1. "What are the core values of the organization?"
2. "How do the organization's core values relate to ethics?"
3. "What does the organization use as an ethical guideline?"
4. "Is there a set of behavioral standards for all employees?"
5. "How do the organization's systems and goals support ethical behavior?"
6. "How do the organization's systems and goals deal with unethical behavior?"
7. "To what extent is the organization dedicated to doing the right thing?"
8. "What kind of support do employees get for doing the right thing?"
9. "How important is winning?"
10. "Is there a written commitment to personal responsibility, and is this carried out consistently?"

Typically, the reasons why ethical issues and dilemmas the PMs face managing project teams occur may be [16]:

1. Data misinterpretation for personal interest
2. Conflict between employee's and corporate moral values
3. Tension between individual and business moral priorities
4. General unprofessional behavior

However, even if all the ethical resolutions are put onto the paper, project managers should keep on constantly brushing up the team in terms of ethicality. Here are some interesting unethical traits, identified by Shouche [17] that may not come to mind in the first place, though they are quite common and may affect the work dramatically:

1. Imprecise reporting;
2. Treating stakeholders unequally;

3. Being not objective;
4. Free riding;
5. Being resistant to changes;
6. Egoistic behavior.

Going virtual also brings numerous challenges to the ethicality of project management, as delivering ethics via leadership becomes quite hard. To cope with it, the project manager should differentiate his or her leadership style, adjusting it to a particular situation. Here are some approaches one could use [6]:

1. Leading by example. This method is hard yet still possible to be used in virtual projects. It is good for setting basic ethical tone for the project;
2. Stewardship. This is leadership by teaching. This style can be used for giving feedback and explaining some complicated cases. This leader helps but does not make decisions instead of employees;
3. Convincing. This is a more persuasive approach when the leader drives the team to the ethical solution, he or she has in mind;
4. Leading by incentives. The style suggests using clear rewards or punishments. This approach is suitable for clear "black and white" situations, which are, obviously, rare;
5. Authoritarianism. Using this style leader commands what employees must do. It is appropriate in extreme situations when ethical rules are to be broken.

When project leaders follow ethical principles and endorse ethicality, trust in them (PMs) is endorsed. On the one hand, employees start trusting their leaders and, consequently, team members begin following them willingly. Hence, project leaders gain authority that improves their management. On the other hand, clients may also gain trust towards ethical project and even become more ethical on their own as well [14]. Hence, when being engaged in ethical decision making (EDM), PMs should consider the code of conduct of their organization, the code of conduct of their profession (e.g., IPMA Competence Baseline) and their own moral values in the hierarchy visualized in Fig. 1 [16].'

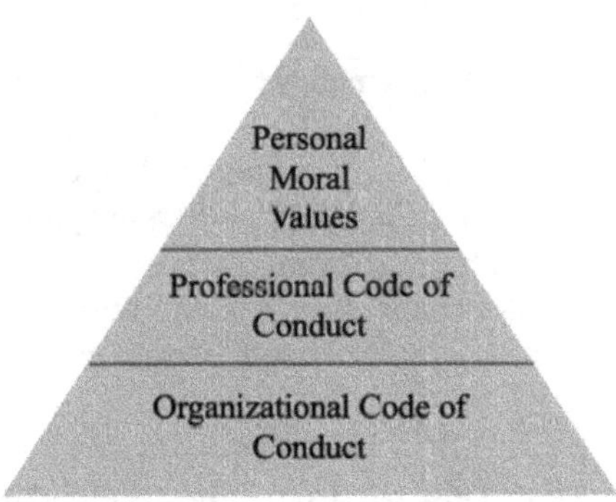

Fig. 1. EDM Hierarchy

When PMs face ethical issues or dilemmas, they can use a five-As PMI EDMF (PMI Ethical Decision-Making Framework) [15] that contains the following steps:

1. *Assessment.* This stage undermines evaluating the case from legislative perspective.
2. *Alternatives.* On this step PM should sort all the adequate possible solutions and list potential outcomes of adapting them.
3. *Analysis.* Analyzing the potential scenarios from the previous stage includes checking, whether a given solution a) maximizes "good-to-bad" ratio and/or prevents negative outcome to happen; b) addresses cultural diversity in the project team; c) is taken under no pressure from other parties or emotional affect.
4. *Application.* On this stage PM should ensure, that the chosen ethical management decision is coherent with adopted ethical norms and fair for all the parties touched.
5. *Action.* When implementing the taken ethical course, PMs should be responsible for their actions and make them visible for the public.

Also, while investigating a case of unethical conduct, it would be helpful for a PM to look at the problem from the perspective of behavioral ethics. Basically, ethically unwanted behavior can be classified in three dimensions [13]:

1. *'Bad Apples'* is an orientation on an individual level that assess form perspective of such qualities like cognitive development, age, education and other aspects that each individual is characterized with.
2. *'Bad Cases'* is an orientation on a situational level. This dimension considers how some cases make employees behave in certain unethical ways
3. *'Bad Barrels'* considers organizational level, how the whole ethical system within a given enterprise (e.g., perceived ethical climate or perceived ethical culture) affect employees' behavior.

Overall, ethicality is an important topic in project management knowledge, mainly covering topics like responsibility, respect, fairness, and professionalism. Even though fundamental topics on project management ethics are provided by professional organizations, project managers should adapt their codes of conduct to their team's specificity in accordance with the global project management standards.

3 Ethicality Management in Project Management Teams

In the year of 2020, the world faced unpreceded pandemics of Covid-19 that affected everyone. To ensure employees' safety and prevent further spread of the virus, businesses had to make their workers work remotely from home, and project teams experienced this shift as well. For many project managers (PMs) fully remotely working project teams were a completely new circumstance. Hence, PMs faced many completely new issues, including ethical ones.

To explore the ethical risks of switching work to a remote style, some forums, blogs, and newspapers have been looked through. It was found that the following corporate ethics topics turned to be relevant: gossiping; overlooking colleagues during meetings and group conversations; being overdemanding; home distractions; ignoring corporate communications; inadequate work leaves policy; excessive control; compensation policy; ecology related issues [2, 3, 18, 19, 21].

Based on this information, five major areas of ethical dilemmas in digital world relevant for project teams have been deduced (see Table 1). Ecological area has been excluded since it is out of scope of this paper.

Table 1. Major Areas of Ethical Dilemmas in Digital World

№	Ethical Area
1	Relations Online
2	Simulations at Work
3	Employees Control
4	Bonuses and Leaves
5	Online Etiquette

4 Methodology

The main methods engaged in this research is surveying project team members. In the survey, employees engaged in project management were asked questions on 1) how regular ethical aspects of their work has changed during pandemics (e.g., gossiping) 2) new ethical issues specific to remote work and 3) their own opinion about "new ethics" in digital world. The questions were deduced from the information covered in the previous paragraph.

The questionnaire also includes control questions on sex, professional organizations, experience etc. These questions were used for segmentation of results. The survey was offered in English and Russian languages in Google Forms and the respondents were guaranteed with full anonymity. The total number of respondents totaled in 91.

All the questions can be sectioned in two parts: 1) the first part (questions II.1-II.13) to indicate changes in ethical aspects during remote work, where respondents showed the degree of their agreement; 2) the second part (questions II.14-II.21) to ask opinions of project team members on probable ethical issues that arise during remote work, where they could agree or disagree. "I do not know" responses were removed from the analysis.

The survey was tested for reliability and calculated Cronbach's alpha, which is $\alpha = 0.7963$ for the main part of the questionnaire. To calculate the alpha, the questions were normalized. Questions of the first part were normalized using the following scale:

- "Absolutely not" = 0;
- "Rather not" = 1;
- "Rather yes" = 2;
- "Absolutely yes" = 3.

In questions of the second part "Yes" = 1 and "No" = 0. Since it is unknown whether the difference between "Absolutely not" and "Rather not" is the same as the one between "Rather yes" and "Absolutely yes", the Likert scale used is ordinal and was indexed accordingly.

To research the difference between samples, based on control questions, chi-square test was used to ensure significance of the difference between groups (alpha level = 5%). The pairs with visible difference are covered in the next chapter.

The results of the survey helped to outline ethical dilemmas that are relevant during distant workstyle. This record of ethical issues was then used as a fundament for in-depth interviews.

5 Questionnaire Results

The results of the questionnaire are presented in Table 2. It was indicated, that out of 13 proposed ethical risks, 6 ones proved to be relevant. Table 3 below illustrates ethical aspects II.1-II.13 that got considerable number of "Absolutely yes" and "Rather yes" responses.

Table 2. Considerable ethical aspects

Ethical aspect (question)	Percentage of positive responses
II.7 (You and/or your teammates started dedicating your work hours to private matters or began doing so more often after you had started working remotely?)	59%
II.5 (Your project manager started expecting you and/or your teammates to be (more) available for communication outside of business hours after you had started working remotely)	54%
II.6 (You and/or your teammates feel that your colleagues started to violate your private virtual space for working purposes (e.g., text you about work tasks in private messengers instead of corporate once) more after you started working remotely?)	37%
II.9 (After you have started working remotely, you and/or your teammates feel that your project manager expects you to work over hours or has started to expect you to work even more over hours than before?)	33%
II.2 (You and/or your teammates started using phones or PCs during meetings for leisure (e.g., checking social media) more often after you had started working remotely)	33%
II.3 (During meetings or common conversations, you and/or your teammates started to be overlooked (e.g., questions or ideas of some employees are ignored in a group chat) more often after you had started working remotely)	25%

As for II.7 [private matters at work], the results may indicate one of the following three:

1. Control from project manager has been decreased, so the tasks can have less strict deadlines now;

2. Project team members started doing tasks faster from home, so they now have spare time during work hours;
3. Project team members started employing different time management (e.g., doing some tasks outside the work hours but playing video games after dinner).

High percentage of positive responses for II.5 [communication after hours] and II.6 says that barriers between private time and work time have not yet been widely established with project teams working remotely. This is not a wanted situation since the consequence of the lack of separation between private and work time may turn into more frequent burnouts and, respectively, higher turnover rates.

33% in II.9 [work after hours] can stem from different reasons. On one hand, project managers could have started demanding more extra work since the overall remote work efficacy is low in their project teams (which may be related with II.7). On the other hand, this can be caused by weak barriers between private time and work time (see II.5 and II.6), so that now project managers just want more work to be done.

Interestingly, results for II.2 [leisure at meetings] and II.7 [private matters at work] are quite different, meaning that project team members are generally more or less involved in meetings. However, the percentage of positive responses for II.2 is still significant. This may indicate that meetings are usually long and not as dense as they should be to keep team involved and interested. Probably, some project managers dedicate too much time to team discussions or include unnecessary topics. Regarding online meetings, not high but yet significant positive rates received II.3, which may mean that project team leaders do not employ involving methods in conversations and discussions.

Furthermore, chi-square test has indicated some statistically significant and visible differences for questions II.1-II.13.

As for dividing the sample by industry, it was found that unlike employees form other industries, IT & Telecom project team members:

1. Have not been subjected more often to be available after work;
2. Have not become more distracted during the working day;
3. Much less often have begun to feel more neglected at meetings;
4. Do not consider it mandatory to turn on the webcam on calls.

In project teams of companies with less than 500 employees:

1. Employees believe that they are personally responsible for the technical difficulties when distant;
2. Most employees want compensation for office space.

When dividing the sample by age, the following features were discovered:

- Respondents 56+ years old started doing private matters when working remotely but did not experience private virtual space violation. They oppose geolocation tracking and food compensation and do not have opinion on electric bills compensation.
- 41–55 years old respondents began doing private matters when working from home but did not experience private virtual space violation as well. They oppose bills and food compensation but support geolocation tracking.

- Employees 31–40 years old neither considered private matters when working distantly nor experienced private virtual space violation. This group is not for bills compensation and geolocation tracking but does support food compensation.
- Respondents 18–30 years old both started doing private matters when working remotely and experienced private virtual space violation. They do not support geolocation tracking and food/bills compensation.

Regarding II.14-II.21, the sample showed to have generally clearly formulated opinions with common-sense results:

- Most respondents believe that it is not necessary to turn on the camera in meetings.
- Most believe that employees are not responsible personally for technical problems during remote work.
- Most consider smoking during online calls unacceptable.
- Most find geolocation tracking unacceptable.
- Most consider the compensation of bills when working from home is not mandatory.
- Most consider the reduction in compensation for lunch unacceptable.
- Most consider it fair to reduce travel compensation.
- Most consider it fair to award compensation for not using office space.

The distribution of opinions are presented in Fig. 2.

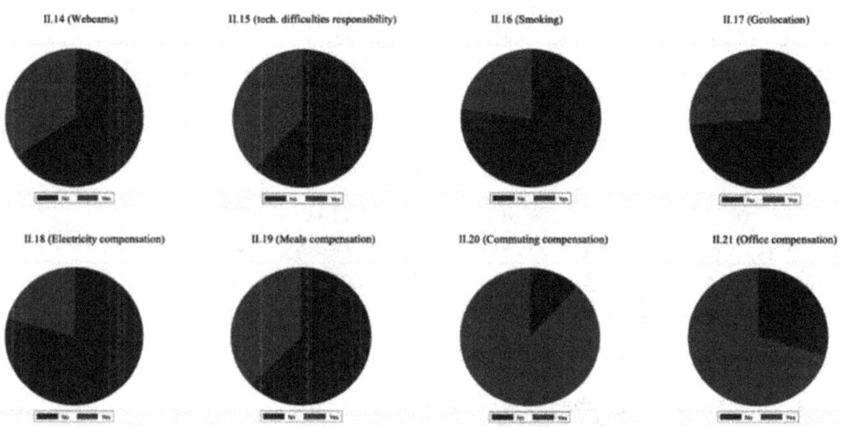

Fig. 2. Opinion distribution on new ethical issues

6 Recommendations

6.1 Questionnaire-based Recommendations

First, project managers would be recommended to address considerable ethical aspects from Table 5 in either code of conduct (corporate or project level), written or oral operational agreement. However, the key feature is not only to declare the wanted behavior but to really cultivate it and incorporate into project team daily behavior. This aim may demand special trainings (e.g., on time management) or special leadership skills from project manager (e.g., leading by example would help set standard for virtual work, since many employees may still not have clear understanding of how remote workstyle should work like).

When addressing II.7 [private matters at work], PMs are not recommended to overuse controlling methods. As the questionnaire has shown, many project team members may be quite annoyed by this. Instead, PMs should employ current tools better. For example, generally, companies track online presence of their workers via MS Teams, Skype, etc. So, PMs could analyze this data and when finding that some employees work considerable time after hours investigate it and interview these workers in friendly and willing to help manner. Probably, these workers will report that they feel more distracted at home, so the PM should prepare a set of solutions beforehand (e.g., trainings, certificates to coworking spaces, solid piece of advice). The bonus of this investigating method is that not only no additional tracking mechanisms are employed but project team members see PMs willingness to help and to mentor rather than to punish.

Another recommendation for II.7, as well as for II.3 [being overlooked], would be to engage some Agile tools as well. For example, daily stand-ups and retrospectives will: 1) keep employees in line in terms of productiveness (they will have to have some progress to report it on stand-ups); 2) let everyone speak out in conversation, give feedback, and feel involved and heard out; 3) help PMs track their employees' work without invading virtual privacy.

Regarding stand-ups or other meetings of any form, PMs should be aware that, according to the survey results of II.2 [leisure at meetings], employees are less attentive in online meetings than in offline ones. This means that PMs should make them more dynamic, informative and involving. For example, moments, when a meeting holder speaks one-to-one with an employee and others neither are involved nor find this information useful, should be avoided. Another recommendation would be not to overuse the privilege of online format and not to schedule meeting too early or too late, substantiating it with the absence of commuting.

As for addressing II.5 [communication after hours] and II.9 [work after hours], there are several moments to be discussed. Firstly, PMs should ask themselves why they are engaging their employees after hours more than usual. If the reason is a cognitive bias, including devaluation remote work in relation to office work, PMs are just recommended to stop doing so after indicating this behavior, since location does not really affect complexity of office-like tasks. If the reason is that employees do not meet deadlines and have started performing poorly, this is the case for II.7 described above. If the reason is the changed client's behavior and their demand for work after hours, PMs would be recommended to develop common practice and standards for the team beforehand (e.g., tasks of which priority must be done even after hours, whether there are any bonuses for such work etc.)

Talking about communication, project team members are highly not recommended to violate private virtual space of their colleagues (II.6), since this may, probably, lead to mental instability and higher burn-out rates. Moreover, this is bad practice in terms of IT security and company's public image (which can be damaged if some messages get resent, leaked or hacked). Instead, PMs would be recommended to use special corporate chats to regulate communication hours, employ tracking methods available and meet corporate privacy standards.

Regarding digital etiquette, it would be recommended to document expected and wanted behavior in corporate virtual space. It is needed because, on the one hand, there is no common practice or standard worldwide and people rarely this kind of education at school or university, and on the other hand, there are visible difference in attitude between people of different ages, industries etc. Also, it was discovered that employees are generally demanding towards compensations for meal and office space, so project managers are advised to think about related issues beforehand. Furthermore, project managers should be careful when engaging tools for tracking their subordinates, since not many are expected to welcome this and perceive it as invasion of privacy.

6.2 Interview-Based Recommendations

Via Zoom and MS Teams, 4 interviews from several Russian project managers were taken. Two of the project managers are engaged in IT industry, and two are of big professional experience. Interviewees were asked whether they faced the ethical challenges from the survey, how they responded to them and how they reflect on the efficacy of their response. The results of these interviews were subsequently used to deduce some generalized recommendations for distant project work best practices in ethical management.

The first project manager "K" is an employee of a big consulting firm and has project management experience of 2 years. Firstly, she recommended introducing daily short reports from each team member to address control and trust issues, arising due to remoteness. Frequent calls also worked, K reported. Secondly, each team should have an agreement on the use of private messengers (which ones, to what extend and on what occasions to use). Since K had noticed that clients had started demanding faster communication during quarantine, she recommended PMs to sometimes "defend" their teams' privacy and not to raise the service expectation bar too high.

The second manager "L" works for venture small firm and has professional project managing experience of 10 years. This interviewee recommended incorporating flexible schedules and assess not on hours worked but on deadlines met, so that there would be no need in extra controlling tools. Separating both offline and online workspaces (e.g., work a desk and rest on couch in different room) would decrease burnouts. The interviewee highlighted that it was important for project managers to consciously plan small talks since occasional conversations disappeared when switching to distant work. The second interviewee expressed an opinion that IT teams were not exposed to a bigger number of tasks after hours when working remotely comparing to pre-pandemics workstyle since they had already had borders between private and office lives blurred. L claimed that usually IT people really live for their work, so they were naturally more involved and actively took on tasks after hours whatever the workstyle was.

The third interviewee, "M", works in IT industry and has got professional experience in managing projects of a year. M recommended to invite the minimal number of staff to calls so that everyone can concentrated and heard out. Practicing "one call – one topic" rule would help keep meeting involving. M also considered that out of all these stakeholders, project managers give team communication the lowest priority, hence, communication after hours increased. So, prioritizing communication with the team would help. Discussing difference between IT and non-IT employees, M proved that in IT teams had a peculiarity that, possibly, others did not. Because developments demanded strict workflow (one task could be done only if the others were), IT employees were used to decomposing the tasks and proactively reporting their work status. Usage of such programs like Jira only helped communicating the workflow and disciplining the employees.

The last manager interviewed, "N", has professional experience of more than 20 years in project management and works in IT industry. N reported that remote style did not really change the game in IT since people in the industry had already adapted their lives to work and were more experienced. As for recommendations, he pointed out the feature of introducing operation agreements between project teams and customers on service level agreed (SLA), to fix, for example, how long it can take for the team to respond. To divide private and corporate virtual spaces N supposed separate work SIM cards to be useful, as the users could have several accounts in messengers.

Overall, project managers are recommended to be more attentive to their subordinates and not controlling when switching to remote workstyle. Managers have been indicated to become more loaded with communication from stakeholders and to experience increased demands from clients but, still, PMs have to value their subordinates' privacy and mental health. Otherwise, managers risk to face higher burnout rates and labor efficacy downfall. Moreover, to keep employees engaged project leaders are recommended to not forget to substitute lost occasional communication (e.g., in hallways or smoking spaces) with online conversations and proactively ask teams' feedback on work regime.

7 Conclusion

Ethicality is a concept highly connected with sustainability. However, even though ethics and sustainability seem to have been in trend only for a while, managers and entrepreneurs have used business ethics in their decisions for a long time. And over the years business ethics get more critical, since people become more and more economically interrelated, changing environment causes new ethical issues and society gets to be more attentive to enterprises' actions in moral dimension.

Corporate ethics is a fundamental area of knowledge since not only it supports and explains the importance of sustainable development but reinforces company's efficiency. Ethicality is a big topic in project management as well and is discussed and regulated by many associations and knowledge books. Ethics in project management teams has become even more relevant due to pandemics of Covid-19 since project team managers encountered new ethical challenges due to changed environment.

It was discovered that ethical aspects emerged the most in during distant work are in comparison to offline work include:

1. Minding private business during office hours;
2. Bigger communication demand from project managers after hours;
3. Violation of private virtual space;

4. More overtime work;
5. Being distracted more during meetings;
6. Being overlooked more.

Regarding respondents' opinion on digital etiquette and courtesy, it was outlined that project teams may have different preferences regarding their age, company size and industry they work in, though, overall, the results are common-sense and expected.

The main recommendations for project managers deducted from the research include:

1. Not to overuse virtual controlling mechanisms and position tracking tools as a way to better project team experience rather than monitor it;
2. To substitute lost occasional communication in office with planned small online conversations or informal chatrooms;
3. To make frequent yet short online calls with minimum people involved;
4. To have an agreement with team on after hours work and communication;
5. To separate private and work messengers.

Nevertheless, pandemics related issues (e.g., wearing a mask) and intercultural problems were excluded out of the scope of the research. Also, respondents and interviewees engaged in the research operate in Russia, which can make results not so relevant for project teams of other cultural descent. Hence, there is more research needed, especially in the international teams since communication in such is usually harder and trickier than in monocultural ones.

References

1. Adams, J.R., Adams, L.L.: The virtual project: managing tomorrow's team today. PM Netw. **11**(1), 37–41 (1997)
2. Bad Remote Work Habits and How to Break Them. Business News Daily. https://www.businessnewsdaily.com/10490-remote-work-habits.html. Accessed 15 Nov 2020
3. BCcampus Open Textbooks. More Telecommuting or Less? https://opentextbc.ca/businessethicsopenstax/chapter/more-telecommuting-or-less/. Accessed 01 Dec 2020
4. Code of Ethics & Professional Conduct. Project Management Institute. https://www.pmi.org/about/ethics/code. Accessed 12 Feb 2021
5. Code of Ethics for the Project Management Profession. PM Netw. **2**(2), 32 (1988). https://www.pmi.org/learning/library/code-ethics-project-management-profession-9111. Accessed 12 Feb 2021
6. Curlee, W.: Ethics and the virtual project manager. PMI. https://www.pmi.org/learning/library/ethics-virtual-project-manager-6709. Accessed 12 Feb 2021
7. Das, A.: Leadership challenges in globally distributed heterogeneous project teams. PMI. https://www.pmi.org/learning/library/leadership-challenges-globally-distributed-project-teams-7321. Accessed 12 Feb 2021
8. Huber, M.: Out of Office: Working Remotely Can Be Isolating; Here's How to Remain Professionally Engaged and Socially Connected. PM Network. https://www.pmi.org/learning/library/remaining-engaged-connected-working-remotely-11082. Accessed 12 Feb 2021
9. Individual Competence Baseline for Project Management. International Project Management Association, The Netherlands (2015)
10. Kostner, J.: Teams without walls: Third Millennium groups. PM Netw. **8**(2), 39–41 (1994). https://www.pmi.org/learning/library/teams-without-walls-third-millennium-4889. Accessed 12 Feb 2021

11. Ljungblom, M., Lennerfors, T.T.: Virtues and vices in project management ethics: an empirical investigation of project managers and project management students. Project Manag. J. **49**(36) (2018)
12. Milhauser, K.L.: Managing distributed project teams. PMI. https://www.pmi.org/learning/library/managing-distributed-project-teams-6035. Accessed 12 Feb 2021
13. Müller, R., Andersen, E.S., Kvalnes, Ø., Shao, J., Sankaran, S., Turner, J.R.: The interrelationship of governance, trust, and ethics in temporary organizations. PMI. https://www.pmi.org/learning/library/interre lationship-governance-trust-ethics-6341. Accessed 12 Feb 2021
14. O'Brochta, M.: Why project ethics matter: Leadership is built on trust. If the foundation is cracked, a project's future is in doubt. PM Network. https://www.pmi.org/learning/library/why-project-ethics-matter-9838. Accessed 12 Feb 2021
15. O'Brochta, M., Meloni, G., Raghupathy, S., Pfeiffer, P., Taylor, M.: The leader's choice: five steps to ethical decision making. PM Netw. **30**(1), 29 (2016). https://www.pmi.org/learning/library/ethical-decision-making-trend-5788. Accessed 12 Feb 2021
16. Raghupathy, S.: Ethics and moral leadership in project management. PMI. https://www.pmi.org/learning/library/ethics-moral-leadership-proj e ct-management-6185. Accessed 12 Feb 2021
17. Shouche, S.: Ethical project management. PMI. https://www.pmi.org/learning/library/ethical-dilemmas-project-management-7084. Accessed 12 Feb 2021
18. The Short- and Long-Term Ethical Issues of Working from Home. The Prindle Post. https://www.prindlepost.org/2020/08/the-short-and-long-term-ethical-issues-of-working-from-home/. Accessed 03 Dec 2020
19. Tips for working remote and some Ethical challenges of telecommuting. Rhode Island CPCU Society Chapter. https://rhodeisland.cpcusociety.org/news/tips-working-remote-and-some-ethical-challenges-telecommuting. Accessed 01 Dec 2020
20. Vergini, S.: Remote Support: Virtual Team Members Deserve the Same Attention as Ones in the Office. PM Network. https://www.pmi.org/learning/library/remote-support-virtual-team-memb ers-deserve-attention-11079. Accessed 11 Apr 2020
21. What happens when toxic office behavior moves online while working from home. CNBC. https://www.cnbc.com/2020/09/24/how-toxic-office-behavior-moves-online-while-working-from-home.html. Accessed 11 Apr 2020
22. Wong, B.J.: You are "IT": how global are your ethics? PMI. https://www.pmi.org/learning/library/global-ethics-business-values-honesty-8276. Accessed 22 Jan 2021

Plan and Evaluate Your PM Effort as Well to Promote the Digitalization of the PM Processes

Morten Fangel(✉)

Fangel Consulting Ltd, Saettedammen 4, Hilleroed, Denmark
morten@fangel.dk

Abstract. When managing projects, the focus is on how to *perform project management* activities as well as on the project execution activities. To *lead the project management* – including *planning* and *evaluation* of the effort itself – is primarily done intuitively and based on years of experience. Or by following a pre-determined standard for project management.

However, a more *conscious* and *systematic* planning and evaluation of the project management effort itself is needed for integrating sustainability concerns into the PM processes as well as other reasons, which will be presented at the congress.

The main steps of a systematic approach to *leading* the project management effort in itself are: *exploring* the project situation and characteristics as well as the project management challenges and potential initiatives; *extracting* activity plans for the project management effort itself; *evaluating* the performed project management activities step-by-step including documenting each activity in a logbook.

At the congress, we aim at in an interactive format to involve the audience during the presentation in planning the management of their current projects.

The concept and methodology have been developed and refined over the last some twenty years. The paper is based on the book [2] and the newly published version in Spanish [3].

Keywords: Planning Project Management · Project Management Complexity · Evaluating Project Management · Project Management Agreement

1 An Example of Planning PM Effort

Let us start with an example of conscious dealing with management effort in a project:

John has just been appointed project manager of a new project. He had expected his new project to be postponed until his current project was closer to completion, but unfortunately, it did not turn out that way. "It is a really exciting project", and "you are just the right man for the task," the project owner had said at the just-terminated meeting that Paul has just left.

S. Bushuyev et al. (Eds.): IPMA 2021, LNNS 704, pp. 50–60, 2023.
https://doi.org/10.1007/978-3-031-34629-3_5

Later that day at lunch, John meets Helen who participates in his current project. "Congratulations with your new job – I just saw on the web-site that you were the lucky guy! How do you intend to approach the management of the new project"? "How"? John looks speculative and replies: "I have not had time to consider it in detail".

While Helen goes on talking about her holiday plans, John wonders why she asked that question. Later, his thoughts return to the start of their common project, into which he had also rushed headlong. It really had taken a considerable amount of time before he succeeded in activating the other participants in the project. During the first months, he had done almost all the work alone – anyhow, that was how he had felt it.

Now John feels very determined. "This project should be different", he thinks. "I must do more to activate the participants – but how? Everybody is so busy". John decides to write an e-mail explaining how he suggests the new project should be approached. Then, when he has discussed the content with the project owner, he will send the e-mail to the future participants.

But what should be included in the plan for his project management? He chooses to play back his preliminary ideas to Helen: "It sounds as if you are going to have a really exciting holiday. But I wonder if you – before you leave – could spend some time with me to discuss my new project. It seems as if you have some ideas – and, in any case, I would like to have your views on my plans to get a better start of the new project. I think you know my points". While Helen considers his question, John reflects – and continues: "And when you come back from the holidays, let's have a talk about the management of my new project". Helen agrees and promises to drop by his office in the afternoon.

Paul was inspired to start on what we call "to plan and evaluate the management effort in a project" – or, in brief, to lead the project management. The leading of the project management takes place consciously or unconsciously in parallel with the process to

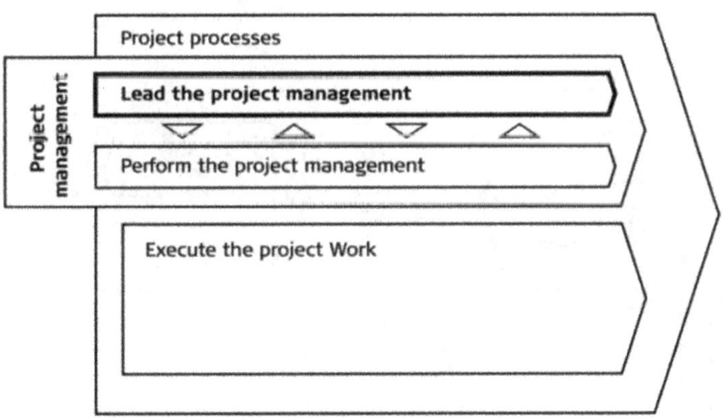

Fig. 1. In the overall project management process, we distinguish between to lead the project management and to perform the project management.

perform the project management, which includes well-known activities such as planning, arranging meetings and sparring with the participants (see Fig. 1).

This paper deals with how you can *more consciously* plan the management effort in the coming period, or in a current situation in your project – how to transfer your knowledge of the project and the situation to concrete managerial initiatives.

2 Three Levels of Guidelines for PM

Before presenting a method for planning PM, let us have a look at a three-level division of guidelines for project management that you may have in a project-based company. The levels of guidelines correspond to the levels of processes in Fig. 1: Execute the project work, perform the project management, and lead the project management.

The relations between the levels of guidelines and the levels of the project processes are illustrated in Fig. 2:

- A classic *project model* – or project execution model – provides an overview of the project processes and accumulated experience from the type of project in focus. The project model includes, for example, project break-down structure, main milestones, organization form, task distribution, and structure of documents.
- One way to use a project model is directly to perceive it as a template for implementing a project process. See Arrow 1 in Fig. 2. This could be realistic in cases where the project model is for a specific type of project – and if the current project is rather repetitive. This corresponds literally to obtaining a "ready-to-wear" project plan – and making minor adjustments to the current project.

  Another way of using the project model is to treat it as a compilation of experiences from which you can pick when planning an actual project. See Arrow 2 in the Figure. In this case, the project plan is "tailored" – but the process could be eased by picking from the model.
- Another level of guideline is a *project management model*, which provides an overview of the project *management* processes and provides generic guidelines for managing several or all types of projects in the company. Typically, the model indicates the management phases marked with main milestones, type of management activities and management roles.

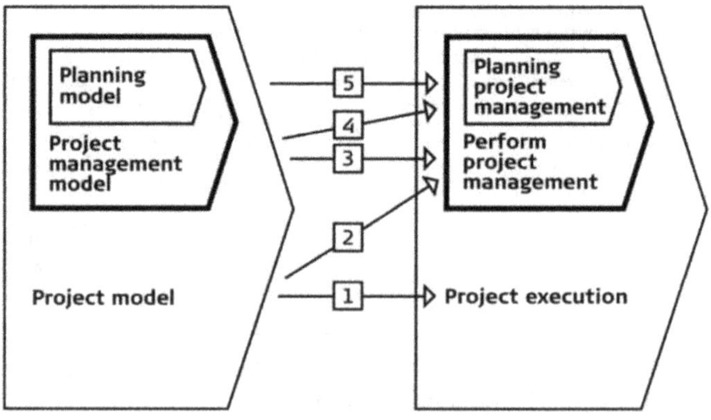

Fig. 2. Coherence between the levels of the models and levels in the project process.

One way of applying a project management model is to perceive it as a template for implementing the project management process. See Arrow 3 in the Figure. Another way is to perceive this model to be a frame of reference and source of inspiration for the planning and evaluation of project management. See Arrow 4.

- A third level consists of a *planning model* with steps and other guidelines for planning and evaluating the management itself. This model is supposed to be so generic that generally it can be used directly as a template. See Arrow 5. This third level of guidelines is the topic of the present article.

In actual practice, you do not have the choice only of one type of guideline or another. The company's project guidelines will typically be a combination. However, the argument of this paper is that the focus of guidelines should be shifted from project models to using PM models – and towards models or methods for planning and evaluating the management.

3 The Need for Planning and Evaluating PM Effort

The core of being a professional project manager is that you know instinctively the amount and type of management effort to be used in a specific project situation. Add to this that we, by and large, always work under time-pressure in projects – so why also spend time for planning the management?

In the area of competence development, we distinguish between four levels: unconscious incompetence; conscious incompetence; conscious competence; and unconscious competence. It might be felt as a constraint – having to fit your own intuition into a "method". However, the best advice is to try to consciously plan project management – not as an alternative, but as a support for the development of your own intuition and solid judgement.

There are many reasons for focusing more consciously on planning and evaluating the management effort in projects:

- It is a tendency of the *natural law* in dealing with projects that we slide into "content mode" – even when it is vital to be in "management mode".
- It serves to promote utilization of the organization's *guidelines for project management* by raising the questions about which parts should be applied in the particular project/situation.
- Limited time and resources for project management make it important to *focus* the management on those efforts that create the most managerial value for the project.
- Increasing *expectations* – both from external and internal parties – regarding the actual management of the project, point at the need for verification of the actual management effort.
- Description of the requested management effort makes it easier to *involve* both project owner and project participants in the management activities.
- Demand is increasing for *ongoing learning* and improvement – also concerning management of projects.

Which of these reasons is especially relevant for you – in your project and your organization?

Planning the project management effort should typically take place at the critical points of the project – such as, for project preparation, project start-up, project evaluation, and project close-down. Of course, it is also relevant during the management of project execution to make a conscious planning and evaluation of the management effort.

4 Method for Planning PM Effort

The central element in planning the project management effort is to consciously realize what the managerial problems or challenges will be during the period in question. In other words: Where could managerial value be created?

In the example from the introduction, John might have a clear option regarding this challenge – to activate the project participants earlier than last time. But who can say that John has faced all the challenges? Let's hope that Helen will help him consider any other managerial challenges when they meet.

When the challenges are clarified, a simple method is to "brainstorm" on possible management initiatives – among which might be to visualize the common objectives for the project and to arrange a project start-up meeting for further clarification.

As a method for the overall planning and evaluation of the management effort in a project, we recommend the 10 steps in Fig. 3. The basic concept is first to clarify the future management effort and to summarize the results in an overall plan for the project management. Based on this, the future management effort is specified as an activity plan for the project management – and perhaps agendas for management meetings. Later, a stepwise evaluation of the performed project management is carried out.

In the example with John and Helen, STEPs 1, 2, 3 and 4 of the method can be utilized as a simple agenda for their discussion. This is not to say that they should accomplish the items step-by-step without reflecting. The steps should rather function as a framework for their discussion, and as an inspiration to allow consideration to move from the first steps toward the last ones – and to a lesser degree vice-versa.

STEP 1: The situation is that Paul has just been appointed project manager. Other things might be added, but let us stick to this information to simplify the example. The situation is that Paul has just been appointed project manager. Other things might be added, but let us stick to this information to simplify the example.
STEP 2: Paul's feeling is that the project is more complex than the one he is just about to finish.
STEP 3: His first reaction to the challenge is that "I must immediately study the new project--
STEP 4: -- by examining the available project description together with the project owner in a separate meeting room".

Based on the inspiring talk at the meeting with Helen, it is recommended that John prepare an activity plan as a diagram covering the STEPs 5, 6 and 7. STEP 5 with activity initiatives will be used to examine the available project description, STEP 6 with the role initiatives indicates that this examination should be made together with the project owner – and STEP 7 with method initiatives suggests that the meeting be organized in a separate room.

Clarify upcoming project management	• STEP 1: Facts of the project and its situation	Where
	• STEP 2: Characteristics of its management complexity	Why
	• STEP 3: Challenges for the project management	What
	• STEP 4: Initiatives for handling the challenges	How
Specify upcoming management activities	• STEP 5: Content and timing of the activities	When
	• STEP 6: Role division of the activities	Who
	• STEP 7: Methods/tools for the activities	How
Evaluate performed project management	• STEP 8: Register performed activities	When
	• STEP 9: Evaluate the specific activities	What
	• STEP 10: Evaluate the entire management effort	How

Fig. 3. Method for planning and evaluating PM effort

The method also implies that John – for example, in the same document as the activity plan – as STEPs 8 and 9 should register and evaluate the actual management activities. And, after the first phase of execution, he should arrange as STEP 10 an evaluation and sparring meeting with Helen.

The advantage of using the 10-step method in Fig. 3 is that it helps you keep focus on planning and evaluating the *management* – and, consequently, prevents distraction into considerations that deal with solving project management tasks – yes, even to entering into project execution.

5 Instruments for Focusing on PM Planning

In practice, three types of instruments have been proven to promote the conscious and methodical leadership of the project management:

- Tools and related documents

Four types of documents can promote both planning and evaluation of the management effort:

1. Initial exploration is documented under the title 'Clarifying up-coming project management'.
2. The clarifying is concluded in an 'Overall plan for project management' to promote focus on the most important aspects. This document can be arranged according to the first four steps of the method.
3. 'Activity plan for the project management' converts the overall plan into detailed plans, including deadlines, distribution of tasks and method proposals.
4. In the 'Logbook for project management', the performed management activities are documented and evaluated according to STEPs 7, 8 and 9.

- Dedicated planning meetings

Some planning meetings should be dedicated to evaluation of the ongoing project management and to planning of the project management for the next period. STEPs 1, 2, 3 and 4 make an appropriate agenda for such a meeting.

- Utilize sparring partner

1. A perfect solution is for the project owner/sponsor to act as a sparring partner for the project manager concerning management of the project.
2. Another possibility is to use colleagues for sparring, either ad hoc (as it was in the case with Helen and John) or in a more formalised network.
3. A third possibility is to hire an internal or external consultant.

Typically, a planning meeting is held at the beginning of those phases that imply important management effort – such as the project preparation, project start-up, project evaluations, and project close-down. Natural participants are the project owner/sponsor, the project manager, and other key persons.

The meeting between Helen and John could be considered as a planning meeting.

6 Describing the Steps of the Method

The following sections specify each step of this conscious planning and evaluation of the project management. The plan should be elaborated for a specific period of the project – because, if the plan is extended to cover the whole project, the contents might be too general and will not promote a suitable management effort.

6.1 Clarifying Up-Coming Project Management

Planning the project management should be initiated by *capturing information* through a brainstorm-like process. This utilizes the participants' intuition and experiences and gives them an opportunity to unfold their creativity. The following scheme can be used to prepare PM planning for a coming period – typically one to two months.
STEP 1: Facts of the project and its situation

- Project definition and stage – in brief.
- Explain the project aim, process, and/or organisation.
- Could be extended by a review of the logbook for project management from the previous period (see STEPs 8 and 9).

STEP 2: Characteristics of its management complexity

- Describe conditions which are determining the management effort in the project. Examples are environment, cross organisation, etc.

STEP 3: Challenges of project management

- The managerial challenges or problems.
- How can the project manager contribute to the project to achieve, for example, a common view or a more positive climate?

STEP 4: Special initiatives for "coping with challenges"

- Management initiatives which make it easier to cope with challenges.

These could be proposals for management activities, such as analyses, planning, communication and follow-up, or proposals for management roles and methods and types of meetings.

All with the focus on what is to be done in addition to what would be done by routine – that is why the title of the step is *special* initiatives.

The first four steps aim at being used as agenda for a planning meeting between the project manager and one or a few others.

6.2 Overall Plan for Project Management

An overall plan for the project management can be established after the creative clarification – and can be structured according to the method's STEPs 1, 2, 3, and 4. The task is to extract the essence of the clarification steps – not to make a complete documentation of the clarification that might have been made earlier in a separate "interim paper".

The plan can also be considered and even named a Project Management Agreement between the project owner and the project manager.

As to the overall plan for project management or PM agreement, I recommend the following:

- Restrict the scope definition to only one page – it is difficult to stick to the plan's intentions during a stressed normal day.
- List the contents as items – typically 3 - 5 items per step in the method.

- Focus on the most meaningful and significant management efforts – and refrain from describing that part of management that takes place irrespective of being listed in the plan.

Please note that such an *overall plan for project management*, both in structure and contents, is similar to an invitation to a project start-up workshop – for example:

1. *Facts of the project and its situation.* A brief introduction to the project and a description of the direct cause for having the workshop now.
2. *Characteristics of its management complexity.* Conditions in and about the project indicating that a workshop is a relevant initiative.
3. *Challenges of project management.* The aims of the workshop and the expected effect.
4. *Special initiatives for coping with challenges.* The agenda of the workshop, including type of analysis, planning and co-operation development. Furthermore, it should be stated who might prepare presentations and whether items in the agenda are purely informative or require a broad involvement. Also, information about facilities and tools for the entire workshop should be given.

A refinement of the planning method includes supplying a checklist for each of the steps. Experience shows that such tools are most helpful for structuring and checking the results from a more open debate. Further, the checklists are useful for editing results from the exploration.

If the checklists are used as a starting point, the output may be too extensive to apply in practical overall plans for the project management.

6.3 Activity Plan for Project Management

The aim of the overall plan for project management is to create an overview and to hold on to important matters concerning interaction in the management effort during the period in question.

Further, a more operational activity plan for the management effort is necessary. We here recommend issuing a separate plan for the management activities. A simple format is just to list activity terms, periods, task distribution, and agreed methods with practical hinds. In small and less complex projects it can be combined with the detailed project plan for the execution.

STEP 5: Content and timing of the activities

- Transform specific management activities from the overall plan into a chronologically arranged list and state the timing of each activity.

STEP 6: Role division of the activities

- Who is responsible for, or who contributes to, these management activities?
- What is the role of the project manager in the specific activity?
- Should the project owners and/or the project participants be involved in the planning?
- Which resources could be implied?

STEP 7: Methods/tools for the activities

- How could the considered management activities be conducted in practice?
- Which methods/tools, examples or practical hints?

6.4 Logbook for Project Management

To promote a conscious follow-up and evaluation of the project management effort, it is valuable as STEPs 8 and 9 to generate a logbook for management of the project – not to be mixed up with a logbook for the entire project.

The aim is:

- to become more conscious regarding the project management roles,
- to plan additional management initiatives, and
- to achieve more learning about project management.

The best way to achieve this aim is to keep the logbook continuously – otherwise you will soon forget the things "that kept you awake" and the considerations that crossed your mind at a particular time.

The logbook could include columns for:

- *Activities of project management,* including meetings and other events.
- Documentation of performed *management roles by activity* – and perhaps the amount of hours spent per activity/event.
- Evaluation of performed management – positive as well as negative.

Practically, the logbook can be combined with the activity plan by stepwise adjusting the activity list to what were performed – and adding a column for evaluation remarks.

7 Implementing Planning and Evaluating of Project Management

As project management consultants, and when acting as coach for project managers, we have applied the described methods in our own projects. This has, on the one hand, indicated how systematic planning and evaluating can promote proactivity of the project management effort.

On the other hand, our applications confirm how easily you can forget all about planning and evaluating the management effort, including your own performance – and instead concentrate on performing the project management activities and events, or even project execution activities.

To promote the planning and evaluation of project management, we recommend that this topic should be included in the guidelines for project management in your organization. Yes – even to re-write your guidelines in order to ensure more focus on demands for this issue and less demand for utilizing specific models and methods.

Initially, project managers may consider this to be more demanding, but the actual essence is that it will ensure more freedom to select what they consider to be the best way of performing project management in their projects.

Latest, the issue of planning and evaluation the PM effort is also included in the new ICB4, IPMA Competence Baseline issued by the International Project Management Association in the competence element Design and in the element Plan and control.

Further, such a focus on the upper, level three in a project model (see Fig. 2) has even been part of the Scandinavian Competence Baseline, NCB3 [1] – based on the IPMA Competence Baseline, ICB3. In this NCB3, we did include planning and evaluation of project management as a major component of project management competences.

The main argument for adding this element to the competence standard is that the capability to reflect on your project management, and to be creative in selecting your project management approach, is one of the key competencies for integrating sustainability concerns into the PM processes.

References

1. Fangel, M., et al.: Competencies in Project Management. Danish Project Management Association, Danmark (2010)
2. Fangel, M.: Proactive Project Management - How to make common sence common practice. Fangel Consulting Ltd. (2013)
3. Fangel, M.: Dirección Proactiva de Proyectos - Coómo hacer del sentido común una práctica común. ETRADETER S.C., Mexico (2019)

Managing Hybrid Teams of Non-commercial Projects

Evgeniy Suslov[1(✉)] and Evgenia Gorn[2]

[1] RANEPA, Sredny prospect VO, 57/43, 199178 Saint-Petersburg, Russia
susloveу@gmail.com
[2] MISIS, Leninskiy prospect, 4, Moscow, Russia
gorn.ea@misis.ru

Abstract. The article summarizes our year experience of analysis, supervision and data collection of hybrid teams functioning in non-commercial projects. The project under consideration included around 50 members and three subdivisions lead by project manager of different managing styles. The presented article states three main problems which all teams faced and a variety of sideway problems. The main three problems can be described as follows, motivational, psychological (e.g. burn-outs), managerial (prioritizing, time-management etc.).

There is an attempt to analyze the reasons of the stated problems, cases of dealing with them by different leaders. In terms of the research, we were also trying to identify correlation between types of leaders and the tools they use to succeed in hybrid team management. Their practices and their implementation outcomes were collected with care, summarized, and described hereby. The article is structured more as a case study and warns that the solutions presented worked for the project and may suit others alike, but they cannot be considered as unique practices and are supposed to be applied with care.

Keywords: Hybrid teams · Non-commercial projects · Team-management · Virtual teamwork

1 Introduction

The recent pandemic, the world has been facing, enhanced the borders of team-management mostly by overcoming the real borders limited by a team member location. Many projects see both the temptation and challenge to invite people to participate online on the regular basis. And that is even more challenging for social project or non-commercial ones. Characterizing the latter, it is worth pointing out that they unite people in the aim of a socially important goal which directs them to creating a socially valuable outcomes used by everybody for free. Such projects do not see any commercial reproduction or marketed purposes like promotion, selling etc. Team building and team management in such projects is not particularly easy both offline or online whereas managing hybrid teams might bring unexpected obstacles which become typical for managing online and offline teams. As Cousins and Robey wrote "In contrast to co-located teams and purely virtual teams, these 'hybrid' teams com-bine face-to-face communication with computer-mediated interaction" [3]. Thus, managing hybrid teams includes issues typical for online and offline teams.

S. Bushuyev et al. (Eds.): IPMA 2021, LNNS 704, pp. 61–67, 2023.
https://doi.org/10.1007/978-3-031-34629-3_6

This paper is structures as follows: first, we present the literature on hybrid and virtual teams, and on problems they might face. Then, we present the methods and project description followed by our finding and discussion.

2 Literature Review

The literature on remote working, virtual and hybrid teams discuss a wide range of issues such as boundaries and control [5, 7], isolation and independence [12], cross-national interaction [2, 13]. Though few studies focus on human factors, such as the way team members cooperate, the way people start suffering from burnouts. This study applies a 'mixed method' approach to research hybrid teams. This study is mostly concentrated on the ability of technologies to control information inside the team which must communicate across large distances or diverse locations [4, 25]. Some researchers emphasized the importance of self-efficacy [21, 22] and knowledge sharing behavior [18] in teamwork as the basis for successful team functioning. However, those studies focused on conventional teams where members work in the same location. Our article goes one step further and addresses a gap by investigating key human factors that influence hybrid team member interaction in non-commercial projects. Using a concurrent 'mixed-methods' approach, our research examines the main human factors affecting teamwork in non-commercial projects. To do this, we first summarized the literature to develop our research hypotheses. We then detail our research methods and discuss the findings. A theoretical framework is then provided and practical recommendations for managing hybrid teamwork are presented.

Our definition of a hybrid team posits that, unlike conventional teams where a group of people work together within a single workplace [11, 17, 20], or unlike the virtual team cooperates across space and time by using information and communication technologies or ICT [1, 9, 13, 14] hybrid teams include the characteristics of both mentioned above. Hybrid teams are teams composed of local and remote employees when working on a non-commercial project, they have had very different work and life experience. At the heart of managing a hybrid team is the awareness that the locals and remotes can invest in the project only time spare from work and, or studies. Locals have an opportunity to meet, have coffee conversations and private talks, can become friends. Remotes do not meet offline, absorbed by their daily routine, can not participate in offline team building meetups.

Communication within a hybrid team plays an important role in influencing team effectiveness because it is a process for sharing personal information to achieve shared mutual knowledge among team members [15]. However, communication between virtual team members is not easy because computer-mediated environment cannot reach 'co-presence of communication' that group members can have with physical contacts with one another at the same location [6]. The global context within which virtual teams are commonly found suggests that cross-cultural and cross-national differences in communication behavior in a virtual team should also be investigated [8]. Although Lee-Kelly and Sankey [12] investigated the impacts of cultural difference and time zone on communication and team relations, there is still a lack of discussion of the influence of communication on virtual teamwork in a cross-national environment. Some researchers have investigated communication media technologies used in virtual teams [14]. However, it appears from the literature review that much less research has been conducted on the way communication between teammates can affect virtual team interaction and member's satisfaction.

Thus, our Hypothesis 1: Hybrid teams in non-commercial projects are difficult to control and motivate and our Hypothesis 2: Communication within a hybrid team plays an important role in team performance.

3 Research Methods

The study utilizes a case study approach which included interviews with volunteers and project managers, briefings observation notes (ON), performance meeting ON and strategic sessions ON together with project reports and messenger's interaction analysis. Relationships among participants were carefully observed even during cor-porate parties and out-of-work activities. Project participants were carefully inter-viewed and these notes on interviews were also analyzed.

The data collection process took nine months and around 120 people were interviewed, the questions mainly focused on 1) the reasons people volunteer 2) why they chose this exact project 3) what problems they faced and what possible solutions they see or can suggest.

4 Results and Discussion

The project under analysis and observation is a youth project directed at making anthropological sciences such as philosophy, social sciences, literature etc. acceptable for everyone. It provides free lectures and workshops free both online and offline.

Within the observation process the project has suffered dramatic changes, such as.

1. Context consumers grew 8 times (from 500 to 4000) which led to the participants growth (from 7 to 55) and subsequently lead to a team managerial crisis.
2. Sponsorship decline which made the project to focus on earning money and provide some paid workshops which needed being managed as well.
3. Answering the demand of the team members who wanted to have scientific, and research stand the project started a scientific laboratory.

The second and the third activities inaugurated a growth of the project into the ecosystem where members-volunteers and shift inside the system obtaining new challenges.

4.1 The Project and Team Description

People involved in the project are aged 18–35 and 79% are female and are 21% male. Most members (37%) are aged 21–24 and around 30% are aged 18 до 21. The location distribution of the project members is also quite wide. Most of them (95%) are from Russia while participants from other countries represent about 1% and include France,

Belarus, Ukraine, Sweden. In terms of Russia, it is worth mentioning that 43% of people are in Moscow, 43% - in Saint-Petersburg, 7% - in Novosibirsk and 4,5% - In Ulyanovsk.

After interviewing project members, we grouped all the problems in three big groups. We actually understand that all of them are usually the reasons and consequences of one another, but they are represented as they were prioritized by the interviewees.

The main three problems can be described as follows, motivational, psychological, managerial.

1. Motivational

   As far as the project is non-commercial motivational problem is top 1 issue in terms of attracting volunteers. In general, they spend about 2–4 h every day to be involved in the project. As we can see from the study group description that most of the members are students with little or no experience in their professional field. Thus, are motivated by

   a. Getting professional experience which is impossible to gain during studies.
   b. Charity. It was quite surprising, but many volunteers consider this project participation as charity especially those who had their degree or professional skills.
   c. Belonging (Maslow's Pyramid). Some participants highlighted that the only reason they wanted to be in the project is literary the wish "to belong to something bigger". It worth mentioning that at least half of these people were not able to provide information about the project goals or mission, or even did not know what the project was doing and who is the project manager. These people were the first to leave the project if something bigger appeared.
   d. Personal advertisement in social media. These people entered the project to get more followers on their profiles on social networking sites and meet new people to build useful connections.
   e. Project mission and values charmed the less part of the participants and. What is more, the close the participant was to the project manager the more they shared the project values.
2. Psychological

   Psychological issues every participant faced are also described and systemized hereby.

   a. Burnouts. Most people suffered from burnouts while the projects take 2–4 h every day. They include meetings and discussions and some technical work itself, most of the people are students and graduates who must spend a lot of time on building their career or studies.
   b. Time difference. The most western location of the project member was Paris (France)/ the most eastern point was Novosibirsk (Russia), time difference is 6 h. All events were held online and offline Moscow time, where time difference is 2 and 6 h accordingly. This time difference might cause many psychological outcomes such as sleep distortion, stress, communication breaks and even depression.
   c. Hybrid communication. Many interviewees noticed that the mixture of online and offline communication bring additional difficulties to the whole process of

communication. The connection between those who practiced offline communication grew stringer though others who could not participate in offline meetups felt isolated, lonely and less important.

d. Social isolation. Many project members felt socially isolated. That might be connected to the whole pandemic period.

3. Managerial

According to the interviewees there are aspects which need managing. They are:

a. Information flow. The project has got too many chats on different platforms which are not being moderated. Information flow is uncontrollable and overwhelming.
b. Information selections. Team leads had problems with understanding what information must be distributed and what information is just for internal use.
c. Team members. Since the project is fully voluntary team members came and left without any notification to the team.
d. Online and offline leaders. Leaders online failed communication in 30% cases in comparison to offline leaders. /1736/

Within three months, all project managers implemented their solutions to improve member performance. We analyzed those solutions in terms of their effectives in non-commercial projects.

1. Consolidated source of information showed 30% team performance improvement and 17% better awareness of the project activities and initiatives. It was organized as a private community at one of the social networking sites which can be joined by invitation. This community site has got all the links for newcomers to get more information about the project. It has got a set of guides for fulfilling all the possible requests on members performance outcomes.
2. Communication rhythm. This practice included regular meetings at least once in two weeks. All the meeting take place the same day and the same time. Regular meeting increased mutual trust in about 5% and increased personal and group performance in 3.8%. One of the project managers also included one-to-one interactions which showed 1.2% increase in group performance.
3. Theme meetings. Though hybrid teams need both online and offline communication management. One of the options which showed a high efficiency was theme meetings. Team members met online to watch and discuss a movie together or the content of lectures or seminars. After these meetings, many friendly chats appeared what helped a lot with adaptation process and added fun to communication. One of managers also included personal updates and small talk at the beginning of every meeting which showed 4.8% conflict reduction.
4. Unification. Whenever templates were introduced team members managed to spend less time on knowledge sharing and tutoring each other.
5. Verbalizing expectation. Entering and leaving interviews which were introduced. At the first interview people could express their expectations and give several suggestions on the performance improvement. What was even better that manager included so called ‘exit’ interviews where people explained the reasons for leaving and improve the whole system based on it.

5 Concluding Remarks

Considering the stated problem, it is worth mentioning that managing teams in non-commercial projects is not easy. Firstly, non-commercial project lacks the man-ager the opportunity to push and control members of the project by increasing or reducing bonuses. Non-commercial projects make managers improve their emotional intelligence. They have to motivate, deal with psychological problems of project members and perform the best as the leaders proving the leadership every minute.

Secondly, managing hybrid teams managers are supposed to be ready for being flexible in terms of communication, online and offline, articulate the tasks, smart the goals and implement digital tools.

Finally, a manager of hybrid teams should create and support trustful atmosphere, open communications style and helpful environment.

To help facilitate this, it's important to build in a good communication rhythm with frequent check-ins, longer deep-dive meetings, and an expectation that all members have brief one-on-ones with each other on a weekly basis.

References

1. Arnison, L., Miller, P.: Virtual teams: a virtue for the conventional team. J. Work. Learn. **14**, 166–173 (2002)
2. Chhay, R.V., Kleiner, B.H.: Effective communication in virtual teams. Industr. Manage. **55**, 27–30 (2013)
3. Cousins, C.K., Robey, D., Zigurs, I.: Managing strategic contradictions in hybrid teams. DBLP. Eur. J. Inf. Syst. Mag. **16**(4), 460–478 (2007)
4. Cummings, J.N.: Economic and business dimensions geography is alive and well in virtual teams. Commun. ACM **54**, 24–26 (2011)
5. Dimitrova, D.: Controlling teleworkers: supervision and flexibility revisited. N. Technol. Work. Employ. **18**(3), 181–195 (2003)
6. Driskell, J.E., Radtke, P.H., Salas, E.: Virtual teams: effects of technological mediation on team performance. Group Dyn. Theory Res. Pract. **7**, 297–323 (2003)
7. Goden, T.D.: Altering the effects of work and family conflict on exhaustion: telework during traditional and nontraditional work hours. J. Bus. Psychol. **27**(3), 255–269 (2012)
8. Jarvenpaa, S.L., Leidner, D.E.: Communication and trust in global virtual teams. Organ. Sci. **10**, 791–815 (1999)
9. Johnson, S.D., Suriya, C., Won Yoon, S., Berrett, J.V., La Fleur, J.: Team development and group processes of virtual learning teams. Comput. Educ. **39**, 379–393 (2002)
10. Kelliher, C., Anderson, D.: For better or for worse? An analysis of how flexible working practices influence employees' perceptions of job quality. Int. J. Hum. Resour. Manag. **19**, 419–431 (2008)
11. Larson, C.E., La Fasto, F.M.J.: Teamwork: What Must Go Right/What Can Go Wrong. Newbury Park, CA (1989)
12. Lee-Kelley, L., Sankey, T.: Global virtual teams for value creation and project success: a case study. Int. J. Proj. Manage. **26**, 175–192 (2008)
13. Lilian, S.C.: Virtual teams: opportunities and challenges for e-leaders. Procedia Soc. Behav. Sci. **110**, 1251–1261 (2014)
14. Lipnack, J., Stamps, J.: Virtual Teams: People Working Across Boundaries with Technology. Wiley, New York (2000)

15. Liu, Y., Yan, X., Sun, Y.A: Meta-analysis of virtual team communication based on IPO model. In: The 2010 International Conference on E-Business Intelligence, ICEBI2010 Organization, Kunming, China, pp. 177–186 (2010)
16. Mazzi, A.: Alternative office structures for telecommuters. In: Proceedings of Telecommuting 1996 Conference (1996)
17. Quick, T.L.: Successful Team Building. AMACOM, a Division of American Management Association, New York (1992)
18. Rice, J.L., Rice, B.S.: The applicability of the SECI model to multi-organizational behaviors: an integrative review. Int. J. Organ. Behav. **9**, 671–682 (2005)
19. Roberts, J.: From know-how to show-how? Questioning the role of information and communication technologies in knowledge transfer. Technol. Anal. Strategic Manage. **12**, 429–443 (2000)
20. Shonk, J.H.: Working in Teams: A Practical Manual for Improving Work Groups. Amacom Books, New York (1982)
21. Stone, R.W., Bailey, J.J.: Team conflict self-efficacy and outcome expectancy of business students. J. Educ. Bus. **82**, 258–266 (2007)
22. Van Mierlo, H., Rutte, C.G., Vermunt, J.K., Kompier, M.A.J., Doorewaard, J.A.M.C.: Individual autonomy in work-teams: the role of team autonomy, self-efficacy, and social support. Eur. J. Work Organ. Psychol. **15**, 281–299 (2006)
23. Purnik, A.A.: Managing non-commercial projects. Moscow, p. 6/17 (2005)
24. Andreev, S.N.: Marketing of non-commercial projects. Moscow, p. 202/320 (2010)
25. Evans, J.R., Berman, B.: Economics of project management. Moscow, p. 192/541 (2019)
26. Orlova, T.E.: Dynamics of virtual teams' development. Saint-Petersburg, p. 22/22 (2006)

Competences for the Management of the Digital Transformation

Carsten Wolff(✉) and Olha Mikhieieva

Fachhochschule Dortmund, Sonnenstr. 96, 44135 Dortmund, Germany
Carsten.wolff@fh-dortmund.de

Abstract. The work environment of many people is developing according to two major trends: the digital transformation and the project-based organisational patterns. Both trends are highly interdependent since projects are a relevant tool for managing the digital transformation and – in addition – projects and project management are digitally transformed. COVID-19 has boosted the shift to virtual collaboration which is a major driving force for the two trends. The competences required for project work and project management are changing therefore, and – consequently – the project management education at universities has to adapt to this change. In this contribution, we analyse which competences have changed, in particular those which affect communication, (virtual) collaboration and the attitude towards change. Based on this we have experimented with project-related virtual learning situations on the level of Master's programmes. The effect of the learning setups is evaluated by students and lecturers and we conclude certain recommendation from this analysis. The main goal of our contribution is to provide ideas and a guideline on how to transform educational programmes in order to deliver competences related to the management of the digital transformation, in particular for virtual project collaboration.

Keywords: Competences · Digital Transformation · EuroMPM · Project Management Education

1 Introduction

Projects are a very relevant tool for the management of the digital transformation. The future isn't only digital – it is also projectized. For the management of the digital transformation with (or by) projects new competences and – as a consequence - new ways of project management education are required. The rise of agile methods and tools and the agile philosophy was perceived as the main answer to the demand for new way of project management. Meanwhile, it becomes clear that the topic is more complex. The digital transformation changes the way we cooperate and communicate. Digital tools are used in project management and virtual cooperation gains relevance.

But this is just one aspect of the digital transformation of project management. Another aspect is the management of the digital transformation with projects. Digital transformation projects are a specific type of projects with specific requirements for management. This goes beyond classical IT projects. Agile was proposed as a way of

S. Bushuyev et al. (Eds.): IPMA 2021, LNNS 704, pp. 68–82, 2023.
https://doi.org/10.1007/978-3-031-34629-3_7

managing (small) software development projects in the first place. It's growing into other domains but at the same time it is combined with (traditional) methods to form new, hybrid approaches. This trend is continuing and the project management methodology evolves with a tremendous pace - challenging project management education and requiring a new view on competences, competence profiles and competence models [14]. Digital competence profile descriptions and data driven approaches for the definition of new competence profiles come up [13]. Digital education methods and tools are a fast-growing segment.

At the same time, new challenges arise. For example, the agile approach has won significant popularity and recognition for its efficiency, offering to achieve better feedback and foster innovation through personal communication (daily stand-ups), but the recent COVID-19 challenges requested everyone to shift to a remote way of working. Which competences are necessary to address such a sudden shift in managing projects and teams remotely and virtually? On the one hand, one needs to deal with cyber security (remote access), data transfer (storing big files in the cloud), video conferencing (web solutions for a big audience without user accounts) and other technologies to deliver results remotely. On the other hand, such soft skills as resilience, emotional intelligence, flexibility, self-organisation, tolerance to ambiguity, readiness to learn, etc. are required more than ever before. Teams are built ad-hoc, for shorter terms and to work mainly remotely and only partially in a project. Team members have to be self-organised. Competence models, competence development and education have to reflect this complexity, the pace of change and the much more individual career paths.

The paper will wrap up the current status regarding competence for the digital transformation, competence management and education. Based on the example of virtual collaboration in self-organizing ad hoc teams, it will elaborate how project managers can develop the competences and how education (especially university education) can be adapted to deliver the competences. The paper will serve as a guideline to develop such educational programmes.

2 Literature Review

This chapter offers an overview of the scientific literature addressing competences for dealing with the sudden shift in managing the digital transformation projects and teams caused by COVID-19 challenges. Furthermore, it elaborates the state-of-the-art in project management education – in particular in higher education – and in the didactic methods for project-based learning (PjBL), problem-based learning (PBL) and digital education.

2.1 Digital Transformation Scope and Challenges

In this subchapter, we define aspects of the digital transformation that can potentially contribute to definitions of the competences needed for successful management of the digital transformation (see Fig. 1).

According to the results of the systematic literature review by Morakanyane [6], digital transformation is "an evolutionary process that leverages digital capabilities and technologies to enable business models, operational processes and customer experiences

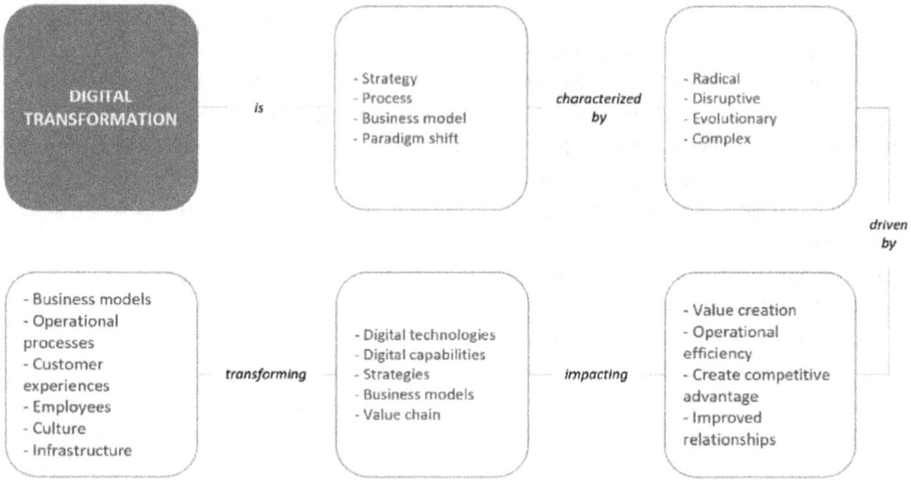

Fig. 1. Digital Transformation Definition Structure [based on 6]

to create value" [6]. The "management of the digital transformation" deals with the planning and execution of the transformation process, leading essentially to change and transformation projects with the digital transformation as the main content. Verhoef et al. [11] state that digital transformation is multidisciplinary by nature, due to the fact that it encompasses changes in strategy, organisation, information technology, supply chains and marketing.

Managing the digital transformation requires an ability to understand some or many technologies, which are applied to execute the digital transformation, e.g. Robotic Process Automation, Internet of Things, artificial intelligence, machine learning, big data, cloud computing, SaaS, IaaS, PaaS, User Interface, etc. [5]. Apart from technical competence, professional and global competence are needed [8] in order to successfully perform the digital transformation. This involves specifically interdisciplinary, international and intercultural competences (3 × "i"). Communication and collaboration in teams are key competences, too, which requires soft skills like resilience and emotional intelligence. The pace and novelty of the transformation journey requires from project managers, team members and also from customers and stakeholders flexibility, self-organisation, tolerance to ambiguity, and the readiness to learn. Verhoef et al. offer their multidisciplinary reflection on the digital transformation and include the insights from such disciplines as information systems, strategic management, marketing, innovation, operations management [11]. Moreover, the research questions for the future research agenda are elaborated such as 'How to construct self-organising teams to attain digital transformation?' [11].

2.2 Competences, Competence Development and Organisational Change

Gollhardt et al. include 'digital transformation management with digital skills' in their explorative digital transformation maturity model as a part of the focus area 'Governance', however, do not elaborate or explain those [2]. Digital competences as 'important criterion in recruiting' are given as one of the metrics for the digital transformation maturity by Berghaus and Back [1]. Also, the importance of innovation and ability to collaborate with IT departments is highlighted [1]. Ivancic et al. point out that "a digital mindset and digital skills have the potential of being an essential mediating capability in determining the success of digital transformation endeavours" [3]. The researchers included direct questions about needed skills for digitization projects in their semi-structured interviews, but did not include any precise results. Still, we see the importance of the mindset that itself belongs to soft skills. Schumacher et al. elaborate maturity items of the Industry 4.0 Maturity Model ICT with the means of the dimension 'People' among other.

Competences of employees, openness of employees to new technology, and autonomy of employees are mentioned here [9]. The typical barriers for an enterprise to unlock the power of digital transformation include complex business processes, highly customized solutions, stringent policies and practices, business units working in silos within the enterprise, lack of proper infrastructure, the gap in skill sets, and multiple and disparate architectures in the ecosystem [5]. In essence, we can conclude that the required competences for managing the digital transformation are not fully known and form a research topic in itself. This is understandable since the digital transformation is just starting. Training and education need to anticipate a continuous change and adaption of the required learning outcomes of educational programmes. Competence models and frameworks and maturity models are required in order to organize the competence development and the emergence and adaptation of competence profiles [14]. Adaptability, innovativeness and agility are therefore quality which educators cannot just teach to students but to have develop for themselves, too.

2.3 Project- and Problem-Based Learning, Work-Integrated Learning

Projects are a didactic tool for learning. This goes beyond learning project management or developing project-related competences. In an educational programme, projects (and typical project situations) can be used and conducted as a didactic format in order to deliver project-related competences [15].

The design of educational programmes – e.g. an academic Master's programme – develop different competence areas in a sequence since they are building on each other (see Fig. 2). Regarding project management competences, students first need to know something about project management, e.g. the terminology and basic concepts. Therefore, programmes build knowledge first. Especially for projects, it is important to practice.

Students develop practical skills by doing projects and doing project management. In this phase it is crucial to bring students into realistic projects or at least project situations (e.g. a weekly meeting or a Scrum stand-up). Only with the knowledge and the practical experience, students can think deeply about project management and reflect on their

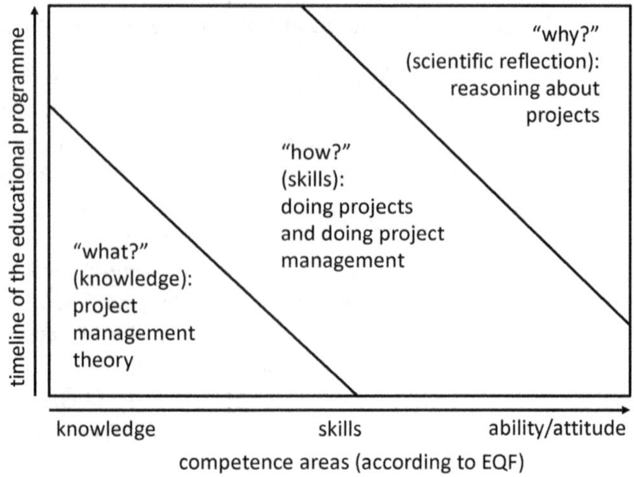

Fig. 2. Development of different competence areas within educational programmes.

experience. This scientific reasoning is especially relevant in academic programmes. The 3 phases and competence areas (knowledge, skills, ability/attitude) do not need to be delivered strictly sequential but can be mixed and iteratively repeated. Nevertheless, they will be clearly visible within a well-designed educational programme (e.g. a Master's) and also within single modules are even learning units. In order to support the "how?" and the practical part, several didactic concepts based on projects have been developed. Prominent ones are Problem-Based Learning (PBL) where students get a complex learning experience (often interdisciplinary) by solving a realistic problem over a longer time period, Project-Based Learning (PjBL) where students conduct complete projects (often in teams) for even more complex problems, and Work-Integrated Learning (WIL) where students are working in companies or conducting company-induced projects [15].

The integration of projects into the curricula can be done in any phase of the programme and they can spread over several phases or semesters, being part of different learning modules which introduce different competences into the project. Such interdisciplinary projects are called Vertically Integrated Modules (VIP) or Capstone Projects. The are specifically useful in order to deliver Overarching Learning Outcomes (OLO), e.g. soft skills, team and project competences, or 3 × "i" competences. Such OLOs are of particular interest for our investigation on the effect of virtual learning due to COVID-19 since they are expected differ a lot if students and teachers don't meet in person.

3 The EuroPIM Cross-Border Master School

The experience presented in this paper is based on the teaching and learning within several interlinked Master's programmes. They are forming the virtual, cross-border Master School of the "European Partnership on Project and Innovation Management (EuroPIM)", a university network of 5 EU universities (UPV/EHU Bilbao, Fachhochschule Dortmund, KTU Kaunas, NTNU Trondheim and KU Leuven) which cooperates with a number of associated universities, e.g. in the Baltics, in Ukraine and in

Kazakhstan. The universities conduct 8 Master programmes which mainly focus on project management, digital transformation, technology innovation and entrepreneurship. In addition, 8 associated Master's programmes are involved. The core element is the European Master in Project Management (EuroMPM) [12] that has more than 15 years' experience in teaching project management and applying the formats of PjBL, PBL and WIL. The EuroMPM programmes at the different universities are interlinked with (currently) 4 double degree agreements.

One core concept of the EuroPIM Virtual Master School (see Fig. 3) is to exchange and mix students and teachers in various didactic formats, starting from lectures and workshops and leading to PjBL and PBL elements, in order to facility the cooperation of the different Master's programmes which are running in parallel at the partner universities.

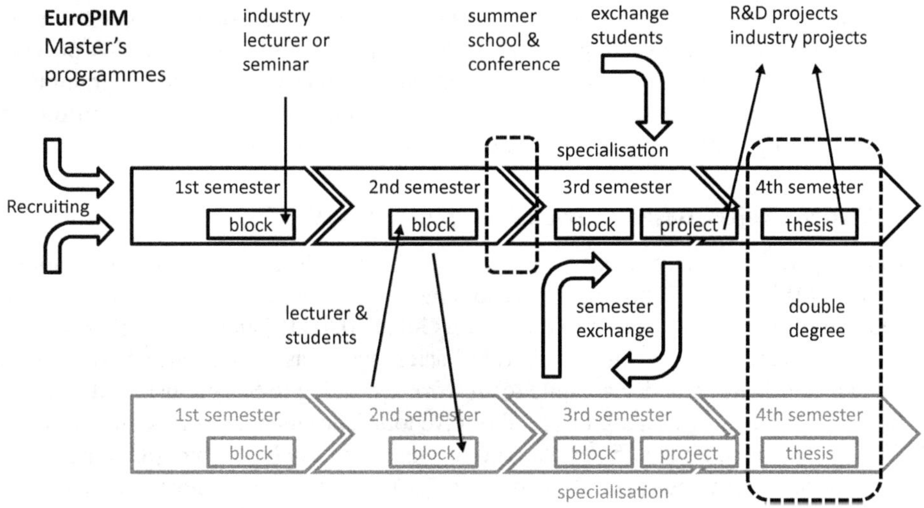

Fig. 3. Cooperation and exchange formats of the EuroPIM Virtual Master School.

This cross-border and cross-programme exchange enables interdisciplinary, international and intercultural (3 × "i") learning experiences and provides realistic project situations of a sufficient complexity. The concept has been tested for many years and evaluated under various aspects [12, 14, 15].

In order to support the cooperation and the development of the EuroPIM Master School, publicly funded projects are conducted on a large scale, including a DAAD Strategic Partnership (EuroPIM), a DAAD Digital Transformation (IMKD) project (ManDEE), a DAAD Ukraine project (ViMaCs), an Erasmus+ Capacity Building project (WORK4CE) and an Erasmus+ Knowledge Alliance (ProDiT) which develops a competence model for the digital transformation (CMDT) and a digital transformation maturity model (DTMM) with partners like IPMA and IEEE. With the beginning of the COVID-19, this cooperation had to be fully virtualized ad hoc.

4 Analysis of Relevant Learning Setups

From summer term 2020 onwards, the EuroPIM Master School had to go fully only, transforming it ad hoc into a Virtual Master School with almost no real mobility. This experience had many similarities with what happened to companies and projects in the same timeframe. Therefore, it is very interesting to investigate how the digital transformation of the Master School influenced the learning and teaching experience. Specifically, it is interesting if the resulting virtual teaching and learning setups are:

- Relevant and meaningful with respect to real work situations (see Sect. 2, WIL).
- Delivering relevant project situation with effective and efficient PjBL.
- Able to deliver the 3 "i" learning experience and the intended OLOs.
- Becoming even more job-relevant since digital cooperation is the standard by today.
- Allow students a more individualized and adapted learning experience.

In the following section, we examine these questions based on examples from the past semesters and based on surveys conducted among students and lecturers. We provide insight on the level of a study module within one singular educational programme, a study module combining students from different educational programmes and on the level of a complete educational programme, all on Master's level.

4.1 Virtual Project-Based Learning on Module Level

Case Study 1. The first case was conducted within the module "Digital Transformation" in the EuroMPM Master's programme (2nd semester) and represented a sub-module that covered Scrum basics according to the Scrum Guide 2020 [10] and was taught 6 weeks, 12 contact hours. The goal of this 'Scrum basics' part was to introduce Scrum as an agile framework and practice an agile team work according to Scrum, using MS Teams with its Kanban feature as a digital collaborative tool. The students worked in 5 teams of 3–6 people each, conducted 5 sprints with Scrum events such as sprint planning, daily scrum, sprint review, and sprint retrospective, formulated own user stories, defined and improved the Definition of Done in teams. The students received 5 topics related to the basics of the digital transformation. Furthermore, the students had to reflect individually on the agile teamwork in their grade course work.

Case Study 2. The second case study was a course 'Introduction to the Scaled Agile Framework' (SAFe) [7] conducted during the summer school (see Fig. 3) and represented a full module on SAFe taught in as an intensive block week course with 40 contact hours. Here, the students received 1,5 days of the theory foundation, studied blended learning materials (theory and videos) and worked in 3 teams of 3–5 people on a real business case on the digital transformation topic during 4 iterations. The students had to work according the SAFe framework and using MS Teams with its Kanban feature as a virtual collaboration tool. The content of the course was about developing new features for a digital application based on the digital value theory. In the end of the course, the students had to submit an individual reflection about working in the agile teamwork, using digital collaboration tools, and the specifics of scaled agile projects.

Evaluation of the Case Studies' Content by Participants. In this subchapter, the evaluation of both study cases (1 & 2) is provided. Both study cases "Scrum Basics" and "Scaled Agile Framework" were evaluated in total by 34 respondents in the end of each course with the means of the online survey. The online survey was conducted using Google Forms and in presence of the teacher (i.e. the respondents could immediately clarify the meaning of the questions, when needed). The survey comprised of two parts. The first part included questions on the learning setup and experiences during the course, where the student had to evaluate the statements according to the Likert scale [4] (from 5 to 1: very much (5) – much (4) – somehow (3) – hardly (2) – not at all (1)). The second part included the evaluation of obtained competences, using the Likert scale in the same format, too. First, we asked the respondents to evaluate, whether the learning situations experienced during the course prepared them to real work situations, a feature known from PjBL and WIL (see Sect. 2) if conducted in presence. The results from Fig. 4 prove that 76,5% of respondents agree that the virtual online version of the course prepared them for work situations properly, too.

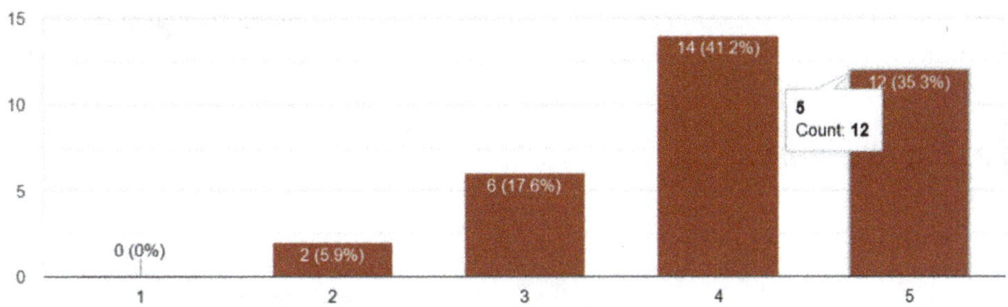

Fig. 4. Do you think the learning situations of this course (theory, teamwork, working according to an agile framework, etc.) has prepared you to real work place situations?

As described above, using MS Teams as a digital collaboration platform was one of the key conceptual settings during the study cases. Thus, the respondents were asked to evaluate, whether the offered learning setup (agile team setup + MS Teams as a collaboration platform) has prepared them to real work place situations. This question is of particular interest since it asks for the technical, digital and tool-related aspect of the learning setup, not so much on the didactics and content of the course. The digital aspect is what distinguishes the new learning situation from the previous ones which were conducted in presence. As presented at Fig. 5, almost 50% of the students reacted to the statement with 'very much', another 35% with 'much', and the rest of 18% with 'somehow'. Moreover, we asked the respondents, if they would use MS Teams to organise their student team work in the future (3rd and 4th semester, see Fig. 3) and received predominantly positive answers, with 65% of participants rated at 5.

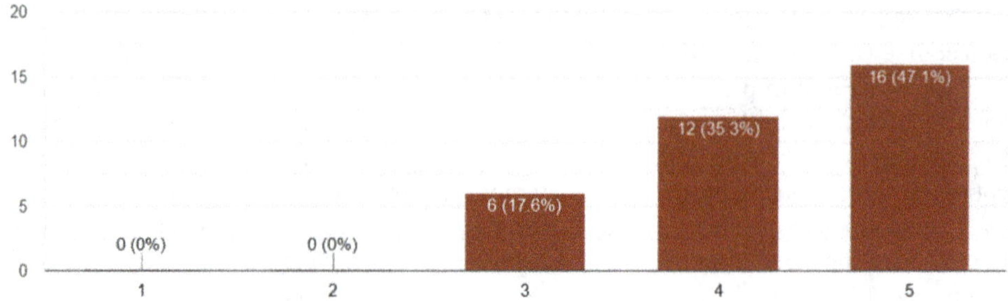

Fig. 5. Do you think the offered learning setup (agile team setup + MS Teams as a collaboration platform) has prepared you to real work place situations?

According to the results shown at Fig. 6, 10 from 34 respondents hardly or not all experienced situations created during the course outside of it. Another 8 students experienced them to somewhat extent, whereas one third of the participants rated this question with 'much' and only four of them with 'very much'. This corresponds with the curricular placing of the two courses in the first half of the study programme where students get acquainted and prepared for digital collaboration situation and virtualized team and project work.

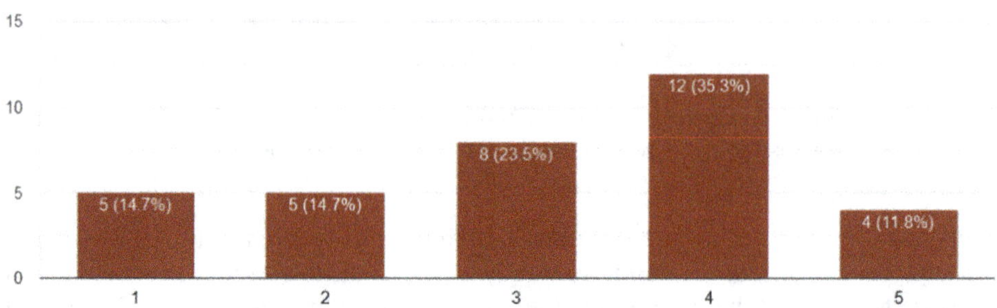

Fig. 6. To which extent did you experience such situations already outside the course?

Evaluation of the Competences Acquired During the Courses. In the second part of the survey, we evaluated the competences obtained by the students, using the Likert scale in the same format, too. The participants had to evaluate to which extent they have developed their competences according to the Likert scale [4] (from 5 to 1: very much (5) – much (4) – somehow (3) – hardly (2) – not at all (1)). We offered for evaluation a list of seven competences (see Table 1). As the subject of the research on the competences for the digital transformation is rather new and the participants of this initial evaluation cannot be considered as subject matter experts to evaluate a sophisticated competence model with 20–40 competences, we kept the list rather short. We aimed to include the competences that are distinguishing the managing of the digital transformation projects and teams under the influence of COVID-19 challenges.

The definitions of competences used for the survey included a short name and a description. The latter is considered to be very important so that respondents have unified understanding of the studied competences. As one can see, most competences are self-explained through the description. The competence 'Dealing with team members' is, however, somewhat outstanding. Under 'dealing', we mean that working in an agile team requires communicating without hierarchies, managing team dynamics, leading fellow team members without official authority to do so: So, how we 'deal' with each other in a self-organised interdisciplinary team in a digital transformation project. In the description, we recognize the fact that, currently, our competence in managing digital transformation projects is just developing, as this is a new trend. It means that in many cases a project team would consist of many unexperienced and, often, not certified team members. 'Learning by doing' is even a part of, for example, the Scaled Agile Framework [7]. Thus, the ability to deal with different levels of competence is important in managing the digital transformation. Similarly, the competence 'Innovation fostering' is included into the studied list for few reasons.

First, digital transformation projects require higher level of innovation due to the uncertainty and rapid pace of information technology development. Secondly, the students receive an assignment that requires their own creativity and is aimed to foster their creativity skills on the master level. Our results show (see Table 1) that the students considered all listed competences as 'much' developed during the courses, scoring between 4,18 and 3,79 out of 5. The most developed competences were 'Working remotely' and 'Soft skills", rated 4,18 and 4,12 respectively. The least developed competence was 'Interdisciplinarity', at 3,79. Remarkably, the authors conducted in 2017–2018 the assessment of the competence development level in the same master programmes [12], including 'Interdisciplinarity'. Although, one cannot directly compare the latter and the current surveys, still, the development of the competence 'Interdisciplinarity' cannot be underestimated especially in the context of the management of the digital transformation.

We can conclude that the virtual learning setup of a module trains relevant, job-related overarching learning outcomes (OLOs) in an effective way. Specifically, the virtualized team cooperation and the usage of a modern collaboration platform contributes positively to the learning experience. This is supported by the application of and agile, project-based learning (PjBL) situation which resembles typical project-oriented work situations. Students rated the soft skills are well-supported by the learning setup. Therefore, the design of project-oriented modules in the first half of the educational programme with agile team work, a practical and realistic task assignment, and supported by a modern IT environment contributes positively to the development of competences required for managing of the digital transformation.

4.2 Virtual Project-Based Learning on Programme Level

The complete Master's programme, not only a single module, should be designed according to the sequence of theory, practice and reflection (see Fig. 2), too. Therefore, the foundations for project work, team cooperation and PjBL/PBL are laid in the first 2 semesters (see Fig. 3), so students are trained in working in teams in virtual collaboration on practice-oriented task assignments. They are familiar with the tools and the relevant project management and communication methods. First, they do this in teams within the cohort of their respective study programme. Then, they are mixed with students from

Table 1. Competences development after a course for managing the digital transformation

Competences	Average
Working remotely through different and changing video conferencing tools	4,18
Soft skills in general (for example, communication, etc.)	4,12
Dealing with team members with different perceptions and competency (knowledge and experience) level in digital transformation projects and agile frameworks	4,03
Virtual leadership as an ability to guide the performance of individuals who telecommute or work in distant locations, dealing with such challenges as navigating time zones, choosing the technology to communicate, defining roles and priorities, resolving conflicts, and celebrating achievements	4,03
Innovation fostering as an ability to contribute to a safe and trustful working environment by allowing everyone to experiment and make mistakes so that each team member can learn and continuously improve the way he or she works	3,97
Tolerance to ambiguity as ability to handle ambiguous situations and even chaos (especially, in agile environments and digital transformation projects with a higher level of uncertainty)	3,91
Interdisciplinarity as ability to understand various concepts and processes (business and technical ones) over a complete landscape of company's handbook	3,79

other study programmes from their own university and from partner universities, delivering a learning setup which is inherently interdisciplinary, international and intercultural (3 × "i"). The positive impact on the learning outcomes is evaluated for the delivery of the learning experience in presence [12, 15], which involves travel and group workshops. In the second half of the educational programme (see Fig. 3) the projects become larger and more complex. Semester mobility, internships and projects in companies combine didactic elements from work-integrated learning (WIL), vertically integrated projects (VIP), and capstone projects. Especially in scientific seminars, project theses and finally in the Master's thesis, the scientific reflection and analysis is becoming more important. While the practical elements become more and more difficult if conducted online, the scientific part is well-suited for individual work from home. Nevertheless, the quality of the digital delivery on the overall educational programme level is very crucial for the learning experience. This has been evaluated for one of the Master's programmes in 2021 [15] in a survey where the answers were collected from 67 (enrolled) Master's student and 30 involved teachers, coordinators and scientific staff from the EuroPIM partner universities (EU + Ukraine) which took part in project-based learning activities in 2020 and 2021.

Figure 7 shows that only a few students were able to travel to partner universities in 2020 and 2021 due to COVID. Still, most of them (37 + 29%) were able to work in cross-programme projects. Both the projects and the 3 × "i" experience are rated important. Figure 8 shows the assessment of the virtual learning situations and their job-field relevance. Students rate the new, virtual learning experience as effective/good and relevant for the job.

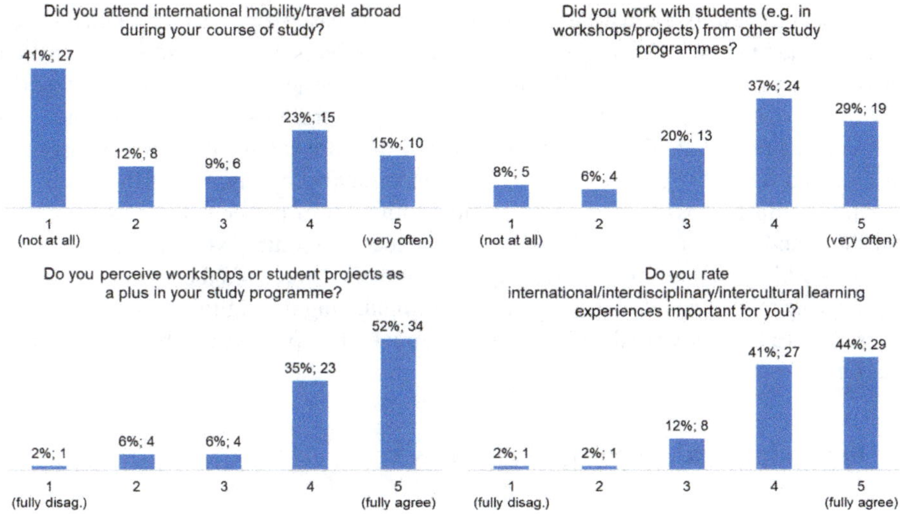

Fig. 7. Effect of missing physical mobility on the learning experience (students) [15]

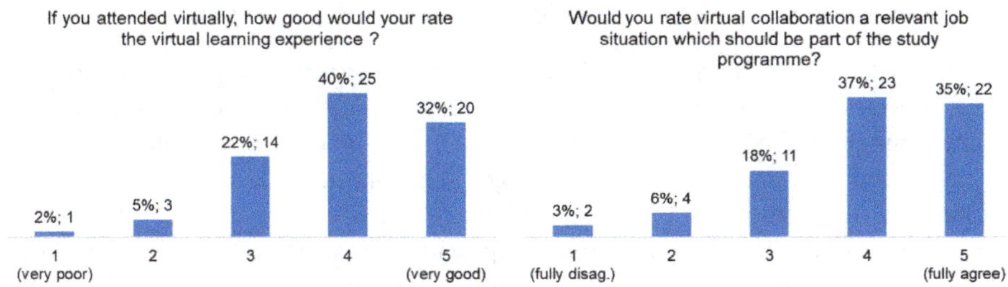

Fig. 8. Quality and (workplace) relevance of the virtual learning experience (students) [15]

In addition, teachers were asked for their assessment and they agree to a very high level with the assumption that the project-based learning situations (55 + 38%) and their digital implementation (27 + 57%) are relevant for the future job-field of students (see Fig. 9).

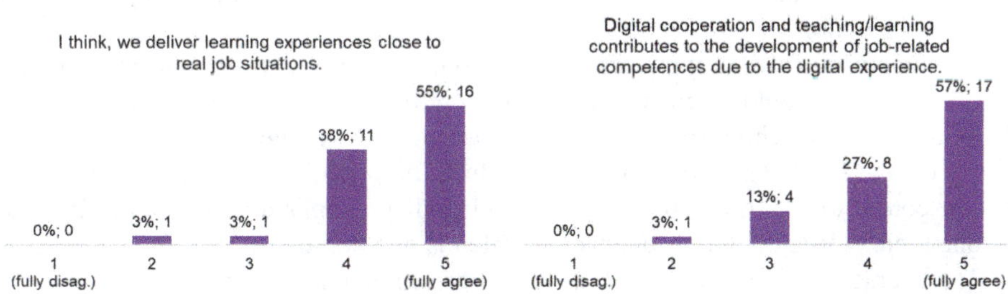

Fig. 9. Quality and (workplace) relevance of the virtual learning experience (teachers) [15]

The survey results indicate that a virtual, project-based approach can deliver most aspects of the intended learning experience since it reflects the reality of many working situations in companies. Especially, it is relevant for project setups in such industries as IT and consulting industries. Such industry sectors are moving to a more and more agile working pattern, mainly based on projects which are conducted cross-organisation and cross-border. This can be reflected by educational programmes, if they deliver a cross-programme, cross-border collaboration with a lot of projects, team assignments, workshops and other project- and problem-based learning setups, which are conducted in an agile working style within a modern IT environment. Educational programmes which are preparing students with the competences for managing the digital transformation are encouraged to build a virtualized learning setup which supports such relevant learning situations.

5 Discussion and Recommendations

In order to provide a guideline for transforming educational programmes to deliver competences related to the management of the digital transformation, in particular for virtual project collaboration, we compiled a list of recommendations.

On the *level of the overall educational programme* (e.g. a full Master's programme):

1. The Master's programme should be embedded into an interdisciplinary and international Master School which serves as a cross-organisational, cross-border learning environment and which is operated as an international project portfolio, generating job-related working situations with a high practical relevance.
2. Students should move on a learning trajectory which starts with theory and smaller project assignments and builds the complexity of real project environments step-by-step. The international and intercultural mix are built gradually throughout the semesters (see Fig. 3), combining different elements of PBL, PjBL, WIL and projects.
3. Throughout the curriculum the bases for scientific reflection and research are laid.
4. Instead of weekly schedules, an organisational pattern with agile teamwork on tasks, assignments and projects should be deployed and agile project work is explicitly trained.
5. A digital education ecosystems (DEE) [15] should serve as a professional virtualized IT environment, using industry standard IT tools, e.g. for collaboration and communication.
6. The delivery of overarching learning outcomes (OLO) and relevant competences (see Table 1) should be considered throughout the whole student journey.
7. Learning and teaching from anywhere at any time should be supported.

    On the *level of a module* within the education programme, we recommend:

8. The course topics should be structured and studied in agile iterations or sprints (i.e. one topic is studied during one week, which is one iteration).
9. The course should be organised in student project teams of 3–6 people.
10. The course should be conducted by using digital collaboration platform, i.e. MS Teams that offers video, chat, and Kanban features.

11. The course can benefit from such Scrum events, as Sprint Planning, Sprint Review, Sprint Retrospective or their alternatives i.e. according to the SAFe.
12. Application of further agile elements, such as Definition of Done, managing tasks in Kanban, acceptance criteria offer additional visualisation of the project goals and reducing miscommunications between students themselves and with a teacher.
13. In the beginning of the course, it is recommended to include basic knowledge on applied agile project management framework (i.e. Scrum, SAFe), at best, using the blended learning approach with self-tests.
14. In the end of the course, it is recommended to request a reflection report on applying the agile framework, using the digital collaboration platform, and collaborating in remote and distributed teams.

Such a guideline is aimed to provide conditions for project-related virtual learning situations on the level of Master's programmes, so that the students acquire not only the technical competences in digital transformation, but also the competences related to the management of the digital transformation and virtual project collaboration.

6 Conclusions

COVID-19 has drastically accelerated the digital transformation, both in project management and in higher education. Being able to manage the digital transformation with projects and to perform a digitally transformed project management becomes a key competence for future graduates, especially from Master's programmes in IT, digital transformation, project management and innovation and entrepreneurship. The EuroPIM consortium with its system of cooperating Master's programmes was confronted with the task to digitally transform the education ad hoc after COVID-19 spread.

We took this as an opportunity to investigate and evaluate how education can stay relevant and deliver relevant learning situations and competences in a fast-changing world. By implementing learning situations which are intentionally similar to work-related situations we provide a setup which allows both learning in a virtualized project-world and learning for a virtualized project-world. With this paper, we want to share the experience and give first indications of the performance and success of the approach. Furthermore, we believe that we delivered a guideline for a transformation of educational programmes in order to deliver job-field related, 3 × "i" and project-oriented competences in an agile and flexible setup with a high resilience against interruptions and changes within the course of the COVID-19 pandemic.

References

1. Berghaus, S., Back, A.: Stages in digital business transformation: results of an empirical maturity study. In: MCIS, p. 22 (2016)
2. Gollhardt, T., Halsbenning, S., Hermann, A., Karsakova, A., Becker, J.: Development of a digital transformation maturity model for IT companies. In: 2020 IEEE 22nd Conference on Business Informatics (CBI), vol. 1. IEEE (2020)
3. Ivančić, L., Vukšić, V.B., Spremić, M.: Mastering the digital transformation process: business practices and lessons learned. Technol. Innov. Manage. Rev. **9**(2) (2019)

4. Likert, R.: A Technique for the Measurement of Attitudes. Arch. Psychol. **140**, 1–55 (1932)
5. Maheshwari, A.: Digital Transformation: Building Intelligent Enterprises. Wiley, Hoboken (2019)
6. Morakanyane, R., Grace, A.A., O'Reilly, P.: Conceptualizing digital transformation in business organizations: a systematic review of literature. Bled eConference **21** (2017)
7. Scaled Agile Framework, Inc. (2021). https://www.scaledagileframework.com/#. Accessed 31 July 2021
8. Rajala, S.A.: Beyond 2020: preparing engineers for the future. Proc. IEEE **100**, 1376–1383 (2012). https://doi.org/10.1109/JPROC.2012.2190169
9. Schumacher, A., Erol, S., Sihn, W.: A maturity model for assessing industry 4.0 readiness and maturity of manufacturing enterprises. Procedia Cirp **52**, 161–166 (2016)
10. Schwaber, K., Sutherland, J.: The scrum guide (2020)
11. Verhoef, P.C., et al.: Digital transformation: a multidisciplinary reflection and research agenda. J. Bus. Res. **122**, 889–901 (2021)
12. Wolff, C., Dechange, A., Mikhieieva, O.: European master in project management (EuroMPM) - joint approach for master education. In: 12th IEEE International Scientific and Technical Conference on Computer Sciences and Information Technologies (CSIT), vol. 2, pp. 115–122 (2017)
13. Wolff, C., Mikhieieva, O., Nuseibah, A.: Competences and the digital transformation. In: Ayuso Muñoz, J.L., Yagüe Blanco, J.L., Capuz-Rizo, S.F. (eds.) Project Management and Engineering Research. Lecture Notes in Management and Industrial Engineering, pp. 221–234. Springer, Cham (2020). https://doi.org/10.1007/978-3-030-54410-2_16
14. Wolff, C., Omar, A., Shildibekov, Y.: How will we build competences for managing the digital transformation? In: 10th IEEE International Conference on Intelligent Data Acquisition and Advanced Computing Systems: Technology and Applications. Metz, France (2019)
15. Wolff, C., et al.: Cross-border projects in digital education ecosystems. In: Auer, M.E., Hortsch, H., Michler, O., Köhler, T. (eds.) ICL 2021. LNNS, vol. 389, pp. 382–394. Springer, Cham (2021). https://doi.org/10.1007/978-3-030-93904-5_39

Risk Simulation Challenges and Success Driven Project Management

Victoria Shavyrina and Vladimir Liberzon(✉)

Spider Project Team, 7 Semenovskaya Sq., Moscow, Russia
smt@spiderproject.com

Abstract. To be reliable, project risk simulation must take into account all existing constraints, simulate risk events that may occur and risk responses that management will apply in case they occur, simulate corrective actions that management will take in case project performance gets worse (or better) than expected.

The main problems with risk simulation tools include modeling the consequences of uncertainty rather than its source, missing correlations between probability distributions of project parameters, missing corrective actions people apply when risks occur and threaten project goals, using different leveling heuristics than those used for managing projects with resource constraints.

A correct project risk model can address the first three of the above problems but the last one is the most serious and cannot be solved if you are using a risk simulation tool which is external to project scheduling software you manage your project with. Probability distributions created by any external risk simulation tool are not reliable when project schedule has resource constraints.

Success probability is the probability of meeting project target. Success probability trends are the best performance indicators. Negative trends show that project contingency reserves are consumed faster than expected and corrective actions should be considered.

Keywords: Risk management · Risk simulation · Success Driven Project Management · Success probability trends · Monte Carlo simulation · Three Scenarios method

1 Introduction

Reliable project delivery is impossible without proper risk management.

Projects always include something unique, the estimates of durations and costs are always uncertain, risk events can happen and lead to changes in project scope, etc. That is why it is necessary to estimate existing uncertainties and create project contingency reserves that are necessary for meeting project targets with sufficient probabilities.

Contingency reserves are determined by risk simulation that is usually done using the Monte Carlo method. Some project management tools like Spider Project and Safran include risk simulation capabilities while the users of others tools rely on external software for risk simulation.

S. Bushuyev et al. (Eds.): IPMA 2021, LNNS 704, pp. 83–91, 2023.
https://doi.org/10.1007/978-3-031-34629-3_8

The Monte Carlo method can be used for simulation of both uncertainties of initial data and risk events.

Risk simulation is becoming popular but most risk simulation and management tools and the ways they are used miss some important functionalities, which makes the results of this simulation unreliable.

In this article we will discuss the most common of these problems.

Risk simulation is used for setting reliable targets and reasonable contingency reserves. The probabilities of meeting project targets are called Success Probabilities. When a project is executed, some risk events occur, new risks are identified, some activities are late, etc. As a result, contingency reserves are consumed and success probabilities change. It is necessary to be able to analyse project performance and take timely corrective actions as needed. This can be done using Success Driven Project Management, which is a proven project management methodology that integrates scope, schedule, cost, and risk management. It is based on the analysis of success probability trends and will be discussed in this paper.

2 Risk Simulation Requirements

To be reliable, Risk Simulation must:

- consider all constraints and restrictions including resource constraints, material supply constraints, funding restrictions, etc.;
- simulate possible risk events and risk responses to be applied by the management in case these events do occur;
- simulate corrective actions to be taken by the management in case project performance gets worse (or better) than expected as a result of risk events.

To meet risk simulation requirements, one must be able to use the GERT network that includes the probabilistic and conditional branches, and simulates potential loops in project execution.

A risk model must simulate initial uncertainty, dependencies between project parameters, and how the project will be managed in real life.

3 Risk Simulation Problems

Major Risk Simulation Problems that will be discussed in this article include:

- simulating the result of uncertainty rather than its source;
- missing correlations between probability distributions of project parameters;
- missing corrective actions to be taken if a risk event occurs and threatens project objectives;
- using leveling heuristics other than those employed for managing projects with resource constraints.

3.1 Simulating the Result of Uncertainty Rather Than Its Source

Now let us consider a simple project consisting of a single activity; the volume of work for this activity is 80 pieces.

Activity duration depends on the assigned resource's productivity with the optimistic estimate of 1.25 pc/hour, the most likely estimate being 1 pc/hour, and the pessimistic estimate being 0.8 pc/hour. The workday is 8 h. Thus, the optimistic duration of the activity/project is 8 days, the most likely duration is 10 days, and the pessimistic duration is 12.5 days.

The source of uncertainty is the assigned resource's productivity – however, most risk simulation tools do not work with resource productivity and only simulate duration uncertainty.

If we simulate productivity uncertainty, the probability of finishing the activity less than in 10 days is 53.7% (see Fig. 1) but if we simulate duration uncertainty, the probability of the same is only 42% (see Fig. 2).

Duration uncertainty is a result of productivity uncertainty. Simulating the result rather than the source of uncertainty leads to wrong estimates.

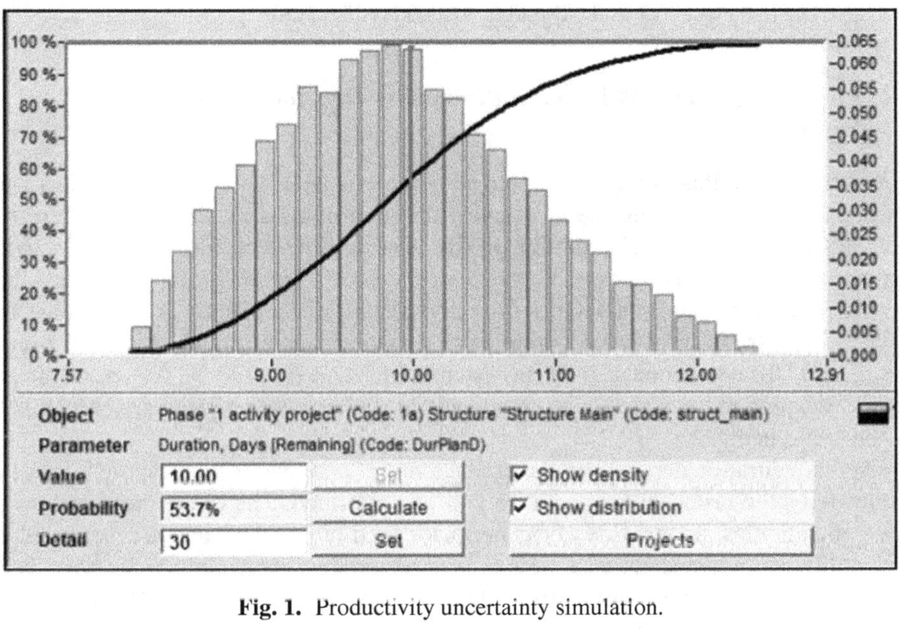

Fig. 1. Productivity uncertainty simulation.

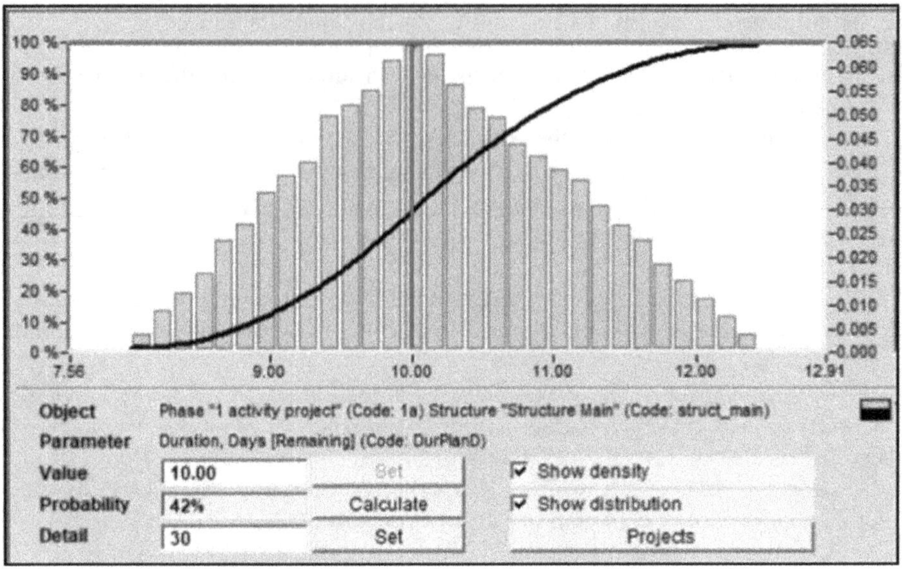

Fig. 2. Duration uncertainty simulation.

3.2 Missing Correlations Between Probability Distributions of Project Parameters

Now let's assume that we decided to create a more detailed model of our project and, instead of one activity with the most likely duration of 10 days, created a path of ten activities with the most likely duration of 1 day each and the same distribution of activity duration (−20%, + 25%).

If durations of these ten activities are considered independently of each other, the probability distribution of project duration will change and the probability of finishing it less than in 10 days becomes 26.4% instead of initial 42% (see Fig. 3). It does not appear logical: the probability distribution should not depend on how activities are presented in the project schedule.

Activity duration depends on the productivity of assigned resources: the higher the productivity, the faster all ten activities will be performed. Therefore, the durations of these 10 activities are strongly correlated with each other. With this provision, initial probability will not change. Had the source of uncertainty (resource productivity) been simulated instead of the result (activity duration), this problem would not exist.

It is not easy to set correlations between durations for all activities executed by the same resources. Therefore, the probabilities calculated using the Monte Carlo method can be incorrect if some dependencies were missed.

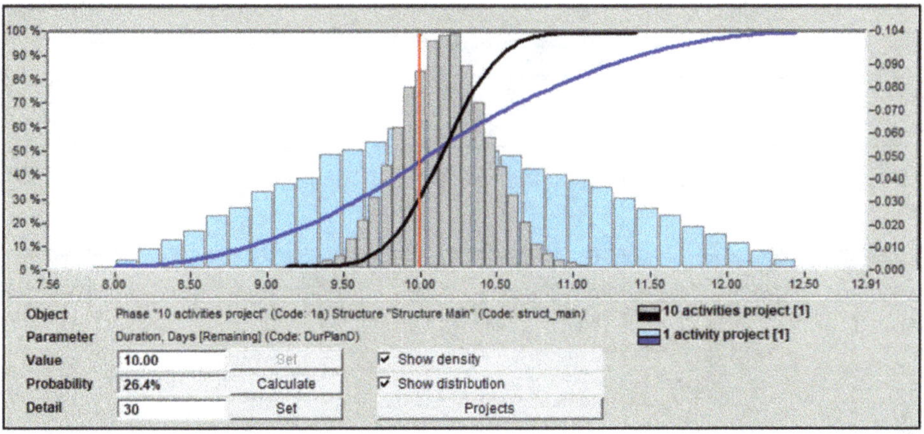

Fig. 3. Probability distribution of 10-activity project duration compared to initial distribution.

3.3 Missing Corrective Actions

When projects are late and over budget, corrective actions are taken. The Monte Carlo model should simulate these corrective actions in the respective iterations; otherwise the results will not be reliable because the model does not reflect the real-life actions. It means that the project model must include conditional branches "if … then" and risk simulation software must automatically select the right branch, based on the predefined conditions.

Besides, risk simulation software must be able to work with risk events and probabilistic branches activated with certain probabilities in case of occurrence of risk events.

If risk responses and corrective actions are not simulated, the probability distributions shown by the Monte Carlo method will be incorrect.

3.4 Missing Corrective Actions

Project constraints can include resource, supply, and funding restrictions to be taken into account in all Monte Carlo iterations. It means that, in each iteration, resources must be leveled using the same leveling algorithms as in project management software used for project resource management. Otherwise, the obtained probability distributions will be incorrect.

Different software programs, however, use their proprietary leveling algorithms and create different resource-constrained schedules for the same projects. This is due to the lack of a mathematical solution for finding the best (shortest) schedule when project resources are constrained.

Let's have a look at the resource-constrained schedules calculated for the same project by Microsoft Project (Fig. 4) and Spider Project (Fig. 5).

The default leveling algorithm used by Oracle Primavera P6 creates the same schedule as created by Microsoft Project for this project (Fig. 6):

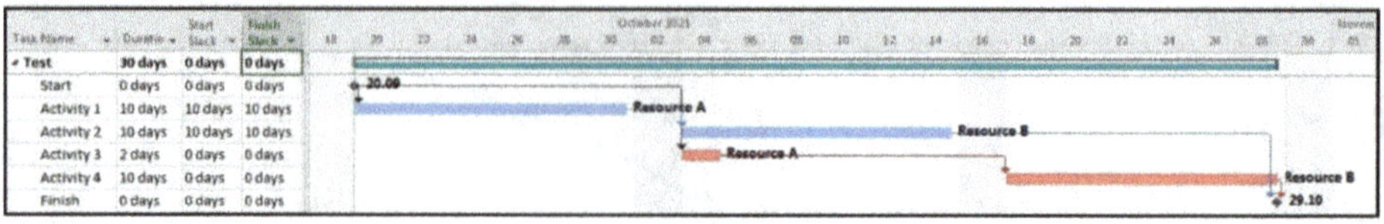

Fig. 4. Microsoft Project schedule of project "Test".

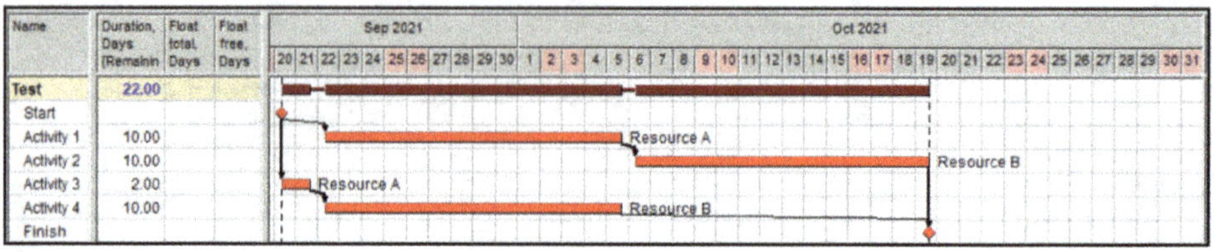

Fig. 5. Spider project schedule of project "Test".

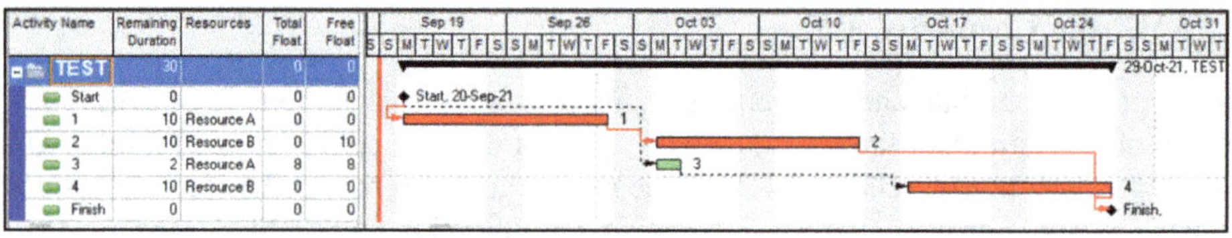

Fig. 6. Oracle Primavera P6 schedule of project "Test".

We have entered the same activity duration ranges (−20%, + 25%) and triangular distributions of activity durations, and obtained different probability distributions for project duration calculated with Spider Project heuristic and heuristics used by other tools.

In Fig. 7 you can see probability curves created for project "Test" duration, scheduled using Spider Project leveling algorithm (left) and the algorithms used by other tools. The probability of finishing this project in less than 26 days is 99.66% if to use the schedule created by Spider Project and only 0.075% if resources are assigned as advised by other tools.

If project targets are set based on one heuristic and another heuristic is used for project resource allocation and management, the Monte Carlo simulation results can be misleading.

Any external risk simulation tool is not reliable when project schedule has resource constraints.

4 Success Driven Project Management

Success Driven Project Management (SDPM) methodology was developed in Russia in the 1990s and has been successfully used in many projects, programs, and organisations since then. SDPM is supported by the Russian project management software

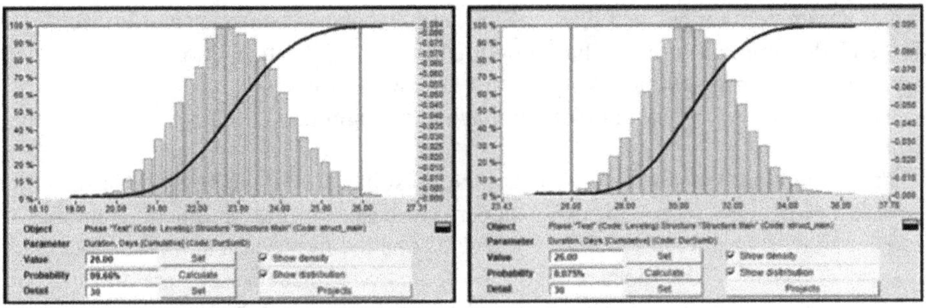

Fig. 7. Probability Distributions of Project Test Duration if to use different leveling algorithms.

Spider Project although its basic approaches can be used with other project management software tools.

The main ideas driving SDPM:

- Project management team must have time and cost buffers (contingency reserves) for managing project risks and uncertainties. These buffers are not linked to any activity sequence. Project buffer is the difference between target value and the value of the same parameter in the working schedule. SDPM recommends to use the optimistic project schedule in project workforce management to avoid the Parkinson's Law effect. The optimistic schedule is based on the optimistic estimates of project parameters (durations, volumes, resource productivity, costs, etc.) and does not include project risk events.
- Project targets must be set using risk simulation. These targets must have reasonable probabilities of being met.
- Project status information is useful but not sufficient for decision making. Decision making must be based on the analysis of project trends. Project buffers are consumed during project execution and project management is about managing these buffers. If they remain positive to the moment of project completion, then project management has been successful and the targets were met. There must be the tools for project performance analysis based on measuring project buffer penetrations. The best indicator of buffer penetration and project performance status is the current probability of meeting the project target. If the probability of meeting the project target is increasing, then the project buffer has been consumed slower than expected; otherwise the project buffer has been consumed too quickly and project success is endangered. Success probability trends are the best integrated performance indicators: they take into consideration project risks; they depend not only on performance results but also on the changes in project environment. Other methods of performance analysis make conclusions, analysing the past performance, but the success probability trend analysis looks into the future.

5 Three Scenarios Method

In the Monte Carlo risk analysis we perform thousands of iterations trying to precisely estimate the probability distributions of project parameters. But what about initial data? Projects are unique, we do not have reliable statistics, we rely on the rough estimates of project parameters ranges and distributions, risk events probabilities and impacts. The Monte Carlo simulation results based on this rough initial information provide us with approximate estimates while taking a lot of time and efforts.

That is why we also use another approach to risk simulation that we call the method of Three Scenarios.

Project planner creates the optimistic, the most likely, and the pessimistic scenarios of project execution. The optimistic scenario is based on the optimistic estimates of project parameters and does include risk events and their consequences; the most likely scenario is based on the most probable estimates of project parameters and includes those risk events that have a probability of at least 50%; and the pessimistic scenario uses the pessimistic estimates of project parameters and includes all of the identified risk events.

After the scheduling, these scenarios give us three values for any project parameter. This allows us to create an approximate distribution of its probability, based on some assumption about the shape of probability curve.

Spider Project software automatically creates these three scenarios based on the data entered for the Monte Carlo simulation and automatically creates the probability curves based on these three scenarios.

Probability curves created using the Monte Carlo and Three Scenarios methods are not the same and show slightly different probabilities of meeting project targets but it is not the values of these probabilities that really matter but their trends. Both methods show the same trends of success probabilities but the calculations based on the Monte Carlo simulation can take hours for large projects with resource constraints while using Three Scenarios to obtain the same information takes mere seconds. An example of success probability trends calculated by both methods is shown in Fig. 8.

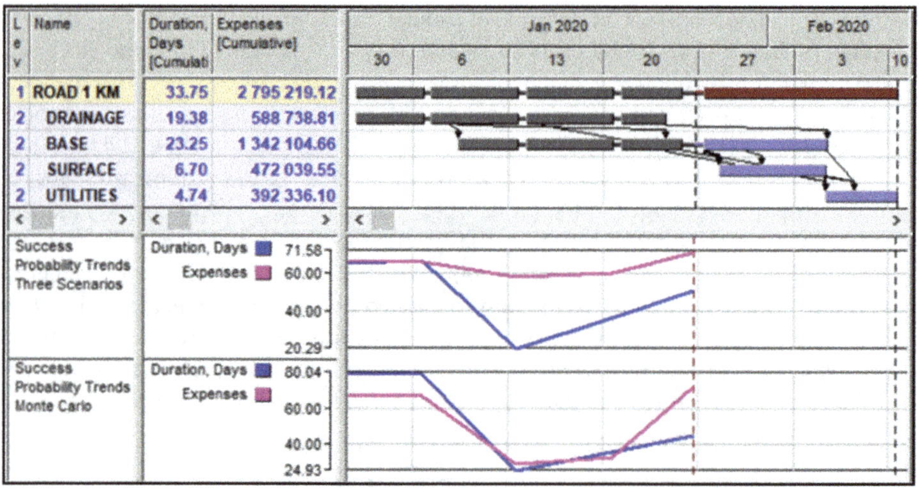

Fig. 8. Success probability trends calculated using the MC simulation and 3 Scenarios method

Spider Project users usually prefer the Three Scenarios method for day-by-day management and, from time to time, verify their estimates using the Monte Carlo simulation.

6 Conclusions

When using the Monte Carlo simulation for quantitative risk analysis:

- Try to set ranges for those parameters that cause uncertainty rather than for the results of uncertainty
- Set correlations between uncertain interdependent parameters
- Include in the risk model corrective actions people take when project targets are jeopardized
- Do not rely on the external risk simulation tools if your projects have resource constraints

Use the optimistic schedule to set tasks to project workforce but set project targets that have reasonable probabilities of being met.

The trends of the probabilities of meeting project targets are the best project performance indicators.

The Three Scenarios method is a simple alternative to the time- and effort-consuming Monte Carlo simulation.

References

1. Archibald, R.D.: Managing High-Technology Programs and Projects, 3rd edn. Wiley, New York (2003)
2. Liberzon, V.I., Shavyrina, V.V.: Methods and tools of success driven project management, Proj. Perspect. **XXXV**, 31–37 (2013)

Schedule Optimisation Methods and Tools

Vladimir Liberzon(✉) and Victoria Shavyrina

Spider Project Team, 7 Semenovskaya Sq., Moscow, Russia
smt@spiderproject.com

Abstract. In this paper we will discuss several methods and tools for schedule optimisation. These tools include activity drags, flexes and schedule adjustment for project schedules, calculated using the Critical Path Method. In the case of resource-constrained schedules, the same methods can be used but most of the popular scheduling tools do not have necessary functionality. It appears that most planners do not load resources into their schedules because automatic resource leveling, performed by popular scheduling tools, provide poor results; risk simulation that takes into account resource constraints creates unreliable probability curves; and software calculates wrong resource constrained activity floats and shows wrong activities as critical. Since most projects have resource and other constraints, using Critical Path Method for project scheduling and risk simulation leads to creating too optimistic schedules. Project targets set ignoring resource constraints have low probabilities of achieving. This is one of the main reasons why many projects are late and over budget despite using contemporary project management software.

Keywords: Drag · Flex · Resource leveling · Risk simulation · Schedule optimisation · Scheduling

1 Introduction

Project schedule created using Critical Path Method does not take into account resource and cost constraints. It assumes that the project has unlimited resources and financing but still can be improved using tools and techniques discussed in this paper. Critical Path Method calculates total and free floats of project activities but these tools are not sufficient for proper schedule analysis and optimisation. Additional tools that can help to improve CPM schedules and will be discussed in this paper are Drags, Flexes and Schedule adjustment.

2 Activity Drag

Activity drag is the amount of time that an activity on the critical path adds to the project duration. Alternatively, it is the maximum amount of time that one can shorten the activity duration by until it is no longer on the critical path or until its duration becomes zero. Activity drag is a metric developed by Stephen Devaux that can be used for finding

S. Bushuyev et al. (Eds.): IPMA 2021, LNNS 704, pp. 92–98, 2023.
https://doi.org/10.1007/978-3-031-34629-3_9

the best way to crash project schedule. Activity drag is the difference between project duration in the initial schedule and in the schedule where activity duration is taken as zero.

For some project activities, drag can be negative. Negative drag shows that increasing an activity duration can decrease the project duration.

3 Activity Flex

Flex is a metric of activity duration flexibility, introduced by Rafael Davila.

Activity Start Flex shows the time difference between the earliest possible activity start that does not violate any constraints or dependencies and the planned activity start in the current schedule. It shows the time amount that can be added to an activity's duration without delaying its finish.

Activity Finish Flex shows the time difference between the latest possible activity finish without violating any constraint and the planned activity finish in the current schedule if activity start is not moved.

Both metrics imply increasing activity duration while keeping activity start or finish intact.

Activity Flex shows whether it is possible to use less assigned resources or less assigned resource workloads without delaying other activities.

4 Sample Project

Let's look at a sample CPM project schedule in Fig. 1.

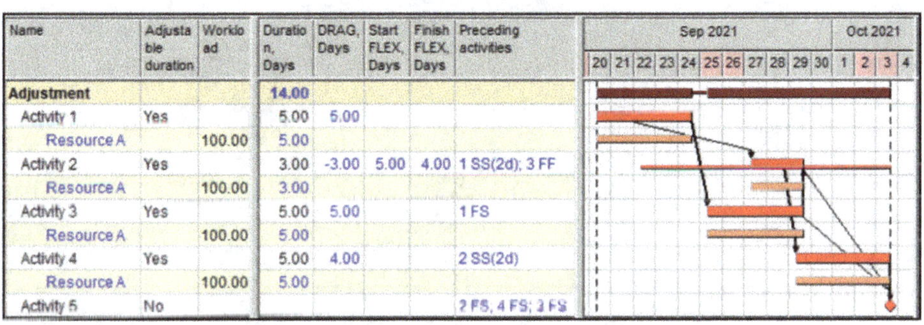

Fig. 1. CPM schedule of Sample Project "Adjustment".

Project "Adjustment" consists of four activities and finish milestone. In the CPM schedule of this project, all activities are critical.

Dependencies between the Sample Project activities are shown in Fig. 2.

Activity 4 has 5-day duration but only 4-day drag because, with less than 1 day duration, this activity will have a positive total float.

In this schedule, activity 3 can start earlier and finish later without delaying any other activity and thus it has positive start and finish flexes.

Preceding activity name	Succeeding activity name	Type	Lag	Lag unit	Lag type	Lag calendar	Driving	Critical
Activity 1	Activity 2	Start-Start	2.00	Day	Time	A	No	No
Activity 2	Activity 4	Start-Start	2.00	Day	Time	A	Yes	Yes
Activity 1	Activity 3	Finish-Start		Hour	Time	A	Yes	Yes
Activity 3	Activity 2	Finish-Finish		Hour	Time	A	Yes	Yes
Activity 2	Activity 5	Finish-Start		Hour	Time	A	No	No
Activity 4	Activity 5	Finish-Start		Hour	Time	A	Yes	Yes
Activity 3	Activity 5	Finish-Start		Hour	Time	A	No	No

Fig. 2. Activity dependencies in the Sample Project "Adjustment".

This activity has a negative drag, which means that increasing its duration we can decrease overall project duration.

It is usually possible to increase activity duration. It can be done by using fewer assigned resources or by using the same resources but with less workload. Fredric Plotnick suggested to automate schedule duration optimisation by increasing durations of some activities. Spider Project is one of the few packages that can do it. We call it schedule adjustment.

Schedule adjustment is a schedule optimisation technique that minimises schedule duration by increasing durations of selected activities, usually those with the negative Drags. Having applied this technique to our Sample Project, we got a schedule created by Spider Project software (see Fig. 3).

Name	Adjustable duration	Workload	Duration, Days	DRAG, Days	Start FLEX, Days	Finish FLEX, Days	Preceding activities
Adjustment			10.00				
Activity 1	Yes		5.00	1.00			
Resource A		100.00	5.00				
Activity 2	Yes		8.00				1 SS(2d); 3 FF
Resource A		37.50	8.00				
Activity 3	Yes		5.00	1.00			1 FS
Resource A		100.00	5.00				
Activity 4	Yes		5.00			1.00	2 SS(2d)
Resource A		100.00	5.00				
Activity 5	No						2 FS; 4 FS; 3 FS

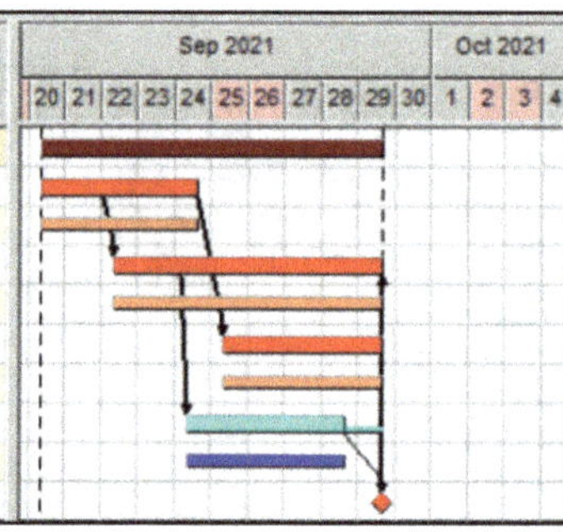

Fig. 3. Adjusted schedule of the Sample Project.

Adjusted schedule became 4 days (almost 30%) shorter when activity 2 duration was increased by 5 days.

Resource A workload on activity 2 is now 37.5% instead of 100% in the original schedule. It means that 62.5% of resource work time can be used on parallel activities.

In Spider Project software, one of the activity properties is its adjustability, i.e., whether its planned duration may be increased if necessary, and only activities with adjustable durations take part in the schedule adjustment.

5 Resource Constrained Scheduling

People can only rely on the schedules created by scheduling software if these schedules take into account all constraints and conditions that are considered by people when they schedule manually. In particular, a schedule must take into account the existing resource constraints. That is why, in most projects, it is necessary to use resource leveling and calculate feasible schedules that take into account the existing constraints.

Resource constrained schedule must be analysed the same way as CPM schedule. Project scheduling software must calculate and show correct resource-constrained floats and Resource Critical Path, also known as Critical Chain, as well as resource dependencies that show why project activities were delayed. A feasible schedule must take into account not only renewable resource constraints but also financial and material delivery restrictions.

6 Sample Project "Test"

Project management tools use different resource-constrained scheduling algorithms and create different schedules for the same projects.

Now let's have a look at another sample project that consists of four activities and the start and finish milestones. This project's CPM schedule is shown in Fig. 4.

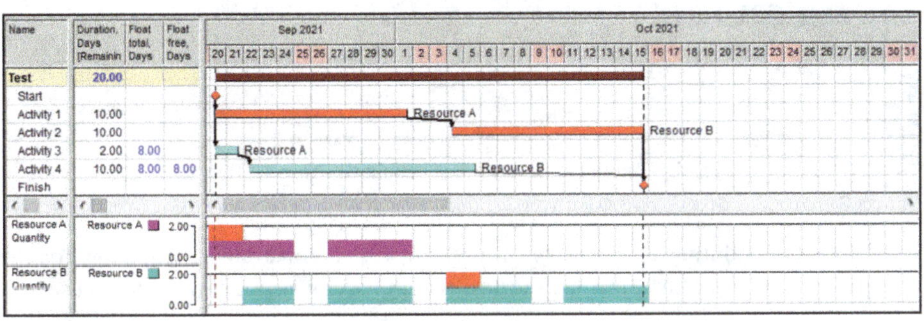

Fig. 4. CPM schedule of the project "Test".

In CPM schedule of the project "Test", resource A assigned to activities 1 and 3 and resource B assigned to activities 2 and 4 are overloaded.

7 Resource Leveling of Project "Test"

CPM schedules created by different scheduling tools are the same but resource leveling can lead to different results.

We have leveled our project using different tools.

Spider Project schedule is presented in Fig. 5. Its duration is 22 days.

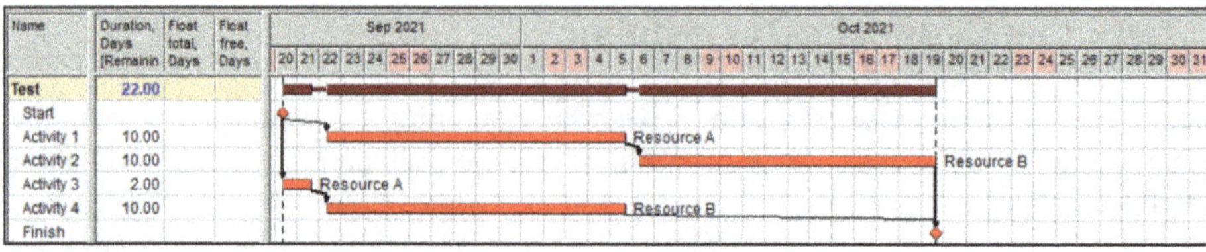

Fig. 5. Spider Project schedule of Project "Test"

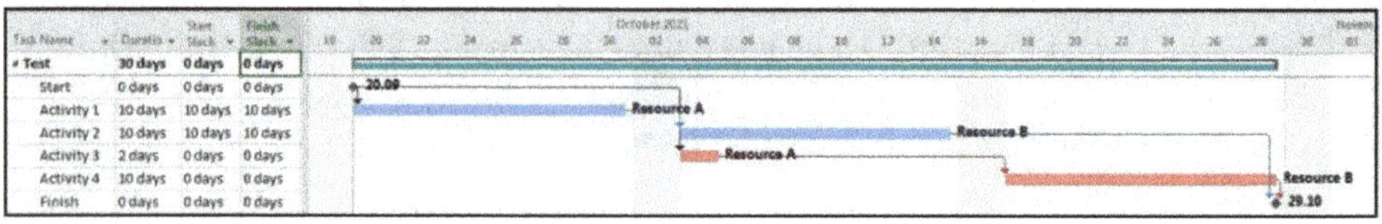

Fig. 6. Microsoft Project schedule of Project "Test"

Microsoft Project schedule of the same project is shown in Fig. 6. Project duration became 30 days.

Oracle Primavera P6 offers its users to select leveling priorities but its default schedule of the project "Test" is the same as the schedule created by Microsoft Project.

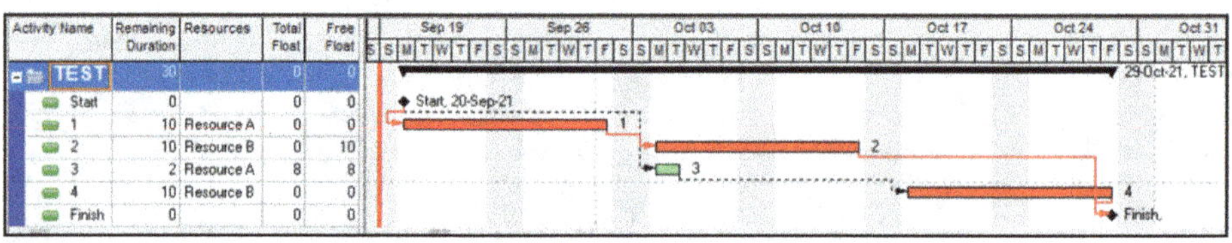

Fig. 7. Oracle Primavera P6 schedule of Project "Test"

Different scheduling tools create different resource constrained schedules for the same projects. The tool that creates shortest schedules can save a fortune to its users.

8 Resource Constrained Floats

An activity's total float in the resource constrained schedule must show the same thing as in the CPM schedule: what delay of the activity start or finish does not delay project finish.

Free float must show what delay of an activity start or finish does not delay other activities.

However, unlike the CPM floats, activity resource constrained floats depend on the project schedule. Activity float can be different in different resource constrained schedules of the same project.

In Spider Project schedule (Fig. 5), all activities are critical and this is correct.

Microsoft Project schedule (Fig. 6) shows that activity 3 is critical but it is obvious that in this schedule it can be delayed by 8 days without delaying project finish and without delaying any other activity. Microsoft Project also shows that activities 1 and 2 have 10 days total and free slack (float) although any delay of these activities delays project finish. Resource constrained floats calculated by Microsoft Project are wrong.

P6 schedule (Fig. 7) shows the right floats of activity 3 but a wrong free float of activity 2.

These and most other tools calculate wrong resource constrained floats, which can be misleading and complicates resource constrained schedule analysis and finding the best ways to improve schedules.

9 Risk Simulation Problems

There is no mathematical solution for finding the shortest resource constrained schedule and scheduling tools use their own proprietary algorithms for resource leveling. As a result, they create different schedules for the same projects. It has an impact on the Monte Carlo risk simulation requirements: it is necessary to use the same leveling algorithm for project scheduling and for risk simulation if the project has resource constraints. Otherwise, the results of risk simulation can be wrong and misleading.

Let's return to our sample project "Test". We set the optimistic durations of project activities at 20% shorter and pessimistic durations at 25% longer than most likely and performed the Monte Carlo simulation, assuming that activity duration distributions are triangular.

Below are the probability curves created for project "Test" duration scheduled using Spider Project leveling algorithm (left) and using the algorithm used by other tools (Fig. 8). You can see that the probability of finishing this project faster than in 26 days is 99.66% with the schedule created by Spider Project and only 0.075% if resources are assigned as advised by other tools.

Today there are many Monte Carlo risk simulation programs that are used with the project models created and managed using other tools. It is necessary to remember that the results of risk simulation of the projects that have resource constraints, obtained using external risk management tools, are not reliable.

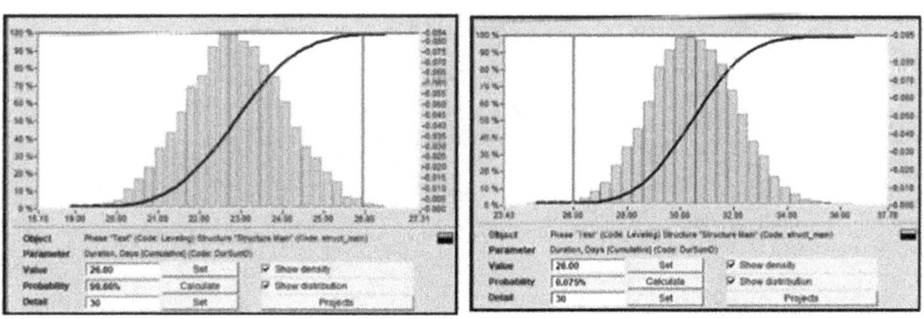

Fig. 8. Probability Distributions of Project "Test" Duration with different leveling algorithms

10 Risk Simulation Problems

Creating CPM schedule is only the first step in finding the best way to execute the project. There are tools that help to find the best way to crash project schedule (activity drag) and to optimise resource requirements using activity flexes.

Some schedules can be improved by increasing duration of certain activities, usually those that have negative drags. Finding the best way to optimise project schedules by increasing activity durations can be automated. In Spider Project, this function is called Schedule Adjustment.

Resource constrained schedule optimisation is a much more complex task. To achieve good results, it is necessary to use scheduling software tools that are able to create good project resource-constrained schedules and to calculate the right resource-constrained activity floats and drags. When you manage projects, you need to know which activities belong to the Resource Critical Path and what contingency reserves must be created to set reliable project targets. The scheduling tools must have this functionality to be used in project resource management.

A Survey of Artificial Intelligence Tools in Project Management

Alexander Mikhaylov(✉)

IT PMO at Nornickel, 1st Krasnogvardeyskiy proezd, 15, Moscow 123100, Russia
almikh@mail.ru

Abstract. Artificial intelligence today is one of the key technological drivers for project management digital transformation and evolution. In this paper we consider three main areas of artificial intelligence applications for project management – digital assistants, intelligent tools for various areas of project management and project management systems with artificial intelligence features. We also discuss the requirements of prospective artificial intelligence system for project management which is called iPMO – the requirements can be used both to develop a new intelligent project management system as well as extend the capabilities of existing solutions. Despite the impressive capabilities of artificial intelligence, project managers are responsible for decision-making. Artificial intelligence plays the roles of assistant and advisor. Artificial intelligence enhances the competence of a project manager, frees up time from routine and allows people to focus on collaborative work and complex management tasks. Project managers also should enhance leadership, communication skills and professional competencies, including managing hybrid project teams using digital assistants and intelligent tools.

Keywords: Artificial Intelligence · Digital Assistants · Robotization

1 Introduction

The first attempts for using of artificial intelligence for project management date back to the 1980s. A widescale research of expert systems for decision-making in project management was done for NASA [1]. In 1987 William Hosley published the famous article «The application of artificial intelligence software to project management» [2]. In the same years one of the first intelligent project management systems called Callisto was proposed [4]. Over the past years significant progress has been achieved in the artificial intelligence methods and techniques such as machine learning, neural networks, speech recognition, machine vision as well as robotizing of routine operations and others. The present wide usage of artificial intelligence is also driven by availability of large computing power to run the applications and big data processing.

Today artificial intelligence is one of the key technological drivers for project management digital transformation and evolution [3]. According to a survey provided by IPMA in association with PwC «Artificial intelligence impact in project management», the most expected artificial intelligence roles are project manager assistant (52%) and

S. Bushuyev et al. (Eds.): IPMA 2021, LNNS 704, pp. 99–105, 2023.
https://doi.org/10.1007/978-3-031-34629-3_10

project management advisor (44%) [6]. PMI also highlights the importance of using artificial intelligence to improve performance of project management [5]. PMO Flashmob has published a report considering of opportunities and threats of artificial intelligence in PMO [8]. In Project Association we completed the research of using artificial intelligence in project management, the summary is presented below and the full reports are available online [7, 9].

2 AI Capabilities in Project Management

There are three main areas of artificial intelligence in project management can be defined:

- Digital assistants to the project manager;
- Intelligent tools for various areas of project management (planning, risk management, contract management, communication management and etc.);
- Project management systems within features of artificial intelligence.

2.1 Digital Assistants

Digital assistants is one of the most rapidly developing and promising areas of artificial intelligence for project management. The digital assistants can help the project manager in different ways - in coordinating projects, monitoring the execution of tasks and assignments, organizing meetings, preparing reports, updating data in the project management system and other tasks. As usual a digital assistant communicates through a messenger like WhatsApp, Telegram or others. Communications are possible in various ways - using menu commands, text messages or even by voice. Communication capabilities depend on the functionality of the assistant and the use of artificial intelligence. An example of such solutions is the digital assistant Dina [16] and Bitrix24.Assistant implemented on the Bitrix24 platform [14]. Popular voice assistants, for example, Yandex.Alisa or Google.Assistant, also can be used for communications. Another application is Stratejos – a digital assistant for the Slack platform that is used by many project teams. Stratejos can process project data and saves time on switching between Slack and external systems such as Jira or Trello [26].

A promising area of work is the use of artificial intelligence for coaching project teams. Digital coaches use machine learning to identify patterns and deviations in the behavior of project participants, to analyze the performance of the project team. A digital coach will help the project manager understand the psychological condition of the project team members, provide feedback on the team's work and options to improve productivity. Examples of such solutions are Isabella AI Coach [17] and AIDA Agile Coach for Jira [11].

2.2 Intelligent Planners

The intelligent planners can help for development of project plans. These tools are especially effective for large projects with complex constraints. Intelligent planners allow us to track many project parameters, resource availability, changing task relationships and priorities. It also can be used to understand how changes in one project will affect related projects in a portfolio. An example is Aurora, originally this tool was developed for NASA, then it was widely used in other organizations and areas [13]. Liquid Planner is another example of an intelligent planner tool [18].

2.3 Risk Management

Risk management is perhaps one of the most demanded direction for the development of intelligent tools. For example, the PPM Insights module connects to project management systems such as Microsoft Project Online or Jira. The intelligent module analyzes project data to identify risks, predict the progress of projects. In case of project plan deviations, the artificial intelligence can help the options for risk mitigation [22].

2.4 Predictive Analytics

Identifying and monitoring project risks is closely related to predictive analytics - may be the most impressive feature of artificial intelligence. Predictive analytics tools answer the question "What will happen?". Such tools are based on machine learning and are able to make conclusions and learn from the projects, analyze projects based on previous experience. This gives us the ability to predict what might happen in projects and the whole portfolio. Such a good example is RISHI-XAI tool, which combines a knowledge base, machine learning system and explanatory artificial intelligence (eXplainable AI) [23].

2.5 Contracts Management and Documents Checking

Intelligent tools can improve the speed and quality of document processing. This is important for project contracts for the performance of works, services, the purchase of licenses and goods. It is also important for regulatory documents on the organization of project activities. AI applications in this area use semantic modeling of contracts and other documents. It is achieved by creation a document model to check it in digital form. Then it becomes possible to establish a links between documents, to identify contradictions and duplicates in documents.

The verification of contracts is accelerated and the risks of their conclusion can be reduced. This is achieved by identifying risky clauses and additional agreements that may violate previously reached terms. Such tools are especially effective in the case of complex project contracts, consisting of tens or even hundreds of pages, with different pricing schemes and terms of delivery of goods and services. Usually intelligent tools support text recognition, semantic processing and information search in documents by meaning. Machine learning allows you to classify documents, automatically extract document details. An example of tools in this area is Naumen Legal Tech [19] and Nia Contracts Analysis, which is part of the intelligent platform Infosys Nia [20].

2.6 Digital HR Partners

The direction of digital HR assistants is rapidly developing, using the capabilities of artificial intelligence to help in hiring project personnel, selecting the most optimal candidates for project managers, and forming project teams. Artificial intelligence can strengthen the capabilities of HR partners, free HR specialists from routine work. Digital HR helpers can undertake the doing of a large number of the same type of calls and the initial check of the candidates' resume. Artificial intelligence applications can even provide an initial interview using instant messengers. Digital HR assistants provide project managers and HR professionals with pre-selected resumes and analytics of candidates. Such services as Sever.AI [24] or PineStem [21] can be useful in this area.

2.7 Advanced Project Management Systems

The functions of artificial intelligence and robotization are used in various project management systems. For example, Autodesk Construction IQ helps to manage construction projects using Autodesk BIM 360 platform. The system receives and analyzes the entire set of data on project implementation, quality and safety of construction sites, and regularly monitors and evaluates risks [15]. The Advanta system provides the customized triggers – active agents performing a large number of routine actions that previously had to be performed manually [10]. Wrike's artificial intelligence can monitor and predict the project risks [27] and be helpful for selection the most critical project tasks for execution [28]. The Aitheon Project Manager system supports project management using artificial intelligence, provides automated formation and assignment of tasks, and risk analysis [12]. The Smart Projects is used multi-agent systems and knowledge bases (ontologies) for adaptive planning and resources management [25]. These are just a few examples of the implementation of artificial intelligence and robotization in project management systems. There is no doubt that in the future the capabilities of artificial intelligence will expand, and new intelligent systems will also appear.

3 IPMO Requirements

The requirements for the artificial intelligence system of PMO were developed in research of Project Association [9]. This promising artificial intelligent system is called iPMO and was designed to help manage projects, programs and project portfolios. Artificial intelligence can help in project planning and monitoring, risk management, issues prediction, problems solving and other areas. However, project managers are responsible for making the final decisions and this is an indispensable condition.

3.1 Conceptual Architecture

iPMO is thought as a virtual member of the project team. The system should interact with project managers, team members and stakeholders. For this it should be integrated with communication applications primarily to e-mail, audio and video conferencing systems, corporate social networks and instant messengers. The system should have access to the projects portfolio, support bi-directional integration with an existing corporate project management system or even be part of it. The access to external databases and files, integration to the corporate information systems and business intelligence modules such as PowerBI or Qlick are also required.

3.2 Key Features

All requirements are belonging for two stages – project planning and execution, as well as for the domains of system usage such as analytics requirements, risk management, communications management and others. iPMO should be available 24/7 through textual and voice communications, provide consulting support for project participants and stakeholders and quick answers to typical questions.

The system should monitor the project risks and tasks progress using the results of semantic analysis of the entire set of project data - reports, documents, communications, meetings records and etc. To monitor the project, the system should support machine vision for using the cameras and drones. It also should be possible to collect the status of project tasks and understand the project progress via sensors and a kind of chips and electronic labels in products and objects - where they are and what about them.

Based on the information collected the system should provide early detection of potential risks, problems and possible changes in projects, inform the project manager about the issues. The system should provide the options to mitigate the detected risks based on knowledge base of previous projects and results if machine learning. It's well to estimate the options using utility function based on quantitative indicators.

An important feature of the system should be to identify the interdependencies between different projects in the same portfolio. This feature is critical in the case of blocking, when the tasks in one project can be completed only after tasks in other projects. This problem may occur is case of project teams have a lack of communications and they don't see interdependencies with each other. Artificial intelligence overseeing the entire portfolio and can greatly help to mitigate such risks.

As a virtual participant of the project iPMO can initiate meetings and discussions of open issues itself. Project participants can use augmented (AR) and virtual (VR) reality devices to visualize the project progress and final products vision and construction objects.

Obviously iPMO should use the project data for self-learning. The project management knowledge base should be for this and support Big Data processing. Knowl edge verifications should be provided by project experts and experience, knowledge and lessons learning from previously completed projects should be used for planning and monitoring of new projects.

3.3 Analytics

An intelligent system should provide four main types of analytics: descriptive analytics – answers the question "what happened?", diagnostic analytics – "why did this happen?", predictive analytics – "what will happen?" and prescriptive analytics – "what can we do?". Predictive and prescriptive analytics are the most expected functions of artificial intelligence, they carry the most value for project management, however they also the most complex for implementation. For example, predictive analytics should provide the believable end date of a project or the current stage of a project. The likelihood of project risks is also a kind of predictive analytics as well as the forecast of the project budget spending. Prescriptive analytics makes it possible to provide options for problems solving in the case of a project plan deviation or risk occurring. The system should provide a quantitative assessment of the proposed options, estimation of their impact on the timing and budget and project risks.

3.4 Requirements Usage

The provided requirements can be used for extending functionality of existing project management systems or for developing a new system. From a practical point of view the effectiveness of artificial intelligence usage in project management should be assessed on standard indicators like increasing the number of successful projects and customer satisfaction. The successful usage of artificial intelligence requires a sufficiently high level of project culture. Basic automation of project management processes should be done first as well as project data should be migrated into digital environment.

4 Conclusion

Artificial intelligence can be defined as amplification and partial imitation of human intellectual abilities. Despite the impressive capabilities of artificial intelligence, project manager is responsible for decision-making. Artificial intelligence plays the roles of assistant and advisor. This ethical issue of responsibility is actual to all areas of artificial intelligence usage. Project managers no need to worry about artificial intelligence taking their jobs. Artificial intelligence enhances the competence of a project manager, frees up time from routine and allows people to focus on collaborative work and complex management tasks.

In the future we will certainly meet the Hybrid Project Teams (HPT) in which humans, digital assistances and other artificial intelligence agents including robots will work together. Insofar as artificial intelligence rises up project managers should enhance leadership, communication skills and professional competencies, including managing hybrid project teams using digital assistants and intelligent tools.

Acknowledgments. I would like to thank Project Association (https://projects.management/) volunteers for their contribution in research of using artificial intelligence in project management and chairman Vadim Bogdanov for organizational support. The full lists of participants are available in online reports [7, 9].

References

1. Harris, R., Shaffer, S., Stokes, J., Goldstein, D.: Application of expert systems in project management decision aiding. NASA, Goddard Space Flight Center, Greenbelt, Md. (1987)
2. Hosley, W.: The application of artificial intelligence software to project management. Proj. Manag. J. **18**(3), 73–75 (1987)
3. Mikhaylov, A.: Using artificial intelligence in project management. Proj. Program Manag. **1**, 6–12 (2021)
4. Sathi, A., Morton, T., Roth, S.: Callisto: an intelligent project management system. AI Mag. **7**(5), 34–52 (1986)

5. AI Innovators: cracking the code on project performance (2019). https://www.pmi.org/learning/thought-leadership/pulse/ai-innovators. Accessed 02 Feb 2023
6. Artificial intelligence impact in project management (2020). https://www.ipma.world/assets/IPMA_PwC_AI_Impact_in_PM_-_the_Survey_Report.pdf. Accessed 02 Feb 2023
7. Artificial intelligence tools for project management (2020). https://projects.management/infopage.html?Page=aicatalog&ReferrerID=375f0ec4a7. Accessed 02 Feb 2023
8. The AI PMO – threat or opportunity? (2020). https://houseofpmo.com/members/library/inside-pmo-ai-pmo-threat-or-opportunity/. Accessed 02 Feb 2023
9. Requirements for the artificial intelligence system of PMO (2020). https://projects.management/infopage.html?Page=aireq&ReferrerID=375f0ec4a7. Accessed 02 Feb 2023
10. Advanta Triggers. https://wiki.a2nta.ru/doku.php/product/triggers/. Accessed 02 Feb 2023
11. AIDA Agile Coach. https://www.digitalmarketplace.service.gov.uk/g-cloud/services/175170253001707. Accessed 02 Feb 2023
12. Aitheon Project Manager. https://aitheon.com/project-management. Accessed 02 Feb 2023
13. Aurora. https://www.stottlerhenke.com/products/aurora/. Accessed 02 Feb 2023
14. Bitrix24.Assistant. https://www.bitrix24.ru/apps/?app=bitrix.assistant. Accessed 02 Feb 2023
15. Construction IQ, https://construction.autodesk.com/tools/construction-iq/, last accessed 2023/02/02
16. Dina. https://dinabot.com/. Accessed 02 Feb 2023
17. Isabella AI Coach. https://www.pr.com/press-release/722469. Accessed 02 Feb 2023
18. Liquid Planner. https://www.liquidplanner.com/. Accessed 02 Feb 2023
19. Naumen Legal Tech. https://www.naumen.ru/products/legal_tech/. Accessed 02 Feb 2023
20. Nia Contracts Analysis, https://www.edgeverve.com/artificial-intelligence/nia/nia-contracts-analysis/, last accessed 2023/02/02
21. PineStem. https://pinestem.com/. Accessed 02 Feb 2023
22. PPM Insights. https://ppm.express/ppm-insights/. Accessed 02 Feb 2023
23. RISHI-XAI. https://www.digite.com/news/the-worlds-first-xai-explainable-ai-product-for-enterprise-project-intelligence/. Accessed 02 Feb 2023
24. Sever.AI. https://sever.ai/. Accessed 02 Feb 2023
25. Smart Projects. http://www.kg.ru/solutions/smart-projects/. Accessed 02 Feb 2023
26. Stratejos. https://stratejos.ai/. Accessed 02 Feb 2023
27. Wrike AI Project Risk Prediction. https://help.wrike.com/hc/en-us/articles/360055046934. Accessed 02 Feb 2023
28. Wrike AI Recommended Tasks. https://help.wrike.com/hc/en-us/community/posts/360052602393. Accessed 02 Feb 2023

Developing Management and Implementation Mechanism for a Digital Supply System Transformation Programme in Construction Company

Diana Hayrapetyan(✉) and Anna Yakovleva

Graduate School of Business, National Research University Higher School of Economics, 20 Myasnitskaya ulitsa, Moscow 101000, Russia
hayrapetyandiana@gmail.com

Abstract. Recent years have been marked by a rapid development and digitalization of many processes, leading to the transformation of traditional ways of doing business. In these increasingly fast-changing times, companies need to adapt to the changes in order to enhance their marketability and, accordingly, digitalize their core business processes. Although a significant number of studies where held, there is a lack of methodological approaches to develop a set of success factors according to which the digital supply system transformation programme in construction is implemented. The aim of this study is to develop a management mechanism for a digital supply system transformation programme and provide recommendations for its implementation in the construction company.

In order to develop a management mechanism for digital supply system transformation programme in a project-oriented construction company, the drivers, as well the key measures and success factors of the digital transformation programme were identified. Afterwards, the significance level of identified measures and success factors of the programme were evaluated through questionnaires among stakeholders that enabled to focus on the most valuable aspects of the programme.

Based on the data obtained, a management mechanism for the digital supply system transformation programme for a construction company was developed. Furthermore, the recommendations for its implementation in the construction company were developed and presented to the management of the company.

Keywords: Digital Supply System · Digital Transformation Programme · Programme Management · Programme Success Factors · Programme · Transformation

1 Introduction

The twenty-first century can be described as an era of globalization and digitalization. The coming years are characterized by the impetuous digitalization of many processes, leading to the transformation of traditional business practices. At these fast-changing

S. Bushuyev et al. (Eds.): IPMA 2021, LNNS 704, pp. 106–120, 2023.
https://doi.org/10.1007/978-3-031-34629-3_11

times, companies need to adapt to the changes to improve their competitiveness and, accordingly, transform their core business processes. This is driving the relevance of this research.

One of the most important processes in construction is the supply system in the company. Successful supply system has a major impact on the overall success of a construction project. Various studies have been conducted to identify the relationship between supply success factors and project success [3].

Digital business process transformation requires an analysis of the current system in a company and the identification of problem areas. Based on the revealed information, it is possible to determine the drivers of the transformation. Furthermore, once the drivers of the program are analyzed, it is necessary to define the aim of the transformation programme.

Moreover, in order to manage the transformation programme more efficiently, we need to identify and rank the success measures and factors of the digital transformation by significance. Hence, studies on the topic include questionnaires or interviews for the significance assessment of the transformation success factors [15]. While there is a considerable number of studies on digital transformation, there is a lack of methodological approaches regarding the analysis of success measures for a digital transformation programme in construction.

The aim of our study is to develop a management mechanism for the digital supply transformation programme and provide recommendations for its implementation in the construction company.

In order to achieve the aim of the study the following objectives were to be accomplished:

1. Analyze the theoretical foundations of transformational program management
2. Identify the drivers and the aim of the transformation
3. Determine the most significant success measures and success factors of the programme
4. Develop a management mechanism for a digital supply transformation program in a construction company

Throughout the study, a detailed research analysis was conducted to define the concept of digital transformation and transformational programme management. Moreover, best practices of digital transformation programme management were analyzed to identify its peculiarities.

The study investigated a digital supply transformation programme by identifying the drivers and defining the aim of the transformation programme. The aim of the programme is to digitally transform the supply system in the construction company, increase transparency of the process, improve planning quality and reduce budget overruns and delays in deadlines by at least 10%.

Through interviews and stakeholder surveys, the most significant success measures and factors for a construction company's digital supply transformation programme were identified. Based on the findings, we developed a management mechanism for a digital supply transformation programme, that consists of 5 steps:

1. Define the purpose of digital transformation;

2. Prepare the organization for transformation;
3. Launch digital transformation;
4. Conduct digital supply transformation;
5. Maintain changes.

The mechanism, as well as recommendations for its implementation, have been handed over to the management of the company and applied to manage the digital supply transformation programme.

2 Literature Review

Over the past few decades, global industries have not only faced technological changes that have led to opportunities such as increased flexibility, reactivity and product customization, but have also posed a variety of challenges such as rapid technological change, increasing levels of complexity and changing consumer preferences and legal requirements [12]. This has led to complex situations in the corporate context: many new technological possibilities are perceived, but people are unsure how to use and implement them simultaneously in terms of product and service offerings [6].

Digitisation (i.e. the process of converting analogue data into digital data sets) is defined as the use of digital capabilities. Digital transformation is defined as a process that is used to restructure the economy, organisation and society at a systemic level [2, 16].

Digital transformation encompasses changes at all levels of society and, by combining different technologies, opens unanticipated opportunities and the potential to create radically new products, services and business models [8]. These innovations can lead to new forms of collaboration between organisations or changes within an organisation [5].

Digitalisation, and subsequently digital transformation, is driving change in the corporate world as it creates new Internet-based technologies with implications for society as a whole [16]. While digitisation describes the process of converting analogue and noisy information into digital data [2], digital transformation is used to describe any change in an organisation due to their increasing use of digital technology to improve both performance and scope.

As the pandemic moves through its inevitable phases, companies that have paused transformation will want to resume or even start at a new pace and accelerate the realisation of a transformational aspect of the organisation. But if they are experiencing a COVID-19-related slowdown and are forced to focus on more pressing issues, they may have to continue working with limited resources. With less margin for error, companies will look for new ways and resources to ensure the success of their transformation programmes.

In the beginning, there was a strong emphasis on the use of digital technology. Then it became clear that digital transformation involves more than just a technological shift. Moreover, it does not require technology, but it is also important to ensure coherence between the organisation's strategy and other external and internal factors. Some definitions focus on the impact of digital transformation, such as improving operational efficiency. Some authors include customer value creation, such as optimising customer needs and experiences [14], in their definitions, while others exclude it. The lack of a coherent definition and underlying elements in the literature remains a huge problem [9].

The speed at which digital technologies can spawn new "smart" products and services is only comparable to their ability to extend the reach and range of social interactions through ubiquitous infrastructure and flexible platforms [10]. Firms are under pressure not only to change existing business models, but also to manage a portfolio of different business models to cope with increasingly fickle customers who demand flexibility and personalisation of products and services [7]. However, radically changing the business model(s) and organisation to exploit new technologies is neither simple nor straightforward. It involves stepping out of your comfort zone and possibly abandoning practices that employees and customers have come to expect or even take for granted.

Even though technical barriers are being erased and the digital landscape is expanding with new opportunities, technology does not automatically bring convenience or value unless companies carefully consider the context in which it is deployed and how to extract practical or monetary benefits. Indeed, some have referred to digitised products and services as 'social cyber-physical offerings', stressing that value is derived only when physical, technical and social systems are intelligently intertwined [11].

A business process transformation programme focuses on the ability of making changes to exactly how the business is run and how repetitive processes in the company are produced, and can include agile transformation. Typically, business process transformation programme involves streamlining and automating repetitive processes so that the organisation can focus on projects of higher importance and cost. It is an ongoing effort that starts with the most common processes and then moves on to those with less impact. The ultimate goal of business process transformation is to free the company from the burden of secondary tasks so that it can innovate or deliver higher value services and offerings to the market [13].

A number of academic publications and studies show the importance of digitalisation for the construction industry to organizations in the sector. A team of authors [1] conducted a literature survey on digitalisation to identify research topics in this area and to propose recommendations for future research.

The researchers linked digitalisation, innovation in the business model and sustainability in the construction industry environment. Their aim is to raise the academic debate on how companies in industry can take advantage of digitalisation to innovate in the business model. In particular, they seek knowledge relating to achieving sustainable benefits for industry that represent the greatest potential for economic, environmental and social impact.

Despite the considerable amount of research, there is still the problem regarding deficient methodological approaches to managing the digital transformation programme in construction.

3 Research Design and Methods

The research consisted of four consecutive phases that are presented in Fig. 1. The first phase of the study analyzed scientific research and best practices for transformational programme and digital transformation programme management. As a result, the main aspects of change in company, as well as a set of success factors for a transformation programme were identified.

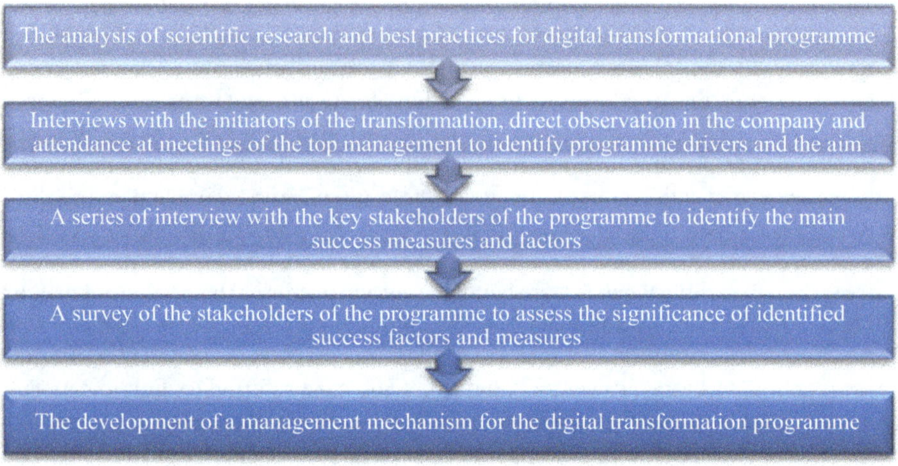

Fig. 1. Research design

The second phase of the study consisted of interviews with the initiators of the transformation, direct observation in the company and attendance at meetings of the top management, including the founders of the company, CEO, Deputy CEO and others. Subsequently, the drivers of the digital supply system transformation of the company, as well as the aim of initiating the transformation were identified.

As the third phase of the study, a number of interviews were conducted with key stakeholders and beneficiaries of the digital supply transformation programme in the construction company. The interviews were conducted with management of the company and the members of the programme team (programme performers).

Key research findings in the third phase included the identification of the measures by which the success of digital supply transformation programme were assessed. Moreover, programme success factors were identified.

The fourth stage of the research included a survey of the stakeholders of the programme in order to assess the significance of the identified success measures and factors. Respondents were asked to evaluate the programme success measures and factors on a Likert scale from 1 to 7, where 1 is not significant and 7 is strongly significant for programme success.

The survey was sent to 55 stakeholders of the transformation programme and 43 relevant responses were received from company employees, whose work is directly or indirectly related to the programme.

About 30% of the respondents were from fundamental area, 21% from a technical area and 16% from a programme team. The percentage of the management among respondents was 33%, including 7% of senior management, 12% of construction project managers, and 14% of business unit and department heads. The percentage distribution of the percentage of respondents by position is presented in Fig. 2.

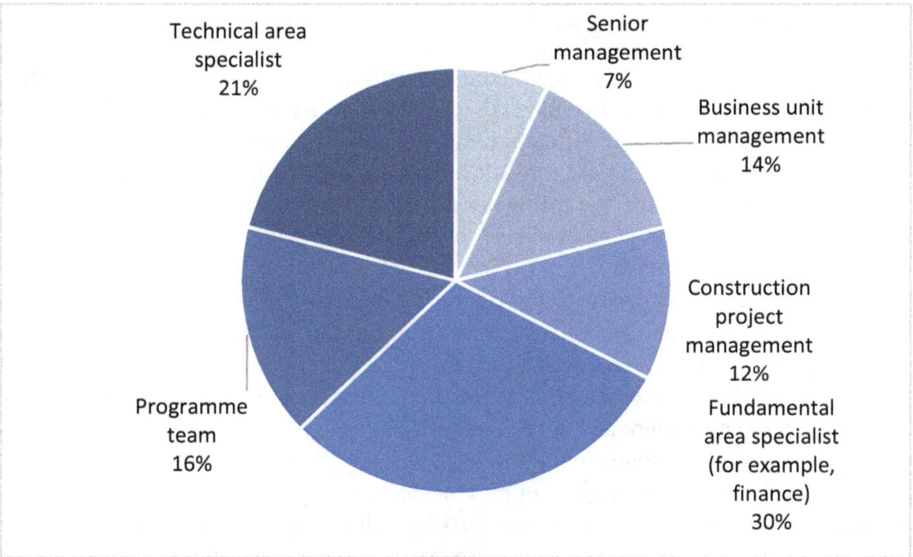

Fig. 2. Distribution of survey responses by position

In the fourth stage, the previously identified measures and success factors were ranked based on the average scores of their significance as per the results of the survey. In addition, we took into account the average scores of importance of the measures and factors received from the management and decision makers of the transformation of the company. As a result of the fourth stage of the research, the most significant measures and success factors of the digital supply transformation programme for the stakeholders of the construction company were identified.

Based on the data obtained during the fourth stage of the study, a management mechanism for the digital supply transformation programme was developed, considering the most significant success factors of the programme identified in the previous stages. Furthermore, the study identified recommendations for the implementation of the management mechanism of the digital transformation programme for the company.

4 Results and Discussion

The key drivers behind the digital supply transformation programme have been identified. By drivers, in this case, we understand the tasks set before the organization, the solution of which is crucial for the further functioning of the company's supply system at a proper level and without disruptions. As a result of processing all the information received at the meetings and interviews, the following drivers of the programme were identified..

Based on all the information obtained about the programme in the study, we developed a management mechanism for digital supply transformation programme in a construction company. The results of the research are presented below.

4.1 Programme Drivers and Aim

The key drivers behind the digital supply transformation programme have been identified. By drivers, in this case, we understand the tasks set before the organization, the solution of which is crucial for the further functioning of the company's supply system at a proper level and without disruptions. As a result of processing all the information received at the meetings and interviews, the following drivers of the programme were identified.

The first driver of the programme is to improve the quality of expense and labor cost calculations for a construction project as well as for specific parts of the project. Actual cost calculation is of great importance to a company for a number of reasons. To begin with, up-to-date and detailed delivery cost information can be used to improve the quality of planning for future work of the construction project. Besides a company will have an opportunity to monitor price trends in the market to increase the efficiency of the procurement department and at the same time alternate the construction technology, using less expensive materials without compromising the quality.

The second driver of the digital supply transformation programme is the reduction of coordination time, as well as the processing of purchase orders. Currently the period of coordination of purchase orders in the company is not regulated. Thus, can often exceed the estimated delivery time for the construction project, delaying not only delivery time, but also the timing of critical works and the entire construction project in the long term. In addition, since the constructors who submit orders are aware of the risk associated with delivery delays, they may order materials much earlier than required, thereby creating the need for extra costs for storage and security of materials in the company's warehouses. Moreover, in case of large-sized materials there are additional expenses for the transportation of materials from the central warehouse of the organization to the target construction facility.

The next driver of the programme is to increase the transparency of coordination and purchase order processing. First, during the coordination of orders, adjustments can be made that are not necessarily conveyed to the initiator of the order. Such adjustments may result in delivery of wrong materials (material characteristics may differ from required) forcing to duplicate the orders with delayed delivery date. Secondly, because of low transparency of order coordination process, the initiator cannot know for sure at what stage of coordination the order is and whether it was not declined or abandoned altogether. Thirdly, when an order is processed, the lack of transparency can lead to poor quality of sourcing by a procurement department. Moreover, it presents difficulties for the analysis and trucking the key performance indicators.

An important driver of the programme is the improvement of the planning quality. This applies both to the planning of supply volumes and to the costs of materials for a construction project. Moreover, the lack of necessary up-to-date information leads to poor quality of analysis and makes it nearly impossible to develop a list of planned procurement volumes for a construction project. Improving the quality of planning will mostly be reflected in analytical work aimed at comparing the planned and actual indicators of costs and volumes of supplies.

Another driver of the transformation programme is the facilitation of real-time interaction between departments. Often the insufficient prompt exchange of information can lead to serious consequences. For example, a supply department employee needs to have up-to-date information about available inventory in the warehouses of the company when processing orders. This will decrease the chance of purchasing materials that are already available in the warehouses. Furthermore, information exchange between the constructors and engineers is very important. This necessity is especially important when there are additional works on a construction facility, which implies making changes in the project documentation.

One of the most important drivers of the transformation programme for the company's senior management is access to up-to-date information about costs and the possibility of generating "instant" real-time reports. In addition, management is considering the need to create an internal audit/technical control department to carry out analytical work and verify the reliability of the information provided.

The last, but not least driver of the programme is the standardization and regulation of all business processes. Some processes are different in various construction projects, which leads to a decrease in management efficiency of some projects due to lack of experience or qualification of the project manager. The regulation of processes common in all projects of the company will enable to exchange experience, lessons learned and use best practices.

Analysis of the main drivers of digital supply transformation in the construction company resulted in identification of the aim of the programmes that was agreed with the initiators. The aim of the programme is to digitally transform the supply system in the construction company, increase transparency of the process, improve planning quality and reduce budget overruns and delays in deadlines by at least 10%.

4.2 Programme Success Factors and Measures

Once the aim of the programme had been identified, we started conducting interviews with key stakeholders of the digital supply transformation programme. We interviewed the heads of various departments, including procurement, accounting, warehouse. Furthermore, a follow-up interviews with the company's senior management were conducted. During the interviews the question of the success measures of the programme was raised. By success measures we refer to measures by which the success of the digital transformation programme is going to be evaluated. As a result, the following programme success measures were identified:

- reduced variance between planned and actual quantities of requirements by 10%;
- reduced variance between estimated and actual costs by 10%;
- pre-agreed whole scope of supply for the construction works of the project;
- bonding the cost of materials in orders to the cost of specific works within the construction project;
- automatic recalculation of the construction project cost when materials are delivered to the target warehouse;
- reduction of order coordination time to 1 day;
- reduction of delivery request processing time by up to 2 working days;
- possibility to track coordination phases in real time;
- automatic data boosting on availability of purchase order's positions at warehouses;
- automatic reservation of warehouse inventory for a particular order;

- generation of “instant” reports for the management in real time;
- regulations of all business processes and clear job descriptions;
- quick access to scanned copies of documentation with reference to the orders for authorized persons.

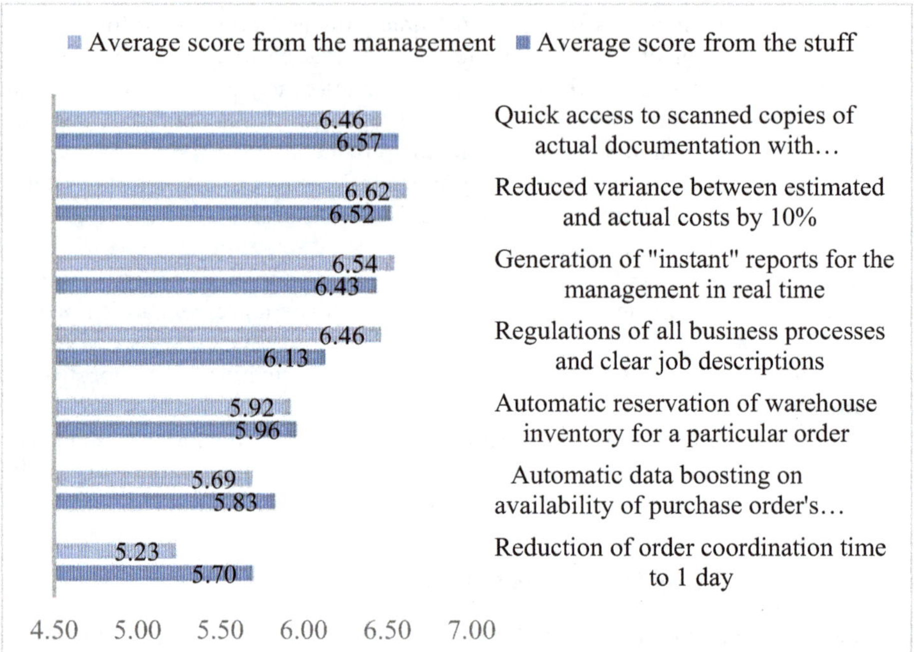

Fig. 3. The average rating of importance for the success measures of a digital supply transformation programme

Figure 3 shows the average score of the importance of the success measures as per survey. According to respondence, the most significant success measure is “Quick access to scanned copies of documentation with reference to the orders for authorized persons” with a standard deviation of 0.51. But it can be noted that for the management “Reduced variance between estimated and actual costs by 10%” and “Generation of “instant” reports for the management in real time” are more significant. “Reduction of order coordination time to 1 day” is characterized by a high value of the standard deviation of estimates, that indicates the divergence of employee opinions.

The study also included the analysis of success factors for the digital supply transformation programme. Initially, a list of digital transformation success factors was formed based on the best practice analysis. The identified factors were adjusted and adapted to the digital transformation programme in company. As a result, the success factors of the programme were divided into three main parts:

1. factors related to the company’s employees;
2. factors related to the vision of the transformation programme;

3 general factors.

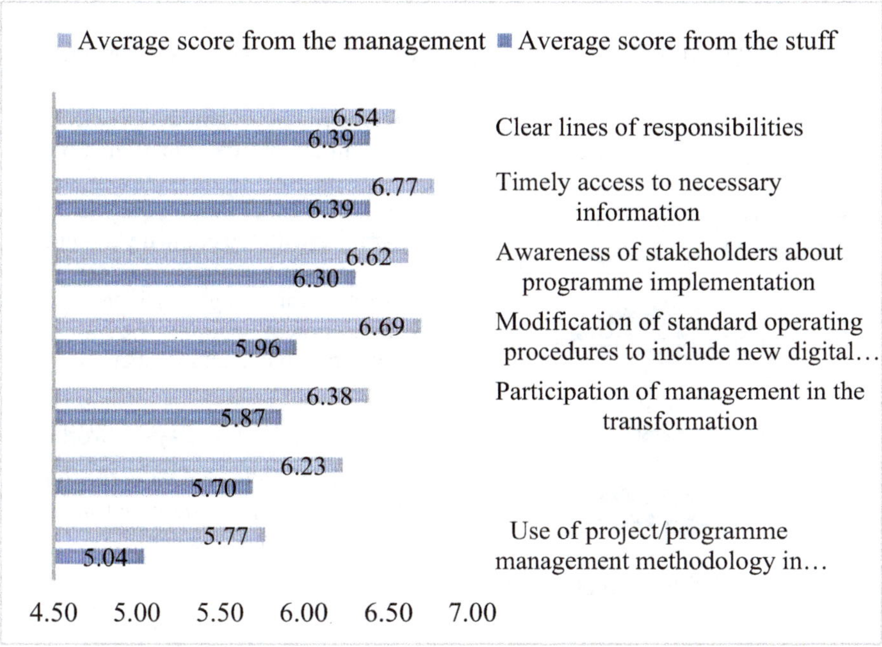

Fig. 4. The average rating of importance for the general success factors of a digital supply transformation programme

The general success factors for the digital supply transformation programme of company are:

- clear lines of responsibilities;
- awareness of stakeholders about programme implementation;
- participation of management in the transformation;
- timely access to necessary information;
- use of project/programme management methodology;
- modification of standard operating procedures to include new digital technologies;
- the strategic correlation between operational activities and digital transformation programme implementation activities.

Figure 4 illustrates the results of the importance scores for the general success factors of the digital supply transformation programme by employees and management of company. The most significant factor based on employee responses is building a clear line of accountability in the management of the program. This factor is very important both during meetings and decision making. The company needs to develop a clear hierarchical system of responsibilities. According to the management of the company, the most significant general factor was the timely availability of information necessary to implement the digital transformation. This factor is especially important given that third parties are supposed to be involved in the implementation of the programme.

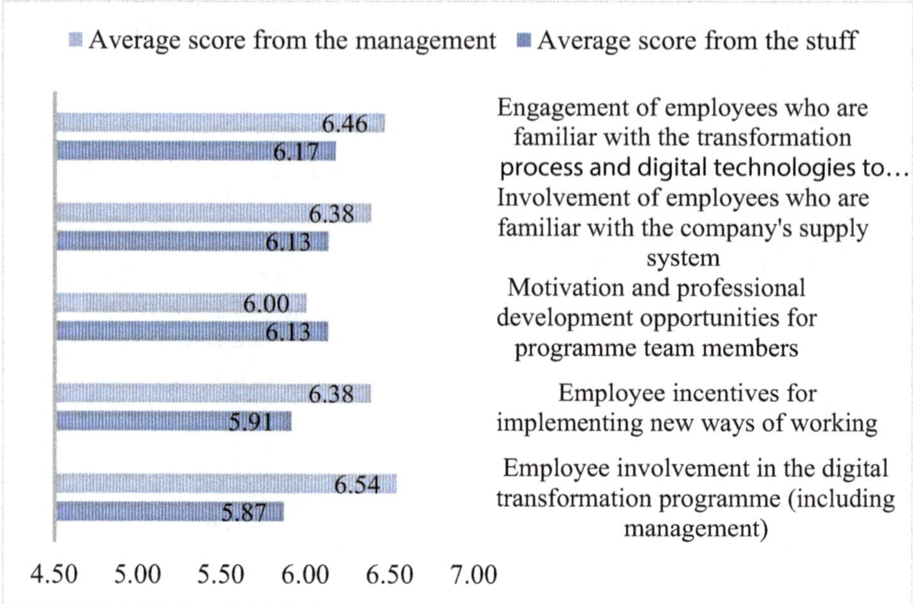

Fig. 5. The average rating of importance for success factors of a digital supply transformation programme related to the employees

The success factors of the digital supply transformation programme related to the employees of the company are:

- motivation and professional development opportunities for programme team members;
- employee incentives for implementing new ways of working;
- employee involvement in the digital transformation programme (including management);
- engagement of employees who are familiar with the transformation process and digital technologies;
- involvement of employees who are familiar with the company's supply system.

Figure 5 highlights the factors, as well as their average scores of importance received from employees and company management. The most significant success factors of programme according to the respondence are the engagement of employees familiar with the process of digital transformation and employees familiar with the current business process in company.

The success factors associated with the vision of a digital supply transformation programme are:

- transparent project management and decision-making processes;

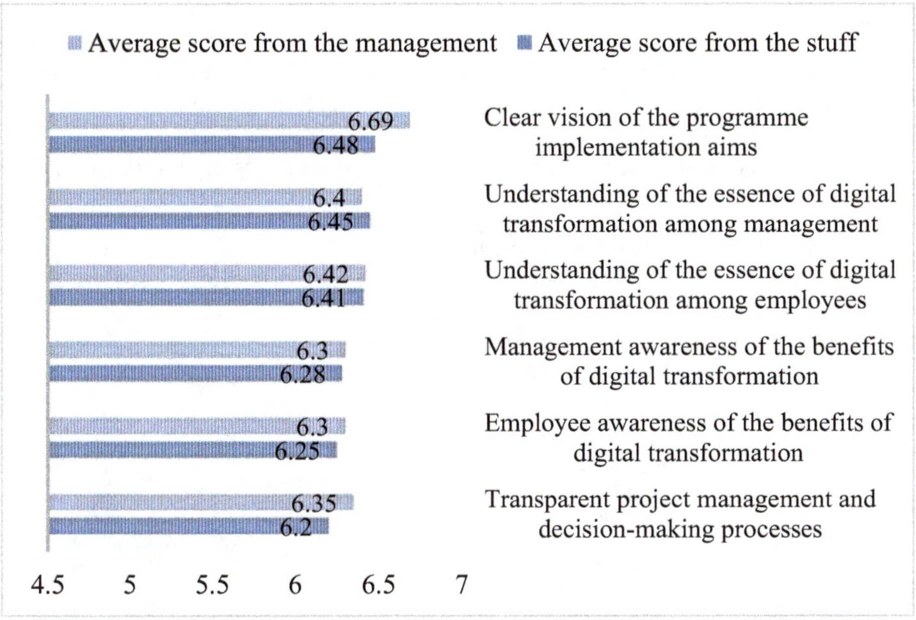

Fig. 6. The average rating of importance for success factors of a digital supply transformation programme related to the vision

- a clear vision of the programme implementation aims;
- employee awareness of digital transformation benefits;
- management awareness of digital transformation benefits;
- management's understanding of digital transformation necessity;
- understanding of the essence of digital transformation among employees;
- understanding of the essence of digital transformation among management.

The importance scores of the programme success factors as per the survey are presented in Fig. 6. The most significant success factor according to both employees and management is a clear vision of digital transformation implementation aims.

4.3 Digital Supply Transformation Programme Management Mechanism

Based on the results of this study a management mechanism the programme in construction company was developed. In order to get to the target state, the company needs to make the following 5 consequent steps:

1 Define the purpose of digital transformation;
2 Prepare the organization for transformation;
3 Launch digital transformation;
4 Conduct digital supply transformation;

5 Maintain changes.

The management mechanism is illustrated in the Fig. 7. It is necessary to consider the factors of each step of the digital supply transformation programme management mechanism separately.

The first stage of the mechanism is to define the purpose of digital transformation that includes:

- analysis of the current state of the company;
- keeping the interview stakeholders and identify the need for digital transformation;
- assessment of the urgency of the transformation;
- revealing of the success criteria for the programme;
- providing a clear vision of the goals of implementing digital transformation.

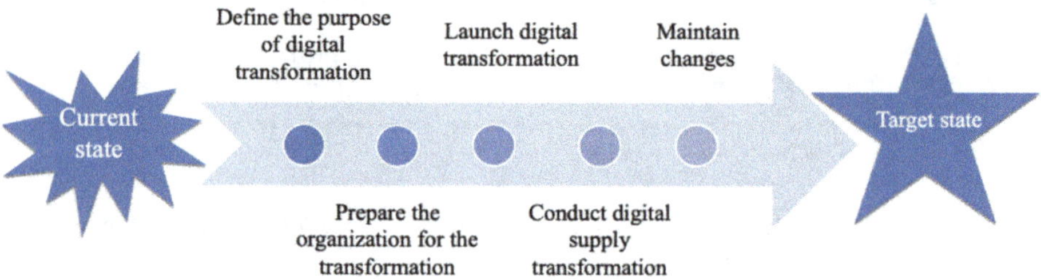

Fig. 7. Management mechanism for digital supple transformation programme

The second step of the programme management mechanism is to prepare the organization for transformation. The following objectives must be accomplished:

- consider the external and internal factors leading to change;
- provide a clear understanding of the transformation on the part of management and employees;
- increase involvement of top management and ensure communication with them via "one window";
- shift to a transformational mindset;
- ensure interactions with employees and their readiness for change;
- increase awareness of digital supply system transformation among external and internal stakeholders.

The third step of the digital transformation management mechanism is the launch of digital transformation. This step contains the following tasks:

- develop a roadmap for error-free transformation initiation and identify programme milestones;
- ensure transparent project management and decision-making processes;
- develop clear lines of responsibilities;
- involve employees who are familiar with the supply system in the company;

- develop a communications plan for the programme and provide channels for timely information sharing.

The next stage of the digital transformation management mechanism is to conduct supply transformation that includes the following tasks:

- ensure delivery of short-term results;
- plan the development and implementation of broader initiatives to achieve results in the medium term;
- provide professional development opportunities for programme team members;
- develop a system to motivate employee engagement in digital supply transformation.

The last stage of the digital transformation management mechanism is the maintenance of change. The following objectives must be accomplished:

- establish new overall strategy and operational models;
- make changes in corporate culture and team structure to ensure sustainable operations.

5 Conclusions

The series of recommendations were identified based on the developed management mechanism for a digital supply transformation programme. Firstly, before starting to preparations for supply transformation in the company, it is necessary to form a clear vision of the aims of digital transformation implementation. Namely, the SMART aim and the measures of success should be defined.

Further, in the initial stages of programme management, it is necessary to obtain a clear understanding of the essence of transformation among management and employees. Not only do they need to understand the purpose of the programme, but they also need to get clear understanding of how to achieve the desired results.

After defining the purpose of digital supply transformation, the company needs to synergize the transition to a transformational mindset and make sure that all employees directly or indirectly related to supply system are aware of the business process transformation. Moreover, it is recommended that the company develops a motivation system for employee engagement in the digital transformation process.

Furthermore, it is necessary to develop clear lines of responsibilities to manage the digital supply transformation programme, as well as the projects included in the programme. Besides, it is very important to ensure transparent processes of project management and decision-making. Maximum openness and transparency across all processes will help to build trust within employees and possibly reduce resistance to change.

The programme should provide opportunities for professional growth and development of new competences for the members of the digital supply transformation team. For example, it is possible to invite experts to provide trainings or share experience in digital supply transformation, it is also possible to promote participation in and fund attendance to conferences where new digital technologies are presented.

Once the transformation has been carried out, the aim of the programme has been achieved, and the desired results have been obtained, it is also important to sustain the changes. Thus, in order to ensure sustainable performance, it is for the company to make changes in corporate culture and adopt transformational thinking.

References

1. Agarwal, R., Shankar, C., Mukund, S.: Imagining construction's digital future. McKinsey & Company (2016)
2. Brennen, J.S., Kreiss, D.: Digitalization. In: Jensen, K., Rothenbuhler, E., Pooley, D., Craig, R. (eds.), The International Encyclopedia of Communication Theory and Philosophy, pp. 556–566. Wiley-Blackwell, Chichester (2016)
3. Rossi, B.R., Bauli, M.R, de Carvalho, M.M.: The key aspects of procurement in project management: investigating the effects of selection criteria, supplier integration and dynamics of acquisitions. Production **30** (2020)
4. Henriette, E., Feki, M., Boughzala, I.: The shape of digital transformation: a systematic literature review. In: MCIS 2015 Proceedings, vol. 10, pp. 431–443 (2015)
5. Kiel, D., Arnold, C., Collisi, M., Voigt, K.-I.: The impact of the industrial internet of things on established business models. In: Proceedings of the International Association for Management of Technology (IAMOT), Orlando, FL, 15–19 May, pp. 673–695 (2016)
6. Lerch, C., Gotsch, M.: Digitalized product-service systems in manufacturing firms: a case study analysis. Res. Technol. Manag. **58**(5), 45–52 (2015)
7. Li, F.: The digital transformation of business models in the creative industries: a holistic framework and emerging trends. Technovation **92**, 102012 (2020)
8. Matzler, K., Bailom, F., von den Eichen, S.F., Anschober, M.: Digital Disruption. Wie Sie Ihr Unternehmen auf das digitale Zeitalter vorbereiten, Vahlen, München (2016)
9. Morakanyane, R., Audrey, A.G., O'Reilly, P.H.: Conceptualizing digital transformation in business organizations: a systematic review of literature. In: Bled eConference, no. 21 (2017)
10. Nambisan, S.: Information technology and product/service innovation: a brief assessment and some suggestions for future research. J. Assoc. Inf. Syst. **14**(4), 1 (2013)
11. Ng, I., Wakenshaw, S.: The Internet-of-Things: review and research directions. Int. J. Res. Mark. **34**(1), 3–21 (2017)
12. Rachinger, M., Rauter, R., Müller, C., Vorraber, W., Schirgi, E.: Digitalization and its influence on business model innovation. J. Manuf. Technol. Manag. **30**(8), 1143–1160 (2019)
13. Messenböck, R., Jens, J., Collie, B., Malby, A., Schuler, F., Schneider, S.: Learn from the Best in Organizational Transformation. Boston Consulting Club. November 2020. https://www.bcg.com/publications/2020/four-phases-and-three-journeys-of-successful-transformation
14. Rogers, D.: The Digital Transformation Playbook. Columbia University Press, New York (2016)
15. Stoyanova, M.: Good practices and recommendations for success in construction digitalization. TEM J. **9**(1), 42–47 (2020)
16. Unruh, G., Kiron, D.: Digital transformation on purpose, MIT Sloan Management Review, 6 November 2017. https://sloanreview.mit.edu/article/digital-transformation-on-purpose

Features and Problems of Forming Teams of Strategic Projects of Russian Universities

Svetlana Apenko(✉) and Anna Breusova

Dostoevsky Omsk State University, Mira ave. 55a, Omsk 644077, Russia
apenkosn@yandex.ru

Abstract. Russian universities are involved in the processes of transformation of higher education, which determines the high importance of developing and implementing transformational strategies. In the strategies, universities lay down priority programs and projects for academic leadership and taking strong competitive positions. To implement these programs and projects, teams are required that are able to accept and implement strategic goals, tasks and challenges of the competitive environment. The formation of such teams for strategic projects occurs spontaneously. There are no effective methodologies and methods for forming university project teams. Our research is aimed at overcoming this problem, the purpose of which is to substantiate the specifics of university project teams, identify problems in the formation of these teams and develop ways to eliminate problems. The study was conducted in 2020–2021 using the case study method. The cases of three universities from different regions of Russia are analyzed. The expert survey method was also used. A total of 75 experts representing the universities selected for the study were interviewed. The results of the study were: a list of specific features of university strategic development project teams; a set of problems of forming strategic project teams, recommendations on ways to overcome these problems, taking into account the specifics of university project teams in the recommendations. The directions of improving the formation of teams of strategic projects of universities have been developed. The novelty lies in the role and competence models of strategic project teams adapted to the conditions of universities.

Keywords: Project Teams · Strategic Projects · Team Formation

1 Introduction

Large-scale and deep processes of transformation of higher education are taking place in Russia, in which all universities are involved. For successful transformation, universities develop and implement a transformational strategy, as well as programs and projects corresponding to it. To implement these programs and projects, teams are required that

The study was carried out at the expense of a grant from the Russian Academy of Sciences (project No. 22-28-20359), https://www.rscf.ru/project/22-28-20359/.

S. Bushuyev et al. (Eds.): IPMA 2021, LNNS 704, pp. 121–131, 2023.
https://doi.org/10.1007/978-3-031-34629-3_12

are able to accept and implement strategic goals, tasks and challenges of the competitive environment. However, the formation of such teams for strategic projects occurs spontaneously. There are no effective methodologies and methods for forming university project teams. This leads to the fact that teams are not effective enough and do not always fully ensure the success of projects and the entire strategy as a whole. This urgent problem requires a solution. Our research is aimed at overcoming this problem.

2 The Degree of Study of the Research Topic

Foreign researchers devote their works to the development of universities, the role and content of strategic management in them. About the strategy of universities, its content and features of development can be found in the publications of the authors: Bouillard P., Peters M. A., Besley T., Orji P. I. [3, 13, 14]. Such authors as: Hietajärvi A-M., Aaltonen K., Ammeter A. P., Dukerich J. M., Thamhain H. J. write about the formation of university project teams, the description of methods and technologies for the formation of these teams, their functions and roles [1, 8, 19]. The researchers Jin C. X., Li F. C., Zhang K., Xu L. D., Chen Y., Cavdur F., Sebatli A., Kose-Kucuk M., Rodoplu C. deal with the issues of team building in various aspects [4, 9].

Russian scientists are also engaged in the transformation of universities and the formation of teams. For example, such authors as Madyarov A. A., Dimitriadi N. A., Glechikova T. O., Sogomonov A. Yu., Gusarova M. S., Korkishko A. N., Kozeeva M. E. describe the mechanisms of strategic development of universities, problems and prospects of university development for the long term, the algorithm for developing a university development strategy and involving teams in this development, interaction between business and education in strategic development and other issues [5, 7, 11, 17].

Latyshev A. S. devotes his works to the development of professional competencies of teams of strategic development projects of universities [10]. His works directly laid the foundations for our research.

Such Russian authors as Gulius N. S., Shepel M. O., Vasiliev A.V., Medvedeva E. V., Nurgalieva A.M., Akhmetshina A. R., Saifudinova N. Z., Solodova E. P., Kalmykova D. A., Polupan K. L., Beilkhanov D. K., Kvyatkovskaya I. Yu., Shkunova A. A., Lebedeva T. E., Mashina A.D. deal with issues of working with teams in the context of corporate culture, modern methods of team formation, competence and role models in team formation [2, 6, 12, 16, 18].

We also used the latest research that is indirectly related to our topic, but their use will allow us to develop the classic issues of forming university project teams. There are developments that do not have a high degree of scientific elaboration, but have value for our research. For example, the "University of the National Technological Initiative 2035" has developments on the formation of cross-functional teams. In particular, within the framework of the activities of this university aimed at implementing a training program for the digital economy, competence models for the digital economy and competence-role models of teams are proposed. The ideas contained in these developments are also used in our research.

3 The Purpose and Methodology of the Study

In accordance with the problem posed, the purpose of the study is to substantiate the specifics of university project teams, identify problems in the formation of these teams and develop ways to eliminate problems.

The research methodology is based on systematic and competence-based approaches. The system approach made it possible to analyze the system complex of methods for forming project teams in the context of the broader organizational environment of the university and in the context of the university's strategy, to analyze the system of competencies and roles in the project teams of the strategic development of the university. The competence-based approach contributed to the analysis of teams from the positions of their competence-role structure.

The study was conducted in 2020–2021 using the case study method. The cases of three major universities from different regions of Russia are analyzed. The universities that are actively engaged in the development of strategic projects and form university transformation teams for this activity have been selected.

The expert survey method was also used. A total of 75 experts representing the universities selected for the study were interviewed. The criteria for selecting experts were: work experience at the university for at least 10 years, experience in participating in strategic development projects of the university for at least 3 years, experience in team management and participation in team building processes for at least 3 years. All the selected experts have these characteristics.

The subject area of the research was: a list of specific features of university strategic development project teams; a set of problems of forming strategic project teams, recommendations on ways to overcome these problems, taking into account the specifics of university project teams in the recommendations.

4 The Results of the Study

On average, about 18–23 projects of strategic development of universities are implemented annually at each university selected for analysis. The implemented projects belong to the following types:

- large (54%), medium – sized (28%), small (18%) projects;
- in the field of education (24%), in the field of science (18%), in the field of university management (40%), in the field of innovation and entrepreneurship (18%).

One university under study has a project office and a vice-rector responsible for project work, while two other universities have departments and positions responsible for project management. At the same time, 68% of experts said that specialists who are officially engaged in project management need advanced training in project management.

The study showed that the teams of strategic projects of universities have their own characteristics that distinguish them, for example, from the teams of organizations and enterprises of other branches of activity. Here is a description of these features.

Teams, as a rule, consist of representatives of the university's management corps, less often include university teachers, and much less often teams include students and teaching and support staff. The number of external stakeholders involved in project teams is less than one percent on average at each university. Table 1 shows the data demonstrating the participation of different groups of staff and students in various types of strategic development projects of the university.

Table 1. Structure of university strategic development project teams

Types of projects	Number of projects participating in teams (%)			
	managers	scientific and pedagogical workers	training and support staff	students
In the field of education	48	36	12	4
In the field of science	42	45	4	9
In the field of management	82	14	2	2
In the field of innovation and entrepreneurship	64	22	2	12
Total	59	29	5	7

There is a tendency to use the same representatives of management personnel in different projects. The rotation of project team members is poorly used. So, when asked how often experts participated in different projects, 78% of them said that they participated in more than 8 projects of the university during 2020, that is, in more than 40% of projects. An additional question about the structure of the teams of strategic development projects and the composition of the teams was answered:

- more than 70% of the team participate in different projects in the same composition-23% of respondents noted;
- approximately 50–70% of the team participate in different projects in the same composition-24% of respondents noted;
- approximately 30–50% of the team participate in different projects in the same composition 41% of respondents;
- less than 30% of the team participates in different projects in the same composition – 12% of respondents.

Given that most of the team members are managers, we can conclude that the same managers are involved in different strategic projects.

There is a different degree of activity of their participants in the teams, some of the participants make a much more significant contribution to the project compared to other participants. So, the question of how evenly the workload of team members is distributed during the implementation of one particular project was answered:

- everyone is approximately equally loaded with project activities – 22% of respondents chose this answer;
- about 50% of team members are more loaded than the other 50% of participants – 48% of respondents;
- about 30% of team members are more loaded than the other 70% of participants – 26% of respondents;
- the team leaders work mainly, of which about 10%, the rest work to a much lesser extent – 4% of respondents.

These features of project teams should be taken into account when analyzing problems and developing directions for improving the process of team formation.

The study revealed typical problems of forming project teams. For example, project roles are not allocated in teams, more often teams are formed and function according to the affiliation of their participants to the official status. The need to include a certain position in the team is taken into account, and the ability of team members to perform certain project roles is taken into account to a lesser extent. When forming a team and its development, competence models of the team are not used. The decision to include a person in the team takes place without analyzing whether he has the required competencies. The formation and development of project teams often occurs spontaneously, there are no long-term plans for the development of teams. There is no strategic project management system and, as a result, the formation of project teams is carried out outside the context of the general project management system. Table 2 demonstrates these problems.

5 Discussion of the Results of the Study

Based on the identified problems and based on expert assessments, we have developed directions for improving the formation of teams of strategic projects of universities. These areas include:

- development of standard role models for strategic project teams;
- development of standard competence models of teams, taking into account their role structure;
- development of methods for diagnosing the competencies of candidates for project teams that are most suitable for the requirements of universities;
- development of mechanisms for the selection of strategic project teams;
- development of a system for the development and motivation of project team members, taking into account the capabilities of Russian universities;
- formation of a system of rotation of team members between projects.

In these areas, we have made appropriate developments, some of which have been tested in practice. In addition, these developments correlate with the research data provided in the description of design practices of leading universities and Skolkovo [15].

Table 2. Expert opinion on the problems in the formation of strategic development project teams

Problem	Severity and frequency of the problem (%)		
	Highly pronounced and frequent	Moderate and sometimes manifested	Low-grade and rare
Project roles are not allocated or poorly allocated	64	29	7
The selection to the team was carried out according to the official affiliation, and not according to the competencies of the candidates	47	40	13
The functions in the team are not clearly fixed, the boundaries of the functions are blurred	20	43	37
There was no long-term development plan for the team, self-development mechanisms prevailed	29	42	29
There was a plan for the development of the team, but it was not related to the project management plan	48	31	21
There were no team-building activities related to cohesion	20	37	43

Let's focus on a brief description of our conclusions and developments based on the results of the study.

The practice of implementing university development projects allows us to offer the following role model of strategic teams. The role model of strategic teams requires taking into account the specifics of the activity and can be based on the typology of projects: educational, scientific, managerial (institutional), in the field of innovation and entrepreneurship, which is reflected in Table 3.

For the successful implementation of strategic projects, one of the important requirements is to know the scope of the project implementation. This requirement affected the proposed model. So, such roles as coordinator and administrator are mandatory for a strategic project. If it is possible for current projects to combine several roles for one

Table 3. Typical role structure of the strategic project management team

Project type	Educational	Scientific	Management	Projects in the field of innovation and entrepreneurship
Curator	+	+	+	+
Project Manager	+	+	+	+
Administrator	+	+	+	+
Coordinator	+	+	+	+
Strategic Analyst	+	+	+	+
Methodologist/Tutor	+			
Organizer			+	+
Entrepreneur-expert		+		+
Marketer/Public Relations Manager	+	+	+	+
Implementers	+	+	+	+

member of the project team, then the importance and volume of work within the framework of a strategic project, the need for a significant number of approvals in related areas may lead to the need to separate these functions. If the administrator maintains project documentation, the coordinator is a project management methodologist and accompanies the team during the development and implementation of the project, being an employee of a specialized project department (project office). Also, one of the important functions of the coordinator is the coordination of the project with other strategic projects, which is very important from the point of view of harmonizing the solution of strategic tasks and avoiding duplication of efforts.

The strategic analyst focuses on evaluating the content of the project to the required changes, on the compliance of the project with the strategic goals of the university, in individual strategic development projects, the analyst can perform the functions of a tester at the intermediate and final stages of implementation. In this case, the role of the analyst is especially important in projects aimed at institutional changes in the university, which can occur within the framework of management projects and projects aimed at the development of innovations and the development of entrepreneurship. In projects of institutional changes, it is important to test the system being created at the pilot launch stages as part of the overall project lifecycle. The role of the organizer is required for management projects and projects in the field of entrepreneurship and innovation, the main functions of the organizer include the construction of a new type of connections between the elements of the educational system and the formats of relationships between the main participants of the designed models, which is expressed in the development of organizational structures, regulations, rules of interaction. The role of an expert entrepreneur in the framework of innovation projects is to evaluate the proposed changes when building a full cycle of work with innovations. The role of the methodologist involves the development of the content part of the changes in the educational program. In strategic projects, a significant role is played by the public relations manager, whose functions include positioning strategic changes in the university in the external environment and promoting the project within the university.

The analysis of cases and expert responses allowed us to determine the main competencies within the selected roles (Table 4). Competencies can be refined with increasing the maturity of project management at the university.

Table 4. Description of the competencies of the participants of the project teams of strategic projects for the development of universities

Role	Competencies
Curator	Strategic thinking Willingness to take risks when making decisions
Project Manager	Ability to work in conditions of change Possession of digital technologies, including communications using digital technologies Focus on achieving results Leadership and responsibility in team management Ability to work in a team with representatives of different countries, cultures, nationalities Ability to work in a virtual team Innovation and entrepreneurship Stress resistance Emotional intelligence
Administrator	Maintaining project documentation Ability to work in a team
Coordinator	Strategic and systems thinking Research skills Flexibility, unconventional thinking and work Technologies of project work
Strategic Analyst	Strategic thinking Critical thinking Research skills
Methodologist	Knowledge of the skills of building educational programs System thinking Critical thinking
Organizer	System thinking Flexibility, unconventional thinking and work Research skills Professionalism in the subject area of the project
Entrepreneur-expert	Creative thinking Professionalism in the subject area of the project

(*continued*)

Table 4. (*continued*)

Role	Competencies
Marketer	Open communication skills in the project team and with partners, including online, in social networks. Networks Creative thinking Marketing technologies
Implementers	Ability to work in a team Result orientation Efficiency

In order to develop methods for diagnosing the competencies of candidates for project teams that are most suitable for the requirements of universities, the following recommendations can be identified. At the first stages of the implementation of project management, you can use a questionnaire to assess the competencies of candidates. In the future, you can switch to professional diagnostic tests for assessing competencies.

As the conducted research has shown, project teams are formed quite spontaneously, which affects the results of project implementation. The selection methods should be based on the definition of roles in the development of each strategic project based on the content of the work performed and taking into account the competencies required to perform the role. In the future, you can conduct a survey of project managers in order to determine the most effective participants in the project teams of already implemented projects.

The development of a motivation system can include both material remuneration based on the results of the project implementation, and obtaining the opportunity for additional training by sending internships to leading universities of project management.

An important element of strategic team management is the system of rotation of project team members. The rotation system can be built on the basis of creating a database of potential participants within the selected standard roles by means of a questionnaire conducted to assess competencies. The use of prioritization when collecting information about roles and projects in which a potential candidate sees himself as a participant allows you to build individual trajectories for the development of project management competencies.

6 Conclusion

Thus, the study revealed the specific features of the teams of strategic projects of universities. The problems of forming these teams are also identified. The developments related to the elimination of problems and aimed at more effective formation of project teams are proposed. The obtained research results are of scientific and practical interest. The novelty lies in the role and competence models of strategic project teams adapted to the conditions of universities. We consider the issues raised to be promising for further study. In particular, it is planned in the future to study such aspects as the construction of different types of competence and role models depending on significant factors (types of projects, type of strategy, situational factors of university development, and others), the development of methods for selecting candidates for teams based on different models of competencies and roles in strategic projects.

References

1. Ammeter, A.P., Dukerich, J.M.: Leadership, team building, and team member characteristics in high performance project teams. Eng. Manag. J. **14**(4), 3–10 (2002). https://doi.org/10.1080/10429247.2002.11415178
2. Beilkhanov, D.K., Kvyatkovskaya, I.Y.: Using the competence model in the process of team building. Technical sciences-from theory to practice. №30 (2014). https://cyberleninka.ru/article/n/ispolzovanie-modeli-kompetentsiy-v-protsesse-komandoobrazovaniya. Accessed 24 July 2021
3. Bouillard, P. A multi-objective method to align human resource allocation with university strategy. Perspect. Policy Pract. High. Educ. **20**(1), 17–23, (2016). https://doi.org/10.1080/13603108.2015.1081303
4. Cavdur, F., Sebatli, A., Kose-Kucuk, M., Rodoplu, C.: A two-phase binary-goal programming-based approach for optimal project-team formation. J. Oper. Res. Soc. **70**(4), 689–706 (2019). https://doi.org/10.1080/01605682.2018.1457480
5. Dimitriadi, N.A., Glechikova, T.O.: Algorithm for developing a University development strategy. Bulletin of the RSEU RINH. No. 1 (65) (2019). https://cyberleninka.ru/article/n/algoritm-razrabotki-strategii-razvitiya-universiteta. Accessed 24 July 2021
6. Gulius, N.S., Shepel, M.O., Vasiliev, A.V., Medvedeva, E.V.: Management of organizational culture and work with teams. University management: practice and analysis. №2 (114) (2018). https://cyberleninka.ru/article/n/upravlenie-organizatsionnoy-kulturoy-i-rabota-s-komandami. Accessed 24 July 2021
7. Gusarova, M.S., Korkishko, A.N., Kozeeva, M.E.: Key factors of successful interaction between education and business. Moscow Economic Journal. No. 4 (2020). https://cyberleninka.ru/article/n/klyuchevye-faktory-uspeshnogo-vzaimodeystviya-obrazovaniya-i-biznesa. Accessed 24 July 2021
8. Hietajärvi, A.-M., Aaltonen, K.: The formation of a collaborative project identity in an infrastructure alliance project. Constr. Manag. Econ. **36**(1), 1–21 (2018). https://doi.org/10.1080/01446193.2017.1315149
9. Jin, C.X., Li, F.C., Zhang, K., Xu, L.D., Chen, Y.: A cooperative effect-based decision support model for team formation. Enterp. Inf. Syst. **14**(1), 110–132 (2020). https://doi.org/10.1080/17517575.2019.1678071
10. Latyshev, A.S.: A new approach to the development of professional competencies to increase the involvement of university staff in strategic development projects. Bulletin of the Surgut State Pedagogical University. No. 2 (65). (2020). https://cyberleninka.ru/article/n/novyy-podhod-k-razvitiyu-professionalnyh-kompetentsiy-dlya-povysheniya-vovlechennosti-personala-universiteta-v-proekty. Accessed 24 July 2021
11. Madyarov, A.A.: Organizational development of universities in a strategic context: problems and prospects of the project "5–100". NC. No. 4 (2020). https://cyberleninka.ru/article/n/organizatsionnoe-razvitie-universitetov-v-strategicheskom-kontekste-problemy-i-perspektivy-proekta-5-100. Accessed 24 July 2021
12. Nurgalieva, A.M., Akhmetshina, A.R., Saifudinova, N.Z.: Modern methods of forming an effective team in an organization. Skif. No. 11 (27) (2018). https://cyberleninka.ru/article/n/sovremennye-metodiki-formirovaniya-effektivnoy-komandy-v-organizatsii. Accessed 24 July 2021

13. Orji, P.I.: Problem-based approach in property law – a university's strategy in focus. Law Teach. **49**(3), 372–387 (2015). https://doi.org/10.1080/03069400.2015.1040295
14. Peters, M.A., Besley, T.: China's double first-class university strategy: 双一流. Educ. Philoso. Theory **50**(12), 1075–1079 (2018). https://doi.org/10.1080/00131857.2018.1438822
15. Evstratova, L.A., Isaeva, N.V., Leshukova, O.V. (eds.): Project-based training: a training manual. (2018)
16. Shkunova, A.A., Lebedeva, T.E., Mashina, A.D.: Studying the role structure of a team in the process of team building. Problems of modern pedagogical education. No. 59–4 (2018). https://cyberleninka.ru/article/n/izuchenie-rolevoy-struktury-kollektiva-v-protsesse-komandoobrazovaniya. Accessed 24 July 2021
17. Sogomonov, A.Y.: Sustainable University (image of the future, current problems and trends). Sheets of Applied Ethics. No. 50 (2017). https://cyberleninka.ru/article/n/ustoychivyy-universitet-obraz-buduschego-aktualnye-problemy-i-trendy. Accessed 24 July 2021
18. Solodova, E.P., Kalmykova, D.A., Polupan, K.L.: Team building as a modern way of human resources management. Bulletin of the Samara University. Economics and management. No. 4 (2018). https://cyberleninka.ru/article/n/komandoobrazovanie-kak-sovremennyy-sposob-upravleniya-chelovecheskimi-resursami. Accessed 24 July 2021
19. Thamhain, H.J.: Leading technology-based project teams. Eng. Manag. J. **16**(2), 35–43 (2004). https://doi.org/10.1080/10429247.2004.11415247

Project Manager Competences and University Educational Programs in Project Management

Olga Peshkova(✉) and Alexander Shavrin

Moscow State Automobile and Road Construction Technical University, Moscow 125319, Russian Federation

o.peshkova@madi.ru

Abstract. Project management is a dynamically developing direction of management in the context of the digital transformation of the economy, both in the private and public sectors. And the need for qualified specialists like project managers is very high. Nowadays Russian universities actively develop and implement project management higher education programs based on the Federal State Education Standards. At the same time there is no a unified educational approach in terms of project manager's knowledge, technical and soft skills. This leads to the situation when graduates of different universities - future project managers - get a different set of competences, which creates certain difficulties in the practice of project management. A substantive analysis of the competences described in the recognized project management standards is carried out; different approaches and models of project management competences are taken in consideration to identify the range of competences recognized by all analyzed sources. On the other hand, based on consideration of the Federal State Educational Standards of Higher Education in combination with a study of some industry project management standards in Russia, it is possible to identify possible essentials of the discussed competences, taking into account the tasks of professional activity.

Keywords: Competences · Digital Transformation · PM Standards · Project Manager · University Program

1 Introduction

It is well known that project manager qualification requires a lot of specific competences to do projects well in dynamic environment with high level of complexity and uncertainty. In our opinion it's critically important to develop plans for academic disciplines in terms of higher education programs based on a deep analysis of the requirements for the future project managers' competences. This will allow to train project managers who speak the same professional language and are in a single information and practical space, which, of course, will positively affect the achievement of projects goals. Nowadays there are a lot of scientific investigations in project management area, specific university educational programs are developed in project management, so this sphere becomes an important direction in science and professional practice [1].

S. Bushuyev et al. (Eds.): IPMA 2021, LNNS 704, pp. 132–147, 2023.
https://doi.org/10.1007/978-3-031-34629-3_13

There is no doubt that system approach and result-oriented Bachelor and Master degrees university education corresponding with future professional activity is a key factor of career growth and success in project management area.

Project manager, being responsible on project goals and success, plays a key role in a project organizational structure. That's why his competences are in focus when investigating in the area of project management [1].The authors [2] define the interrelation of project manager competences and project success. Research of the literature [3] was made to create classification of project manager competences and all those competences were compared with job requirements to potential specialists in project management.

The authors of [3] conducted a literature study to form a classification of the project manager competences; and also carried out a comparison of these competences with the requirements of the employer to the applicant for the position in project management based on the analysis of job offers.

The publication [4] provides a detailed consideration of the project manager competences divided into 11 blocks.

Other authors pay attention to changes in project manager competences associated with Industry 4.0 [5]: an environment of a high level of complexity and uncertainty leads to the needs to digitize a bunch of project management processes, and, therefore, requires updating the traditional range of competences.

In accordance with [6], project managers are involved in this area of activity due to their own business interests, which allows us to speak about the business competences necessary for project managers.

It is critically important to analyze the experience of international professional organizations in the development of competence models for project managers [7–9] as well as Russian industry standards for project management [10–12].

In addition, it must be remembered that the Bachelor Program in the project management profile at Russian universities and, accordingly, the competences of graduates, must meet the requirements [13].

2 Methodology and Result

The problem of the research is the lack of information about the features, requirements and results regarding the formation of necessary competences in project management at the university Bachelor's programs in terms of the digital transformation of the economy. The purpose of the study is to use methods of system analysis and synthesis, content analysis, statistical analysis, to study the available sources of information on the problem field of research: scientific publications, conference materials, project management standards, official sources, - to form the necessary pool of project manager competences in terms of bachelor university programs.

There are many approaches to defining the competences of a project management specialist. Let's consider some of them.

The report [14] points to digital transformation as the main trend influencing on activities in modern conditions.

Other authors [5] talk about the digitalization trend: projects have to be managed in a more unpredictable environment with a high level of complexity. This leads to changes in the traditional project management processes, which consequently requires the formation of new technical, contextual and behavioral competences of project managers.

This is confirmed by the data of the report [14], which indicates the need for modern project managers to use cloud solutions, Artificial Intelligence (AI), 5G-mobile Internet, Internet of Things and other modern tools and technologies.

Research [5] presents 2 groups of Hard Skills, and Soft Skills are divided into 8 groups.

The study [15] identifies intellectual, emotional and group management skills.

According to the authors [4], the project manager requires 81 competences, combined into the following blocks: Influencing Skills, Communication Skills, Management Skills, Team Working Skills, Emotional Skills, Contextual Skills, Cognitive Skills, Personal skills, and Professionalism, Knowledge and Experience, Project management Knowledge.

In the study [16], based on the self-assessment of project management specialists, the most important competences were identified like Leadership, Conflict Management and Initiative, Analytical Thinking, Stress Management, Decision Quality and Achievement Drive competences.

To identify the proximity of theory and practice, do Vale etc. [3] conducted a study of the employer's requirements for a project manager position, and the competences necessary for a project manager, which are presented in 4 categories: contextual, managerial, technical and behavioral competences.

Akkermans etc. [1] provides data from a study demonstrating how close the bachelor's degree programs are to the requirements of employers: the attributes of bachelor's programs are compared with the attributes required to the applicant for a vacancy by employers. In the results of training under bachelor's programs, the following ones are most often indicated: Acquired Knowledge, Work Ethics (Value Competences); Global Awareness, Cooperativeness and Self-Awareness (Behavioral Competences); Critical Thinking and Problem Solving (Skills).

We suppose that business skills after completing a bachelor's degree program seem to be very important - this is confirmed by the results of the study [6], showing that it is the business interest that contributes to attracting specialists in this area.

We propose to take into consideration the most used and well-known models, developed by International Project Management Association and Project Management Institute and described in IPMA International Competence Baseline 4.0 [7], PMI Project Manager Competency Development Framework (Third Edition) [8]. In our opinion it is also important to take into account PMI Talent Triangle [9] and knowledge and skills needed in digital environment described in PMI's Pulse of the Profession® In-Depth Report [17].

IPMA ICB 4.0 describes PM competencies as 29 elements, distributed on 3 spheres [7]: People, Project, Perspective.

PMI Project Manager Competency Development Framework (Third Edition) considers PM competencies in three separate dimensions [10]: Knowledge Competence (not detailed in PMCDF); Performance Competence; Personal Competence.

At the same time PMI recommends to use PMI Talent Triangle in which PM competences consists of three groups [9]: Technical Project Management; Leadership; Strategic and Business Management.

Report [11] directly indicates the need Top Six Digital-Age Skills.

All these competences from 4 references [7–9, 17] mentioned above are collected in Table 1.

Table 1. Project manager competences in accordance to the different approaches

ICB 4.0 [7]	PMI PMCDF [8]	PMI Talent Triangle [9]	4. PMI's Pulse of the Profession® In-Depth Report [17]
1.1. People	2.1. Personal	3.1 Leadership	4.1. Top Six Digital-Age Skills
1.1.1. Self-reflection and self-management	2.1.1.Communicating	3.1.1.Brainstorming	4.1.1. Data Science Skills
1.1.2. Personal integrity and reliability	2.1.2. Leading	3.1.2.Coaching and Mentoring	4.1.2. Innovative Mindset
1.1.3. Personal communication	2.1.3. Managing	3.1.3. Conflict Management	4.1.3. Security and Privacy Knowledge
1.1.4. Relations and engagement	2.1.4. Cognitive ability	3.1.4. Emotional Intelligence	4.1.4. Legal and Regulatory Compliance Knowledge
1.1.5.Leadership	2.1.5. Effectiveness	3.1.5. Influencing	4.1.5. Ability to Make Data-Driven Decisions
1.1.6.Teamwork	2.1.6. Professionalism	3.1.6. Interpersonal Skills	4.1.6.Collaborative Leadership Skills
1.1.7.Conflict and crisis		3.1.7. Listening	
1.1.8. Resourcefulness		3.1.8. Negotiation	
1.1.9. Negotiation		3.1.9. Problem-Solving	
1.1.10.Result orientation		3.1.10. Team Building	
1.2. Project	2.2. Performance	3.2.Technical Project Management	

(*continued*)

Table 1. *(continued)*

ICB 4.0 [7]	PMI PMCDF [8]	PMI Talent Triangle [9]	4. PMI's Pulse of the Profession® In-Depth Report [17]
1.2.1. Project design	2.2.1. Project Integration Management	3.2.1. Agile Practices	
1.2.2. Requirements and objections	2.2.2. Project Scope Management	3.2.2. Data Gathering and Management	
1.2.3. Scope	2.2.3. Project Time Management	3.2.3. Earned Value Management	
1.2.4. Time	2.2.4. Project Cost Management	3.2.4. Governance (Project, Program, Portfolio)	
1.2.5. Organization and information	2.2.5. Project Quality Management	3.2.5. Lifecycle Management (Project, Program, Portfolio, Product)	
1.2.6. Quality	2.2.6. Project Human Resource Management	3.2.6. Performance Management (Project, Program, Portfolio)	
1.2.7. Finance	2.2.7. Project Communications Management	3.2.7. Requirements Management and Traceability	
1.2.8. Resources	2.2.8. Project Risk Management	3.2.8. Risk Management	
1.2.9. Procurement	2.2.9. Project Procurement Management	3.2.9. Schedule Management	
1.2.10. Plan and control	2.2.10. Project Stakeholder Management	3.2.10. Scope Management (Project, Program, Portfolio, Product)	
1.2.11. Risk and opportunity		3.2.11. Time, Budget and Cost Estimation	
1.2.12. Stakeholder			
1.2.13. Changes and transformation			
1.3.Perspective		3.3.Strategic and Business Management	

(continued)

Table 1. (*continued*)

ICB 4.0 [7]	PMI PMCDF [8]	PMI Talent Triangle [9]	4. PMI's Pulse of the Profession® In-Depth Report [17]
1.3.1. Strategy		3.3.1. Benefits Management and Realization	
1.3.2. Governance, Structures and Processes		3.3.2. Business Acumen	
1.3.3. Compliance, Standards and Regulations		3.3.3. Business Models and Structures	
1.3.4. Power and Interest		3.3.4. Competitive Analysis	
1.3.5. Culture and Values		3.3.5. Customer Relationship and Satisfaction	
		3.3.6. Industry Knowledge and Standards	
		3.3.7. Legal and Regulatory Compliance	
		3.3.8. Market Awareness and Conditions	
		3.3.9. Operational Functions (e.g. Finance, Marketing)	
		3.3.10. Strategic Planning, Analysis, and Alignment	

Looking at the Table 1 one can see the areas of intersection and it gives us an opportunity to make a conclusion about nominal competence set of modern project manager.

Let's collect similar competences from [7–9, 17] into Table 2 and show those ones mentioned at least 2 times in Table 1.

So, there are 19 competences, mentioned in several references simultaneously. Saying about bachelor educational programs we don't take in consideration competences mentioned once. It doesn't mean that these competences are not important, but they

Table 2. Similar competences, mentioned at several references

1. ICB 4.0 [7]	2.PMI PMCDF [8]	3. PMI Talent Triangle [9]	4 PMI's Pulse of the Profession® In-Depth Report [17]	5.Frequency of mention
1.1. People	2.1. Personal	3.1. Leadership	4.1. Top Six Digital-Age Skills	
1.1.5. Leadership	2.1.2. Leading		4.1.6. Collaborative Leadership Skills	3
1.1.3. Personal communication	2.1.1. Communicating			2
1.1.7. Conflict and crisis		3.1.3. Conflict Management		2
1.1.9. Negotiation		3.1.8. Negotiation		2
1.1.6. Teamwork		3.1.10. Team Building		2
1.1.4. Relations and engagement		3.1.6. Interpersonal Skills		2
1.2. Project	2.2. Performance	3.2. Technical Project Management		
1.2.4. Time	2.2.3. Project Time Management	3.2.9. Schedule management 3.2.11. Time, Budget and Cost Estimation 3.2.3. Earned Value Management		5
1.2.7. Finance	2.2.4. Project Cost Management	3.2.11. Time, Budget and Cost Estimation 3.2.3. Earned Value Management		4

(*continued*)

Table 2. *(continued)*

1. ICB 4.0 [7]	2.PMI PMCDF [8]	3. PMI Talent Triangle [9]	4 PMI's Pulse of the Profession® In-Depth Report [17]	5.Frequency of mention
1.2.3. Scope	2.2.2. Project Scope Management	3.2.10. Scope Management (Project, Program, Portfolio, Product)		3
1.2.11. Risk and opportunity	2.2.8. Project Risk Management	3.2.8. Risk Management		3
1.2.2. Requirements and objections		3.2.7. Requirements Management and Traceability		2
1.2.12. Stakeholder	2.2.10. Project Stakeholder Management			2
1.2.6. Quality	2.2.5. Project Quality Management			2
1.2.8. Resources	2.2.6. Project Human Resource Management			2
1.2.9. Procurement	2.2.9. Project Procurement Management			2
1.2.1. Project design	2.2.1. Project Integration Management			2
1.3. Perspective		3.3. Strategic and Business Management		
1.3.3. Compliance, Standards and Regulations		3.3.6. Industry Knowledge and Standards 3.3.7. Legal and Regulatory Compliance	4.1.4. Legal and Regulatory Compliance Knowledge	4

(continued)

Table 2. (*continued*)

1. ICB 4.0 [7]	2.PMI PMCDF [8]	3. PMI Talent Triangle [9]	4 PMI's Pulse of the Profession® In-Depth Report [17]	5.Frequency of mention
1.3.1. Strategy		3.3.10. Strategic Planning Analysis and Alignment		2
1.3.2. Governance, Structures and Processes		3.3.3. Business Models and Structures		2

could be included in master educational program. At the same time the growth of complex projects requires to use agile approach more widely and agile practices should be included in university educational programs as well.

Now let's take a look at some Russian professional standards in project management. For example, Professional Standard "Space Industry Project and Program Management Specialist" [10], Professional Standard "Information Technology Project Manager" [11], Professional Standard "Fleet Project and Program Management Specialist" [12] and some others. All these standards use concept of job activities, skills and knowledge instead of competence approach. In terms of bachelor university educational programs let's analyze knowledge requirement first. PM knowledge requirements of some Russian professional standards are presented in Table 3.

Comparing Table 3 with Table 1 one can see that the data from these tables are very similar, but the Table 1 describes specific PM competences more detailed. At the same time Russian professional standards directly say about project management software and documenting of project information as a necessary elements of project manager education despite the fact that these standards were developed in according with specific industry requirements.

As a result the following project manager competences can be recognized as a key ones when developing bachelor university educational programs in project management (in alphabetical order):

1. Agile Practices;
2. Compliance, Standards and Regulations;
3. Conflict Management;
4. Documenting of Project Information;
5. Governance, Structures and Processes;
6. Interpersonal skills;
7. Leadership;
8. Negotiation;
9. Personal communication;
10. Project Cost Management;

Table 3. PM knowledge requirements in accordance with industry standards for the project management

1. Space Industry Project and Program Management Specialist [10]	2. Information Technology Project Manager [11]	3. Fleet Project and Program Management Specialist [12]
1.1. Risk Analysis	2.1. Project Risk Management	3.1. Project Risk Management
1.2. Project Quality Standards	2.2. Project Quality Management	3.2. Project Quality Management
1.3. Documenting of Project Information	2.3. Documenting of Project Information	3.3. Documenting of Project Information
1.4. Strategic goals of a Company	2.4. Project Configuration Management	3.4. Strategic goals of a Company
1.5. Deviation Analysis	2.5. Project Change Management	3.5. Project Change Management
1.6. Project Planning	2.6. Communication Management	3.6. Communication Management
1.7. WBS Development	2.7. Conflict Management	3.7. Conflict Management
1.8. Project Management Standards	2.8. Project Management Subjects	3.8. Project Management Subjects
1.9. Project Milestones	2.9. Project Stakeholder Management	3.9. Project Time Management
1.10. Project Management Software	2.10. Project Requirements Management	3.10. Project Management Software
1.11. Modeling and Optimization		3.11. Project Cost Management
1.12. Project Management System		3.12. Project Knowledge Management
1.13. Project Charter Development		

11. Project Human Resource Management;
12. Project Integration Management;
13. Project Management Software;
14. Project Procurement Management;
15. Project Quality Management;
16. Project Requirements Management;
17. Project Risk Management;
18. Project Scope Management;
19. Project Stakeholder Management;
20. Project Time Management;
21. Strategy;

22. Teamwork.

Students of bachelor program in project management are on their way to their future profession, they will have to learn in practice many aspects of professional activity in project management. It should be borne in mind that bachelor graduates are more likely to work in the field of operational project management as members of the project management team, as opposed to master graduates who are faced with strategic management tasks.

According to the Federal State Standard of Higher Education of the Russian Federation in the field of study "Management" [13], students of bachelor program in project management should master and develop general professional competences, universal competences and professional competences.

As for universal and general professional competences, they are clearly formulated in the Federal Standard [13], but professional competences, the list of which specifies project management profile (distinguishing it, for example, from logistics), is developed by each university independently.

It is on the professional competences necessary to successfully start working in project management after bachelor graduation that, in our opinion, should be focused in a special way.

Authors [5] fix the changes in project manager competences according to Industry 4.0: dynamic environment with high level of complexity and uncertainty requires a high level of digitalization regarding a set of the project management processes and, thus, requires an update of traditional competence set. In digital economy competences needed on project initiation and planning are of particular importance since performance, monitoring and control processes could be entirely automated or transferred to robots and digital assistants.

The source [4] indicates not only competences that can be developed in the learning process, but also personal qualities, such as empathy, reliability, adaptability, creativity, innovation, and intuition.

The increasing complexity of projects requires a wider application of agile project management methodologies, which should be reflected in the University Program.

The Federal State Educational Standard of Higher Education - Bachelor's Degree Program 03.03.02 Management [13] defines several types of professional tasks. To our opinion, the most suitable for the "Project Management" profile are: Information and Analytical, Organizational and Management, Business types of professional tasks.

We propose to apply an approach based on the connotation of professional competences developed at Bachelor's Degree program in "Project Management" with the task types of professional activity (see Table 4).

A number of competences in this case are integrated, complex, may include technical, behavioral, contextual, managerial competencies, hard and soft skills.

To our point, the Integrated Professional Competence represents an integral system based on interrelated contextual, managerial, technical, and behavioral competences necessary for the activities of a specialist in project management. Thus, the competences distributed according to the task types of professional activity can be presented in Table 4.

Table 4. Distribution of professional competences in accordance with the task types of professional activity (Bachelor University Program)

Task type of professional activity	Indicated Competences	Competences
1.Information and analytical	1.1. Able to analyze the impact of stakeholders on project management processes	Negotiation Personal Communication Project Stakeholder Management
	1.2. Able to collect, critically analyze and synthesize the data required to implement various project management processes, applying the necessary set of quantitative and qualitative methods	Documenting of Project Information Project Management Software Project Quality Management
	1.3. Able to document information relevant to various project management processes	Compliance, Standards and Regulations Documenting of Project Information Project Management Software
	1.4. Able to apply various informational project management tools for different subject areas at different stages of the project management life cycle	Documenting of Project Information; Project Management Software Project Integration Management
2.Organizational and management	2.1. Able to organize a project based on the organization's management strategy and/or corporate project management strategy	Compliance, Standards and Regulations Governance, Structures and Processes Project Integration Management Strategy Teamwork
	2.2. Able to initiate a project by organizing the development of relevant documents	Compliance, Standards and Regulations Documenting of Project Information Project Integration Management Project Management Software Project Requirements Management Teamwork

(*continued*)

Table 4. (*continued*)

Task type of professional activity	Indicated Competences	Competences
	2.3. Able to carry out project planning for various project management processes	Documenting of Project Information Project Cost Management Project Human Resource Management Project Integration Management Project Management Software Project Procurement Management Project Quality Management Project Risk Management Project Scope Management Project Stakeholder Management Project Time Management
	2.4. Able to carry out operational management of various functional areas of project management, applying the principle of integration	Documenting of Project Information Personal communication Project Cost Management Project Human Resource Management Project Integration Management Project Management Software Project Procurement Management Project Quality Management Project Requirements Management Project Risk Management Project Risk Management Project Stakeholder Management Project Time Management Teamwork
	2.5. Able to apply various tools to manage the project team; capable of analyzing and resolving conflicts in a team	Conflict Management Interpersonal Skills Leadership Negotiation Personal Communication Project Human Resource Management Teamwork

(*continued*)

Table 4. (*continued*)

Task type of professional activity	Indicated Competences	Competences
	2.6. Able to apply the necessary project management tools for various project management processes	Agile Practices Project Integration Management
3. Business	3.1. Able to consolidate various resources to optimize project management results based on economic and financial assessment	Governance, Structures and Processes Project Cost Management Project Human Resource Management Project Integration Management Project Procurement Management Project Quality Management Strategy
	3.2. Able to assess the legal, technical and technological, social, economic, financial, environmental conditions for carrying out entrepreneurial activities in the field of project management	Compliance, Standards and Regulations Governance, Structures and Processes Project Integration Management
	3.3. Able to identify, assess risks, manage project risks based on risk management techniques	Governance, Structures and Processes Project Risk Management Strategy

3 Discussion

Talking about the difference between competences formed in training at the Master's and Bachelor's programs, we should consider the following nuances: graduate students prepare for senior management positions, therefore, the main competences for mastering are not technical, as it usually happens when studying in bachelor programs, but contextual, managerial and behavioral. If we talk about managerial competences, then in the course of training in the master's program, strategic management skills are developed, and in the bachelor's program, mainly operational management skills.

Of course, there are a set of competences which is critical for a specialist in project management, including, for example, critical thinking or problem solving. But this study deals specifically with professional competences in project management.

The above analysis of the requirements for specialists in connection with digital transformation showed that digital competences are in demand and therefore they should be added to the register of mandatory competences for a specialist in project management. However, as pointed out by the authors [17], digital skills are no longer an additional advantage despite their necessity. Therefore, digital skills also do not belong to the pool of specific competences of a bachelor's degree graduate - they are a given today.

4 Conclusion

This study is a look at the nominal set of competences that should be formed in the course of bachelors training in the project management profile, taking into account the trends of digital transformation of the economy. Based on the analysis of project management professional standards, some sources of scientific literature, official documents, a register of professional competences of a bachelor's graduate - a future specialist in the area of project management - was formed. More and more digital projects are being initiated that use various digital technologies, including AI, which will increasingly be entrusted with routine work. It should be noted that when applying digital technologies based on artificial intelligence in a project, it is necessary not only to strictly adhere to the legal requirements regarding the processing, storage and distribution of data, but also to develop a concept of corporate digital responsibility within the corporate social responsibility of the company [18] to prevent data leakage. According to the study [17], most organizations pay great attention to the development of not only technical competencies (68 percent of the surveyed companies), but also leadership skills (65 percent). Thus, the role of leadership and communicate competences should be taken into account when forming a University Bachelor's program.

References

1. Akkermans, J., Chipulu, M., Ojiako, U., Williams, T.: Bridging the fields of careers and project management. Proj. Manag. J. **51**, 123–134 (2020)
2. Alvarenga, J.C., Branco, R.R., Guedes, A.L.A., Soares, C.A.P., Silva, W.D.S.E.: The project manager core competences to project success. Int. J. Manag. Proj. Bus. **13**(2), 277–292 (2020)
3. do Vale, J.W.S.P., Nunes, B., de Carvalho, M.M.: Project managers' competences: what do job advertisements and the academic literature say? Proj. Manag. J. **49**(3), 82–97 (2018)
4. Bolzan de Rezende, L., Blackwell, P.: Project management competency framework. Iberoamerican J. Proj. Manag. **10**, 34–59 (2019)
5. Ribeiro, A., Amaral, A., Barros, T.: Project manager competencies in the context of the industry 4.0. Procedia Comput. Sci. **181**, 803–810 (2021)
6. Havermans, L., Van der Heijden, B.I.J.M., Savelsbergh, C., Storm, P.: Rolling into the profession: exploring the motivation and experience of becoming a project manager. Proj. Manag. J. **50**(3), 346–360 (2019)
7. International Project Management Association (IPMA): ICB - IPMA Individual Competence Baseline for Project, Programme and Portfolio Management, Version 4.0, Nijkerk, the Netherlands (2015)
8. Project Management Institute (PMI): Project Manager Competency Development (PMCDF) Framework – Third edition, Newtown Square, PA (2017)
9. Project Management Institute (PMI): The PMI Talent TriangleTM. https://www.pmi.org/-/media/pmi/documents/public/pdf/certifications/talent-triangle-flyer.pdf. Accessed 21 Nov 2022

10. Professional Standard "Space Industry Project and Program Management Specialist": Approved by the order №486n of Ministry of Labor and Social Protection of the Russian Federation (2018)
11. Professional Standard "Information Technology Project Manager": Approved by the order №893n of Ministry of Labor and Social Protection of the Russian Federation (2014)
12. Professional Standard "Fleet Project and Program Management Specialist": Approved by the order №190n of Ministry of Labor and Social Protection of the Russian Federation (2014)
13. Order of the Ministry of Education and Science of Russia of 12.08.2020 N 970 "On the Approval of the Federal State Educational Standard of Higher Education - Bachelor's Degree Program 03.03.02 Management". https://base.garant.ru/74561312/. Accessed 21 Nov 2022
14. Project Management Institute (PMI): Pulse of the Profession© 2021 (2021). https://www.pmi.org/-/media/pmi/documents/public/pdf/learning/thought-leadership/pulse/pmi_pulse_2021.pdf. Accessed 21 Nov 2022
15. Oh, M., Choi, S.: The Competence of Project Team Members and Success Factors with Open Innovation. J. Open Innov. Technol. Mark. Complex. **6**(51) (2020)
16. Liikamaa, K.: Developing a project manager's competencies: a collective view of the most important competencies. Procedia Manuf. **3**, 681–687 (2015)
17. Project Management Institute (PMI): Pulse of the Profession® In-Depth Report "The Project Manager of the Future. Developing Digital-Age Project Management Skills to Thrive in Disruptive Times", Newtown Square, PA (2020)
18. Herden, C.J., Alliu, E., Cakici, A.: Corporate Digital Responsibility. Nachhaltigkeits Manag. Forum (2021)

Integration of Sustainable Development into International Projects in the FMCG Industry

Timur Bessolitsyn(✉)

National Research University Higher School of Economics, Myasnitskaya str., 20, 101000 Moscow, Russia
t.bessolitsyn@mail.ru

Abstract. At this moment, the pursuit of sustainable development (further – SD) has become one of the most important aims of large international companies and, accordingly, their international projects. In the field of FMCG, where the quality of products and the speed of their use are very important, sustainability is given great attention, being one of the main success factors for many international FMCG projects. The purpose of the research is to identify the key success factors as well as to develop guidelines for integrating SD into international projects in the FMCG industry. To conduct the research, the following research methods were used: case studies, semi-structured in-depth interviews, online survey. As a result of the study, key success factors of integrating SD into international FMCG projects were identified, propositions were tested and guidelines and recommendations were developed. The results of the research might be useful for FMCG companies and their international projects.

Keywords: International Project · FMCG · Sustainable Development

1 Introduction

Sustainable development (SD) is currently one of the most serious competitive advantages, without which many companies would be unable to evolve, compete on international markets, introduce innovations, and follow the main trends. Sustainable development helps companies, and their international projects grow, boost their performance, and have a resilient and long-term strategy for crises. SD also enables companies to improve the environment and socio-economic sphere, laying a solid foundation for the life and development of the next generations.

This work focuses on the topic of integration of sustainable development into international projects in the fast-moving consumer goods (FMCG) industry. This topic is crucial in the development of modern business and management. The small number of academic studies on the topic, the unexplored practices and processes of companies applying and integrating this concept, as well as the social and environmental significance of sustainable development require urgent attention to this issue.

S. Bushuyev et al. (Eds.): IPMA 2021, LNNS 704, pp. 148–161, 2023.
https://doi.org/10.1007/978-3-031-34629-3_14

The international aspect of projects is no less important when integrating SD, as it simultaneously creates both many advantages and difficulties. It is easier to integrate sustainability into traditional local projects since all the processes in such projects take place in the same institutional and cultural environment. It is important to consider the cultural differences in the multinational project team, the complicated structure of the project, the large number of locations, different time zones, complex communication, as well as political, economic, social and legislative factors in different countries. International projects require more knowledge, resources, and understanding of different approaches to managing these types of projects, as they are more complex than local ones. However, international projects have greater advantages than local ones, such as a large number of sales markets and buyers in different countries, the best specialists from different countries in the project team, advanced technical support, cheaper resources, greater innovation, flexibility, productivity and efficiency, The integration of sustainable development will be much more useful and effective in international projects due to more opportunities to solve the environmental, social and economic problems of the world.

The coronavirus pandemic has posed new challenges and created a new reality for integrating SD into international projects. Kumar et al. (2020) states that companies have to adapt because of the crisis in order to avoid losses [4]. The coronavirus pandemic has had a major impact on integration of SD in the whole world and has brought to light problems such as health issues, inequality, environmental pollution and, most importantly, the vulnerability of humanity to such crises. Many FMCG companies have switched to buying resources in the local market, for example, having abandoned cheap purchases in China, which will also reduce carbon emissions due to reduced traffic and modified supply chain and logistics systems. Hakovirta and Denuwara (2020) suggest that the pandemic will also worsen problems such as poverty and hunger due to the large number of layoffs and lower salaries for employees of companies caused by their difficult financial situation [2].

Currently, there are almost no studies in academic literature on the integration of sustainable development into international projects, taking into account the factors of industry specificity, internationality and the coronavirus pandemic, which creates a scientific gap in this area. The scientific novelty of this study lies in identifying the degree of importance of the factors for successful integration of sustainable development into international projects in the FMCG industry, as well as creating guidelines. The scientific novelty of this work also consists in considering the international aspect of FMCG projects, as well as the new factor of the coronavirus pandemic. The guidelines and recommendations will be applicable and specific and can be useful for international FMCG companies and their projects in integrating sustainable development, which gives this work practical significance. Thus, this work will close the gap in the academic literature and approaches and contribute to the study and development of recommendations for the integration of sustainable development in international FMCG projects.

The main goal of this work is to identify the key success factors as well as to develop guidelines for integrating sustainable development into international projects in the FMCG industry. This study consists of three chapters. The first chapter (Literature Review) focuses on the systematization and critical analysis of academic literature. This chapter covers the theory of the problem. The first chapter is divided into 3 parts: the first part examines SD in FMCG companies; the second part focuses on the specifics of international projects wherein SD is integrated; the third part considers SD in project management. The second chapter describes an empirical study. The second chapter is divided into 2 parts: the first part of the chapter presents research design; the second part analyzes cases on the integration of SD international projects of one of the largest FMCG companies (Unilever, L'Oréal and IFFCO beauty) and in-depth interviews with representatives of FMCG companies. The third chapter focuses on describing the results obtained, on analyzing the online survey, checking propositions and on developing guidelines.

2 Literature Review

2.1 Sustainable Development in FMCG Companies

Mensah (2019) state that Sustainable development (SD) is the concept that describes the process of life, activities, and consumption of the modern society, which does not harm future generations and does not compromise their ability to meet their needs [9]. This is a development that helps society in the long-term period, considering the environment, the social sphere, and economic development. Sustainable development is the idea of growth that contributes to the progress of the entire society, without harming the planet, while solving social and economic problems and finding a balance between these areas.

SD is now being integrated into an increasing number of companies in the FMCG industry. FMCG companies are trying to achieve not only economic but also environmental and social goals. Bashir et al. (2020) state that at present, FMCG companies, due to their large social and environmental footprint, are under strong pressure from society and international organizations to encourage them to change and transform their business models into more sustainable ones. FMCG companies' products mainly use plastic as packaging [1].

Companies are trying to change the design of their packages so that they can be well designed for transport, storage and use, and be attractive to the consumer. Some research has also been devoted to finding out what FMCG products people buy. For example, Vethirajan and Ramu (2019) state that the environmental protection factor in personal care products is as important as the price [12]. Makina and Luthuli (2014) emphasize that consumers are willing to pay more for FMCG products that are sustainable [8]. Supply chain, production and waste disposal are also of great concern to many FMCG companies, because these aspects of business are more entrenched in developing countries, where there are very serious environmental problems.

Leal Filho et al. (2020) The coronavirus pandemic both stopped the integration of sustainable development in FMCG companies and convinced them of the need for sustainability as means of resistance and protection against such crises [5]. The coronavirus pandemic took many companies by surprise; many FMCG organizations had to change their strategies and business operations to cover short-term losses. Nevertheless, it is the principles of sustainable development that could help many companies avoid difficulties, losses from lockdowns and economic problems in their countries.

2.2 Features of International Projects

Koster (2009) state that international projects are projects that go beyond the national borders of one country, that is, they may concern stakeholders of different nationalities, branches or companies from different countries, a large number of locations, divisions, and large-scale goals of the international project itself [3]. An international project differs from a local one in some aspects. It is necessary to make a comparison according to some criteria.

1. **The number of locations.** An international project can involve many locations and operate in different countries, unlike a local one.
2. **Range of the project.** Lientz and Rea (2003) state that international projects are usually much larger in structure and size than local ones. An international project can involve a big project team, departments, and even several companies and states [6].
3. **Risk.** International projects operate in an international environment with a high level of uncertainty and a large number of complexities, which increases the degree of risk.
4. **Organization and management.** Managing an international project is much more difficult due to the presence of a large number of external factors.
5. **Culture.** Lückmann & Färber (2016) argue that in an international project, there may be confusion and misunderstandings between participants due to cultural differences, traditions, customs, as well as differences in corporate cultures [7].
6. Many FMCG companies should understand why they need international projects at all. It is also crucial to consider the main advantages and reasons for success of international projects, taking into account the industry specifics of FMCG.
7. **The most qualified foreign specialists.** Welch & Welch (2012) state that an international project can attract the best specialists from all over the world [13]. This allows international projects to exchange the best ideas and practices, use foreign experience and predict mistakes. In FMCG companies, horizontal mobility between departments and project teams is very common.
8. **Increased flexibility and adaptation.** International projects have to constantly adapt to business conditions in different countries, which increases their flexibility and adaptation. FMCG projects also have to take into account certain aspects of internationalization, such as the requirements for the composition of the product or the ban on testing the product on animals.
9. **Innovations and cheap production.** International projects enable project managers to introduce innovations from different countries, as well as significantly save money by placing cheap production in developing countries. This is what many FMCG companies do by collecting the necessary resources and building factories in low-cost countries.

2.3 Sustainable Project Management

Sustainable project management (SPD) can be defined as the planning, monitoring and control of a project that take into account the three areas of sustainability (environmental, social and economic) in the life cycle of various project elements, such as resources, processes and effects, in order to add value to the project for different stakeholders, presenting to them the process and result in an ethical and transparent way. Toljaga-Nikolić et al. (2020) state that sustainable development in project management helps projects achieve positive results in various elements: logistics, operational efficiency, reputation, investment [11]. Sustainable project management also requires having a specific strategy, understanding the necessary management practices, and taking into account the interests and requirements of stakeholders. This is all very important because the role and number of projects in the world has increased significantly. Thus, it is on the project managers and their approaches to the management board that it will largely depend on whether global environmental and social problems will be solved.

Some researchers have considered how to integrate sustainability into project management. Silvius et al. (2012) underlined the possibilities of introducing SD into the five process groups of project management: initiating, planning, executing, monitoring, and closing [10]. The first two processes allow project managers to integrate sustainability into the goals and objectives of the project itself. The following two execution-related processes allow project managers to implement sustainability already in the process of achieving goals and creating the final result. The last process is related to the transfer of the project itself to the customer; it is worth noting here that if the project is not adopted, then such a project will immediately be considered unsustainable, since resources and energy were wasted in the process for nothing, which is contrary to the principles of sustainable development.

3 Analysis of Practices of Integrating Sustainable Development into FMCG International Projects

3.1 Research Design

In order to address the identified problem of lack of academic approaches to the integration of sustainable development into international projects in the FMCG industry, it was necessary to conduct a study using both quantitative and qualitative methods. The quantitative methods in this work helped to determine the degree of importance of success factors and other elements of integration of sustainable development in international projects in numerical statistical values, while the qualitative methods helped to study the various aspects of the problem and processes in practice in a more profound and complex way. This work used qualitative methods such as **case studies** of international FMCG projects of L'Oréal, Unilever, and IFFCO Beauty companies, as well as **semi-structured in-depth interviews** with three representatives of these companies. **An online survey** with FMCG managers of companies and their project teams was used as a quantitative method. The survey consisted of 12 questions. As a result, it was completed by 16 employees of L'Oréal, 16 employees of Unilever, and 15 employees of IFFCO Beauty.

The above-mentioned qualitative and quantitative methods were also necessary to confirm the **propositions** given below.

1. Use of recycled or biodegradable resources and packaging, human resources, availability of common company guidelines and tools for integrating SD and involvement in the sustainability of all elements of international project might be the most important factors of success of the integration of SD into international projects in the FMCG industry.

2. Sustainable development in international projects might help FMCG companies in internationalization: receive tax and financial incentives and get the approval of local stakeholders when entering foreign markets from the states.

Qualitative methods included composing questions and conducting interviews with three representatives of three FMCG companies; case studies and interviews were used to identify the main aspects and elements of successful integration of sustainable development into international FMCG projects; case studies, online survey and interviews were used to confirm propositions. The quantitative method was applied as an online survey to identify the key factors of sustainable development integration, as well as to confirm the propositions. In the third chapter, based on the research, the main conclusions were made, an online survey was analyzed, propositions were tested, and guidelines and recommendations were compiled, which include the model for determining the success of integration of sustainable development into an international FMCG project, levels of integration of SD into FMCG companies and their international projects and guidelines to them, guidelines for SPM processes in the FMCG industry, guidelines for increasing the engagement of company employees and project teams in SD as well as recommendations on areas and elements for integrating SD into international FMCG projects.

3.2 Analysis of the Integration of Sustainable Development into International Projects in the FMCG Industry (L'Oréal, Unilever, IFFCO Beauty)

Analysis of the Integration of Sustainable Development into International Projects of L'Oréal

L'Oréal is a company that prioritizes sustainable development in its strategy, actions, and international projects. L'Oréal has a major commitment program by 2030 to the planet, society, and products called "Sharing Beauty With All". This program is aimed at reducing environmental damage by L'Oréal, by also addressing social issues such as inequality and poverty. The company began integrating sustainability back in 2013 with the first version of the "Sharing Beauty With All" program and the goals aimed at achieving by 2020. The company has already achieved many positive results, for example, reducing greenhouse gas emissions by 78% compared to 2005; 35 production sites also have become completely carbon neutral.

One of the most important international projects of L'Oréal was the creation of a unique tool for measuring the environmental and social environment of cosmetic products. It is called the Sustainable Product Optimization Tool (SPOT). This tool is implemented in the development of all new products of the company and the transformation of old brands towards sustainable development. SPOT takes into account and calculates the indicators of packaging, formula footprint, ingredient sources, and social benefits of the product.

Successful examples of integrating sustainability into international projects include the examples of the brands ELSEVE, Redken, Yves Saint Laurent, and many others. In the international projects of L'Oréal, there may be problems related to cultural and language barriers due to the multiculturalism of the teams and foreign specialists, but the company tries to actively solve them. The company is also actively engaged in projects in the social sphere and environmental protection. L'Oréal was able to adapt to the coronavirus pandemic thanks to its sustainable strategy.

Analysis of the Integration of Sustainable Development into International Projects of Unilever

Unilever is constantly launching sustainable projects to transform and create new brands. Such well-known brands of the company as Dove and Camay already use eco-friendly packaging created from recycled materials. These international brands constantly hold various social events. The Dove soap shows a girl with special needs, which proves that the company pays great attention to the social sphere. Another example of an international project is the creation of the brand Love Beauty & Planet. This brand includes shampoos, shower gels, and other similar products. This is one of the most sustainable brands of Unilever; all products are made from natural ingredients, and the packaging is made from recycled materials. This brand is certified by PETA and V-Label, which is a very important indicator. As a result of the international project, it was possible to find the necessary set of sustainable ingredients and reduce the emissions of garbage and carbon footprint. The brand is produced only in the greenest production facilities using 100% wind energy worldwide.

Fig. 1. Steps of integration of sustainable development on the example the Clean Line international project

For this study, the international project "Integration of the principles of sustainable development in the Clean Line brand". This project will be analyzed based on a case study and an interview with the brand director of Clean Line. As a result of the analysis, it is possible to highlight the steps taken by the project team of Unilever to introduce sustainability in the international Clean Line project (Fig. 1).

Analysis of the Integration of Sustainable Development into International Projects of IFFCO Beauty

IFFCO Beauty has been one of the main business segments of the IFFCO Group of Companies since 1994. This company operates in the field of FMCG and is the leader of the Arab region. This company is active in 80 countries. The team is multicultural; it employs representatives of 76 nationalities. The company's top sales are beauty products and various personal hygiene products. Integration of sustainable development into the Savannah brand. The headliner of this project is toilet soap, it is in this direction that all the forces for the integration of sustainability were concentrated. Raw materials and production technology of Savannah are certified according to the Halal standard. Obtaining this certificate was one of the main goals of integrating sustainable development into this project. The Halal standard is a model of sustainability for products that are consumed by the Muslim world. As a result of the analysis, it was highlighted that many of the principles of Halal and the principles of sustainable development coincide. The common elements and aspects linking Halal and sustainable development standards are given below:

- The brand's product now does not contain animal products, which is very important for both sustainability and compliance with Halal standards.
- There are no alcohols or parabens in Savannah soap. Hygiene products must not contain substances that can be potentially dangerous to humans. So, for example, parabens found in cosmetics are allergenic and carcinogenic, and alcohol is strictly prohibited in the Muslim world.
- The soap packaging is biodegradable, which has a positive effect on the brand's sustainability.

4 Results

One of the most important tasks of this study was to identify the key success factors for integrating sustainable development into international projects. Based on the analysis of the academic literature, case study, and interviews, the 16 most important factors were revealed (Fig. 2). During the online survey, respondents were asked to distribute 100 points between these factors according to the degree of importance (what factors are the most important and ensure successful integration of sustainable development into international projects in the FMCG industry). The respondents distributed 100 points for all or certain factors while determining the most important of them. Thus, the crucial factors and their weights were identified for further model construction.

Sustainable development helped projects and companies to internationalize. For example, this is reflected in receiving tax benefits, subsidies, or approval from stakeholders in different countries. More than 78% of respondents confirmed that this is true.

Table 1. Key success factors for integrating sustainable development into international FMCG projects.

Success factors	Minimum	Maximum	Average	Median
use of recycled or biodegradable resources and packaging and their possible reuse or recycling	0	28	10,11	12
sustainable development specialists and expatriates with experience in this field	0	28	9,89	10
competent and engaged project teams and company employees	0	27	9,68	8
availability of common company guidelines and tools for integrating sustainable development into international projects	0	27	9,11	8
involvement in the sustainability of all elements and divisions of the organization and the international project (HR, logistics, production, etc.)	0	20	8,53	10
developed infrastructure for sustainable development in countries	0	20	8,38	8
contracts with local ethical responsible resource suppliers and recycling partners in different countries	0	20	7,57	8
a large potential profit from the project and the financial condition of the company that is working on the project	0	21	7,11	6
top management support	0	30	7,09	6
sustainable development as the main strategic direction of the company	0	17	5,89	4
certification and standards of sustainable development project processes and products	0	32	4,51	4
concern for the interests of stakeholders of an international project	0	23	3,3	3
favorable, ethical, and equal working conditions for the project team	0	10	3,25	2
political, economic, and social situation in the countries of the international project	0	19	2,15	1
minimizing harm to the environment and society throughout the project's value chain	0	9	1,77	2
availability of necessary energy-saving and innovative technologies and equipment	0	10	1,66	1

This leads to the conclusion that sustainable development in projects is a very important tool for achieving international goals (Fig. 2).

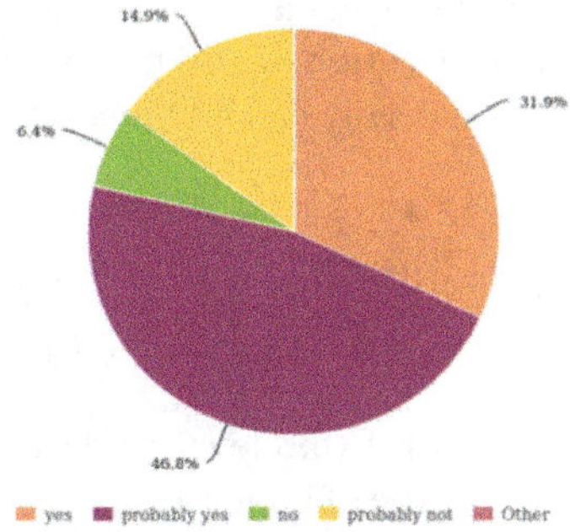

Fig. 2. Sustainable development assistance in the internationalization of companies and their international projects

Sustainable development helped projects and companies to internationalize. For example, this is reflected in receiving tax benefits, subsidies, or approval from stakeholders in different countries. More than 78% of respondents confirmed that this is true. This leads to the conclusion that sustainable development in projects is a very important tool for achieving international goals (Fig. 2).

1. **The first proposition** is confirmed by the following conclusions and findings. First, as a result of an online survey, it turned out that these factors are the most significant and have the greatest weight of importance (average scores). Use of recycled or biodegradable resources and packaging weights 10.11; (human resources) SD specialists and project teams weight 9.89 and 9.68, respectively; common company guidelines and tools weight 9.11 and involvement in the sustainability of all elements of an international project weights 8.53 (Table 1). These are the largest values that were obtained from an online survey of 47 FMCG employees of companies; therefore, these are the most important success factors. Second, in the case studies of all three companies, these elements were found: the rejection of conventional plastic packaging in all projects, the importance of the human factor in strategies and plans, as well as the SPOT tool from L'Oréal and the involvement of all areas of companies proves this proposition. Thus, this proposition must be confirmed.
2. **The second proposition** can be confirmed by the following conclusions. In an online survey, more than 78% of respondents confirmed that this is true. The case studies of all three companies have also identified the assistance of sustainability in internationalization by receiving financial benefits and meeting the requirements of stakeholders around the world. The importance of this aspect was also discussed in two interviews. Thus, this proposition should be confirmed.

In order to help the FMCG company more effectively and correctly integrate sustainable development into international projects, the author has developed **guidelines** and recommendations. The first one is a **model** for determining the success of integration of sustainable development into an international FMCG project. The model will determine the success of the integration of SD into international FMCG project and is based on two values (indicators): “f” (the level of development of the success factor in the international project) and w (the weight, the degree of imptance of this factor in the overall project model). The “w” indicator was already determined in advance above based on the results of the survey on the distribution of 100 points among the factors; that is, the weight of each factor was a certain average and divided by 100 (1).

$$\begin{aligned} SoI &= f1 * w1 + f2 * w2 + f3 * w3 + f4 * w4 + f5 * w5 \\ &+ f6 * w6 + f7 * w7 + f8 * w8 + f9 * w9 + f10 * w10 + f11 * w11 + f12 \\ &* w12 + f13 * w13 + f14 * w14 + f15 * w15 + f16 * w16 \end{aligned} \tag{1}$$

where “SoI” (success of integration), f (the level of development of the factor in the international project), w (weight, degree of importance of each factor). The boundaries of SoI: 0–20: negative forecast; 21–40: probably negative forecast; 41–60: neutral forecast; 61–80: probably positive forecast; 81–100: positive forecast.

FMCG companies are at different levels of integrating the principles of sustainable development into their international operations. Therefore, there is a need to determine the number of these **levels** and develop a description of each of them, as well as create guidelines for achieving each subsequent one. This division of levels was selected (1-Beginner; 2-Simple; 3-Intermediate; 4-Advanced; 5-Higher). To sum up, there are 5 levels of integration and guidelines for the three components of sustainability: environmental, social, and economic. These guidelines will help companies navigate in which directions they need to move.

As the online survey showed, human resources and staff involvement in the process of integrating sustainable development into international projects are one of the most important factors of success. In this regard, the author decided to develop guidelines to increase the **involvement and interest** of the FMCG company’s staff and project teams in sustainable development (Fig. 3).

When deciding to initiate a sustainability integration project, FMCG companies should consider various aspects and elements of sustainable project management. In this regard, the author decided to develop guidelines for sustainable project management processes in the FMCG industry (Fig. 4).

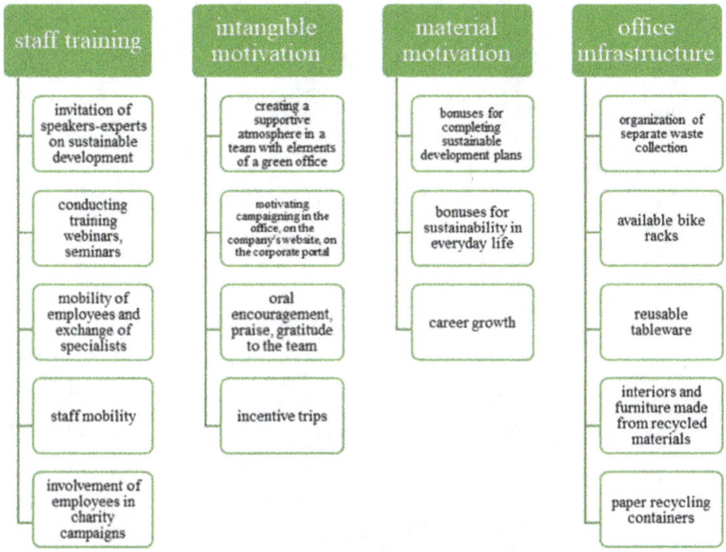

Fig. 3. Guidelines for increasing the engagement of company employees and project teams in sustainable development

The author also developed recommendations for integrating the sustainable development of FMCG companies and their international projects based on a comprehensive analysis and highlighting the key factors of integration success.

1. **Interaction with stakeholders.** Openness, attention, and motivation.
2. **Sustainable resources and packaging.** Recycled or biodegradable resources and packaging and their reuse or recycling.
3. **Internationality of sustainable projects.** Increasing the advantages and reducing the disadvantages of the international aspect of sustainable FMCG projects.
4. **Sustainable and green production.** CO_2, waste, energy.
5. **Sustainable innovation and technology.** Ingenuity, enterprise, and improvement.
6. **Green marketing and sustainable public relations.** New positioning, consumers, and quality.
7. **Ethical and responsible HR.** Equality, inclusiveness and tolerance.
8. **Supply chain and logistics.** Ethics, locality, minimum transport, and CO_2.
9. **Certifications and standards of SD project processes and products.** Confirmation, success, and labeling.
10. **Unified SD integration tools.** Universality, common methods and indicators.
11. **Sustainable development is the priority of the strategy.** Engagement of all elements, unified plans, KPIs, internal policies, awards.
12. **Coronavirus pandemic and adaptation.** Digital, immunity, and transformation.
13. **Green project management.** Standard, methodology, and evaluation.

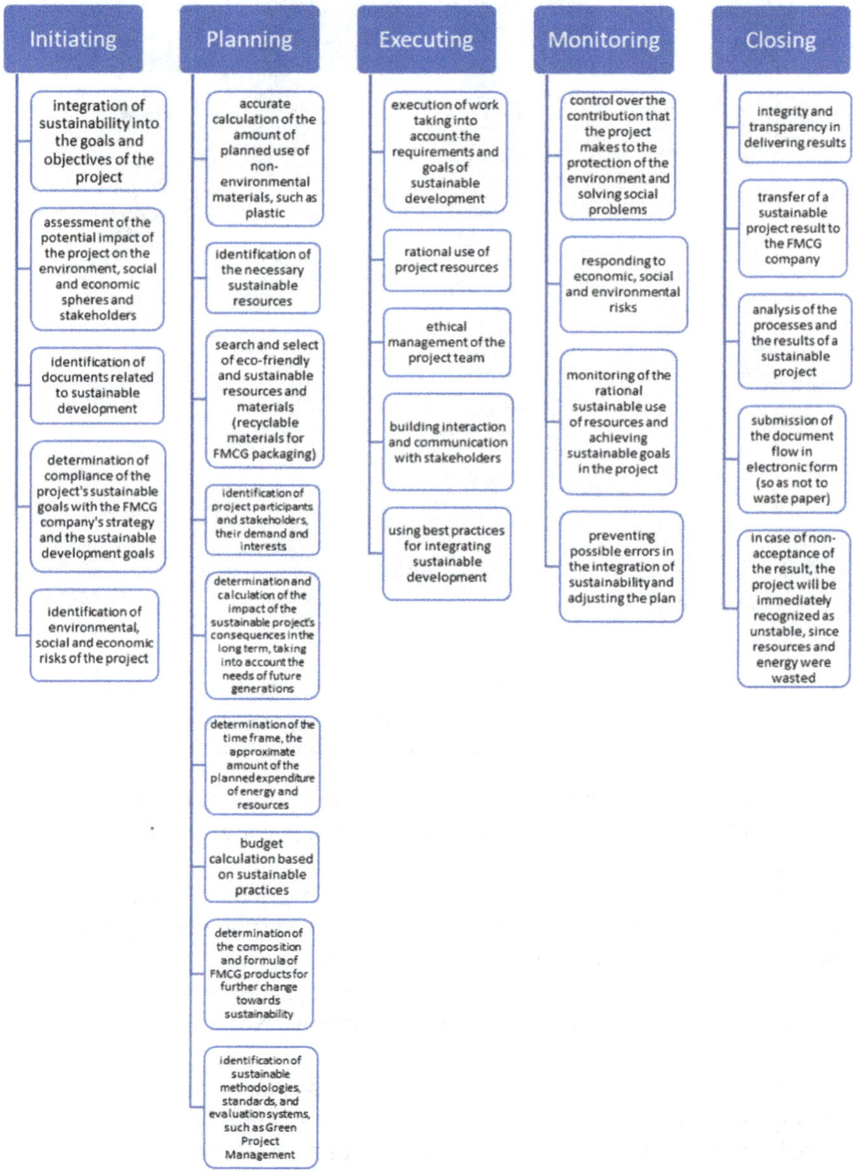

Fig. 4. Guidelines for sustainable project management processes in the FMCG industry

5 Conclusion

Integration of sustainable development is one of the most important components of the success of any company and, accordingly, its projects in the modern world. The management of each business, especially in the field of FMCG, needs to integrate sustainable development, observing its environmental, social and economic principles at every stage of doing business and managing international projects. This study has some limitations, such as the FMCG industry, the size of the companies considered, and the number of respondents and interviewees. At the same time, this work sets the directions for further research aimed at studying the integration of sustainable development into international projects in other FMCG companies or even industries.

References

1. Bashir, H., Jørgensen, S., Pedersen, L.J.T., Skard, S.: Experimenting with sustainable business models in fast moving consumer goods. J. Clean. Prod. **270**(1), 1–122 (2020)
2. Hakovirta, M., Denuwara, N.: How COVID-19 redefines the concept of sustainability. Sustainability **12**(9), 27–37 (2020)
3. Koster, K.: International Project Management, 1st edn. SAGE Publications Ltd. (2009)
4. Kumar, A., Luthra, S., Mangla, S.K., Kazançoğlu, Y.: COVID-19 impact on sustainable production and operations management. Sustain. Oper. Comput. **1**, 1–7 (2020)
5. Leal Filho, W., Brandli, L.L., Lange Salvia, A., Rayman-Bacchus, L., Platje, J.: COVID-19 and the UN sustainable development goals: threat to solidarity or an opportunity? Sustainability **12**(13), 43–53 (2020)
6. Lientz, B.P., Rea, K.P.: International Project Management, 1st edn. Academic Press/Elsevier Science (2021)
7. Lückmann, P., Färber, K.: The impact of cultural differences on project stakeholder engagement: a review of case study research in international project management. Procedia Comput. Sci. **100**, 85–94 (2016)
8. Makina, A., Luthuli, A.: Corporate South Africa and biodiversity in a green economy. Int. J. Afr. Renaissance Stud. - Multi- Inter- Transdisciplinarity **9**(2), 197–212 (2014)
9. Mensah, J.: Sustainable development: meaning, history, principles, pillars, and implications for human action: Literature review. Cogent Soc. Sci. **5**(1), 1–21 (2019)
10. Silvius, G., Schipper, R., Planko, J., Brink, J.V.D.: Sustainability in Project Management (Routledge Frontiers in Project Management), 1st edn. Routledge (2012)
11. Toljaga-Nikolić, D., Todorović, M., Dobrota, M., Obradović, T., Obradović, V.: Project management and sustainability: playing trick or treat with the planet. Sustainability **12**(20), 1–20 (2020)
12. Vethirajan, C., Ramu, C.: Consumers' knowledge on corporate social responsibility of select FMCG companies in Chennai district. J. Int. Bus. Econ. **12**(11), 82–103 (2019)
13. Welch, C.L., Welch, D.E.: What do HR managers really do? Manag. Int. Rev. **52**(4), 597–617 (2012)

Measuring the Agile Mindset on Individual, Team and Organizational Levels - Results of an Empirical Study

Helge Nuhn(✉)

Wilhelm Büchner Hochschule, Hilpertstr. 31, 64295 Darmstadt, Germany
helge.nuhn@wb-fernstudium.de

Abstract. In recent years, agility has become an increasingly relevant aspect of practitioner approaches as well as research on organizational endeavours. While techniques are often in focus, governance aspects as well as mindsets are hypothesized to be more effective aspects or agile organizing, however. As of today, there is little quantitative empirical research aimed at illuminating agile mindsets, and non that aims at mindsets of temporary teams or entire companies. This article presents the findings of a study conducted in a major German IT services company. 79 valid results obtained from 500 polled operative employees allow for the quantitative analysis of agile mindset concepts introduced by Miler and Gaida. Multiple methods like PCA, cluster analysis and SEM were used to explore into the potential structure of the agile mindset on individual, team and organizational levels. This article adds knowledge to the still much unexplored concept of organizational agility and concludes to propose future research for making use of the obtained insights in longitudinal, temporary settings. It is the first study to present factor as well as principal component analysis, and structural equation modelling for a concept that was up to then largely unexplored and subject to qualitative research at most.

Keywords: Agile · Mindset · operationalisation · items · confirmatory · explorative · FA · PCA · cluster analysis · SEM

1 Introduction

In recent years, agility has become an increasingly relevant aspect of practitioner approaches as well as research on organizational endeavours. Disagreement prevails what makes organizations, especially temporary ones, agile. While techniques are often in focus, governance aspects as well as mindsets are hypothesized to be more effective aspects or agile organizing, however [1]. As of today, there is little quantitative empirical research aimed at illuminating agile mindsets. Also, research work on the nature of organizing for agility is still scarce [2].

Yamal LNG project has realized such project management and trade innovations as conquest of harsh conditions of construction and logistics, e.g. the remote site on the Arctic Sea, winter temperature of −50 °C that the site construction must face, full use of

S. Bushuyev et al. (Eds.): IPMA 2021, LNNS 704, pp. 162–179, 2023.
https://doi.org/10.1007/978-3-031-34629-3_15

modules fabricated in Far East and Europe and transported to the site by specially built ships qualified for Arctic Sea navigation; LNG shipment to Far East and Europe via the Arctic Sea even during winter; and marking the first benchmark of China's "One belt, one zone" program realized (idem).

In this EPC industry, similarly, a series of huge natural gas projects such as US$25 billion PEARL GTL – natural gas to liquid – in Qatar, US$40 billion Ichthys LNG in Australia as well as complete refinery projects such as Nghi Son Refinery Project in Vietnam, all exceeding US$10 billion in investment costs according to economic journals, have been completed successfully [Tanaka, 2018]. We can infer from these cases that investment costs, sizes and complexity of projects do not affect linearly probability of success or otherwise of projects. Mediating variables should include an application area (industry branch) of the project in question.

International Energy Agency' World Energy Outlook [IEA, 2015] estimated that the global oil and gas industry, the most representative EPC project arena, was expected to invest US$ 25 trillion from 2015 to 2040 in development projects which stands at US$ 1.02 trillion average per year. This expected oil and gas development investment volume has been drastically reduced in 2020 and 2021 due to two reasons: the drastically reduced demand for fuel caused by the prolonged COVID-19 Pandemic and accelerated global policy on energy transition forcing a shift of gravity of oil and gas companies' investment from oil and natural gas development to that of renewable energy or its derivatives such as green hydrogen and bio fuels by which international oil and gas companies have frozen a considerable part of oil and gas development spending, as of May 2021. This current situation is drastically reducing the size of EPC market [Tanaka, 2021].

A strong warning is that loosing or weakening EPC project management capability will be a serious harm to industrial project management base as this is the most rigorous project management capability, and it would cause a huge loss to our society. Owner companies, EPC contractors and concerned governments should stage concerted efforts to conserve the EPC operations and management capability in facing the 'existential crisis' of capital program delivery [ECI, 2018]; such efforts include closer owner and contractor alignment initiatives for sound growth of the EPC industry and particularly an industry initiative to renovate the structures of capital program (FEED + EPC) delivery for reducing transaction costs in the EPC chain [CII, 2018] while continuing prudent investment to safeguard supply of fuel.

2 Theoretical Basis

Companies these days experience increasing market pressure, uncertainty and high development speeds and many more aspects that force upon them in addition to former paradigms like rigorous exploitation of production processes the need to explore novel products and services [4]. This approach causes companies to organize differently as a rule, often transforming into project-based, temporary organizations [5].

Theory of temporary organizing [6] in its development relies on the presence of collective minds in order to quickly gain robustness in organizations even though they are only in the process of forming, and potentially of a short lifetime anyhow [7]. But also for longer-lasting, yet still temporary organizations, familiarity within a team with regard to shared values and beliefs is important for team integrity in organizations [8].

Various researchers and practitioners have attempted to capture the essence of agile mindsets in the past. Early contributions remark a lack of scientific approach into the matter, like that of Dikert et al. [9]. They also mention "mindsets" as success factors for large-scale agile transformations within extant companies. On an even larger scale, Santistevan and Josserand argue "mind-set" to be necessary precursors for successful "agile" and "fluid" parts of organizing, like with "meta-teams" within multi-national entities [10].

Denning [1] suggests a 10-stage-approach for agile transformations, in which forming agile mindsets is one element. However, there is not operationalization visible that aims at thoroughly understanding the depths of the concept. Gregory and Taylor report the results of a conceptual elaboration on a "view of agile culture" [11]. Their resulting matrix however is mostly formulated in an organization-centric way of assessing "what is the case" and does not focus on values and belief systems as they would be expected to describe a mindset.

Miler and Gaida (2019) [3] presented the most fitting approach from this paper's perspective. Its qualitative, grounded approach with numerous interviews conducted among IT professionals, and also quantitative empirical testing (n = 51). This gives a good basis for analyses into potential underlying structures of mindsets on different organizational levels.

In their paper, the authors ended up with 26 items that were supposed to capture the relative importance of each item for teamwork efficacy. Their paper was taken as a basis for this contribution.

While Miler and Gaida's approach of measuring what people deem as important for successful teamwork, it can also be considered slightly different, namely in a way that asks for "what is important to you?", hence capturing values and belief systems in general, not goal related. What this measuring of agile mindsets still leaves open is that the individual's mindset is embedded in a social system, i.e. in a system of mindsets. Base hypothesis for the research described in the following is that mindsets can be measured and have a specific structure for individuals. But also, that individuals have a perception of others' mindsets, specifically groups of others. Therefore, other useful foci of investigation are the perceived mindsets of the team that the employee is part of, as well as the perceived mindset of the entire organization/company. The specific research question that this paper seeks to answer is for one whether the categorizations of items by Miler and Gaida can be recovered in an additional empirical study, for another if the found conceptualization of agile mindsets can also be transferred to perceptions of team-mindsets and organizational mindsets. Therefore, the specific hypothesis are:

H1: Miler and Gaida's approach can be transferred to measuring individual mindsets.
H2: Team mindsets are measurable and constructed similar to the individual mindsets.
H3: Organizational mindsets are measurable and constructed similar to individual mindsets.

3 Method

3.1 Participants

A total of approximately 500 employees were contacted with electronic communication, inviting them to participate in a study regarding organizational agility. 79 valid response data sets were received, meaning a reasonable response rate of ca. 16 percent. Most participants were male (58), fewer female (18), one was of diverse gender and one did not respond to this question. Age distribution shows a more life-experienced workforce: 25–29: 6; 30–34: 8; 35–39: 9; 40–44: 5; 45–49: 10; 50–54: 17; 55–59: 19; 4 were older than 60 and only one younger than 25. Similarly, 66 of 79 have a work experience of more than ten years. The majority (46) obtained degrees of higher education. Regarding tenure, there are two peaks: greater than 20 years of working with the company (22) and five to ten years with the company (17). Eight had been less than two years, twelve two to five years, 9 10–15 years and 11 between 15 and 20 years members of the organization. Asked for agile roles, 52 responded to be regular team members, eleven were agile masters and 16 product owners. Also surveyed was the state of the agile transition each individual was in with their team: five consequtive milestones, ranging from "not started" (2), to "transition ended" (39), the rest being in between.

3.2 Procedure

The online survey was administered in the time of May and June 2021 to collect data from the employees mentioned above. The link was sent via email through one of the mailboxes of one group member. The survey, its purpose and its approximate duration of 30 min were described on a landing page. Data privacy specifics were addressed, and the purpose and processes of data collection and handling made clear. Care was taken to design the questionnaire in a way that would not allow for identification of individuals once the data collection had been over. This was aligned with internal governing bodies of the organisation.

The data showed no signs of extraordinary premature termination behaviour, nor of extreme cases of outliers. Missing values were filled with column means for further data usage. The questionnaire was entirely in German, consequently the items had to be translated into German language. This was done by three researchers as a group effort, in order to assure correct translation.

3.3 Measurements

Miler and Gaida had researched both qualitatively and quantitatively concepts that they checked for importance regarding effective teamwork. These concepts were taken and turned into statements that allowed polling for acceptance or rejection and that aimed at measuring opinions on individual, team and organizational level. Example: "To me, openness to change is important", "To my team, openness to change is important", or "To my organization, openness to change is important". These statements could then be rated on a Likert-type scale ranging from 1 to 7, combined in a set of three times 26 items. Means and standard deviations are listed in the appendix in Table 2. For composite constructs, see the subsequent tables and their Cronbach Alpha, Rho and AVE values. More quality measures are described in the following sections.

4 Exploratory Factor Analysis

Bartlett's test of sphericity was significant $\chi 2(325) = 3133.63$, $p < 0.001$, the Kaiser-Meyer-Olkin measure also showed adequate results (0.9) for mindset-related items on all three levels. Parallel analysis, scree plots and VSS plots were used to identify the number of components or factors within the mindset-related data of 26 variables per level. Kolmogorov-Smirnoff as well as Shapio-Wilk tests indicated significant deviations from normal distributions, which was considered in later methodological decisions. Consequently, CFA, using the lavaan package, was attempted but could not confirm the factors suggested by Miler and Gaida. Principal component analysis was conducted using R's prcomp function of the stats standard library. Also, exploratory factor analysis was conducted using R's (Rstudio.cloud) psych package, namely the fa function. The factor analysis configuration included a maximum likelihood approach, varimax rotation and 100 iterations.

In an additional step, cluster analysis was conducted further analyse clusters of items. This added some more interesting insights, leading to the contemplation more fine-grained factor-solutions than parallel analysis and scree-plot initially suggested. This was done using the *psych* package's *iclust* function.

Analysis was started with a focus on the individual level. Parallel analysis only suggested two different factors to be considered. In brief, factor 1 is comprised of items that are in a sense "value-oriented", while factor 2 is rather "action/solution-oriented". Examples for the first factor: "sincerity", "openness", "equality", "listening", "positive attitude", "respect". Examples for the second factor include: "continuous delivery", finishing instead of multitasking, "constant work-speed", "customer satisfaction", "customer cooperation", "transparent decision-making". Results are displayed in Table 1 and apart from the mentioned items, it seems that the two factors blend into one another

Table 1. Factor loadings of the FA yielding two factors on individual level.

Item *(To me <X> is important.)*		Factor 1 – "value-oriented"	Factor 2 – "action/solution-oriented
1	Continuous delivery of a valuable product in short intervals	0,90	0,18
2	Cooperation with the customer based on partnership	0,85	0,36
3	Attitude towards customer satisfaction and needs	0,82	0,20
4	Mutual trust	0,79	0,23

(*continued*)

Table 1. (*continued*)

Item *(To me <X> is important.)*		Factor 1 – "value-oriented"	Factor 2 – "action/solution-oriented
5	Direct communication face to face conversations	0,79	
6	Focus on achieving common goal	0,78	0,32
7	Helping each other	0,77	0,47
8	Sincerity	0,77	0,38
9	Mutual respect	0,72	0,15
10	Mutual listening	0,71	0,31
11	Equality in the team	0,67	0,43
12	Searching for a solution to the problem instead of finding the guilty	0,66	0,46
13	Team responsibility	0,64	0,49
14	Continuous improvement and learning	0,63	0,60
15	Openness to change	0,53	0,50
16	Being motivated	0,51	0,63
17	Positive attitude	0,44	0,68
18	Openness to criticism and feedback	0,42	0,46
19	Openness to others	0,42	0,64
20	Self-organization	0,39	0,63
21	Maintain a steady pace of work	0,37	0,29
22	Ability to collaborate	0,29	0,69
23	Sharing knowledge and results	0,24	0,56
24	Asking questions in case of insufficient knowledge	0,16	0,50
25	Finishing the current task before taking the next one	0,13	0,51
26	Transparency in decision-making and actions	−0,12	0,64

Table 2. Means and standard deviations.

Item (To me/To my team/To my organization <X> is important.)		Individual μ (σ)	Team μ (σ)	Organization μ (σ)
1	Continuous delivery of a valuable product in short intervals	4,81 (1,31)	4,74 (1,25)	4,93 (1,15)
2	Cooperation with the customer based on partnership	5,56 (0,75)	5,45 (0,74)	5,32 (1,03)
3	Attitude towards customer satisfaction and needs	5,64 (0,58)	5,43 (0,74)	5,22 (1,10)
4	Mutual trust	5,71 (0,72)	5,55 (0,80)	5,18 (1,13)
5	Direct communication face to face conversations	5,05 (1,29)	4,96 (1,12)	4,65 (1,36)
6	Focus on achieving common goal	5,52 (0,86)	5,34 (0,91)	5,15 (1,17)
7	Helping each other	5,66 (0,68)	5,52 (0,82)	5,01 (1,19)
8	Sincerity	5,63 (0,88)	5,49 (0,87)	4,93 (1,43)
9	Mutual respect	5,78 (0,59)	5,57 (0,75)	5,04 (1,24)
10	Mutual listening	5,72 (0,70)	5,48 (0,80)	5,00 (1,32)
11	Equality in the team	5,51 (0,90)	5,33 (0,90)	5,03 (1,26)
12	Searching for a solution to the problem instead of finding the guilty	5,70 (0,63)	5,27 (1,11)	4,87 (1,43)
13	Team responsibility	5,51 (0,92)	5,19 (1,08)	5,14 (1,05)
14	Continuous improvement and learning	5,62 (0,76)	5,22 (1,00)	4,94 (1,26)
15	Openness to change	5,49 (0,77)	5,22 (0,97)	5,09 (1,26)
16	Being motivated	5,39 (0,87)	5,25 (0,83)	5,04 (1,30)
17	Positive attitude	5,39 (0,94)	5,16 (1,10)	5,06 (1,24)
18	Openness to criticism and feedback	5,54 (0,83)	5,18 (1,13)	4,87 (1,44)
19	Openness to others	5,54 (0,78)	5,18 (1,13)	4,88 (1,51)
20	Self-organization	5,10 (1,08)	5,06 (1,14)	5,11 (1,13)
21	Maintain a steady pace of work	4,35 (1,34)	4,39 (1,42)	4,69 (1,35)

(*continued*)

Table 2. (*continued*)

Item (To me/To my team/To my organization <X> is important.)		Individual μ (σ)	Team μ (σ)	Organization μ (σ)
22	Ability to collaborate	5,54 (0,76)	5,32 (0,82)	5,12 (1,11)
23	Sharing knowledge and results	5,53 (0,73)	5,23 (0,93)	5,03 (1,24)
24	Asking questions in case of insufficient knowledge	5,67 (0,63)	5,36 (0,79)	4,94 (1,29)
25	Finishing the current task before taking the next one	4,41 (1,37)	4,41 (1,43)	4,27 (1,62)
26	Transparency in decision-making and actions	5,34 (0,90)	5,16 (1,11)	4,56 (1,61)

Table 3. Items and Component Loadings – Individual Mindset.

Item (To me/To my team/To my organization <X> is important.) "org"-prefix relates to model 2a		*Value-oriented*		*Action/solution-oriented*	
		Teamwork	Attitude	Organisation	Customer Centrism
1/ CUCE1/orgSDLN3	Continuous delivery of a valuable product in short intervals			0,21	**0,69**
2/ CUCE2/orgCUFO1	Cooperation with the customer based on partnership	0,14	0,32		**0,71**
3/ CUCE3/orgCUFO4	Attitude towards customer satisfaction and needs	0,44	0,14		**0,76**
4/ TEWO1/orgSDLN1	Mutual trust	**0,68**		0,35	0,31
5/ orgSDLN2	Direct communication face to face conversations	0,32	0,29	0,48	0,16

(*continued*)

Table 3. (*continued*)

Item (To me/To my team/To my organization <X> is important.) "org"-prefix relates to model 2a		*Value-oriented*		*Action/solution-oriented*	
		Teamwork	Attitude	Organisation	Customer Centrism
6	Focus on achieving common goal	0,67		0,26	0,48
7	Helping each other	0,72	0,51		0,13
8	Sincerity	0,66	0,63	0,17	
9/ TEWO2	Mutual respect	**0,83**	0,25		
10/ TEWO3	Mutual listening	**0,84**	0,33	0,27	0,15
11/ orgBINS2	Equality in the team	0,65	0,47	0,29	
12	Searching for a solution to the problem instead of finding the guilty		0,55		0,49
13/ orgBINS3	Team responsibility	0,44	0,33	0,56	0,29
14	Continuous improvement and learning	0,51	0,44	0,43	0,35
15/ orgCUFO2	Openness to change	0,55	0,46	0,17	0,37
16/ ATTI1	Being motivated	0,29	**0,79**	0,12	
17/ ATTI2/orgCUFO3	Positive attitude	0,18	**0,81**	0,20	0,12
18	Openness to criticism and feedback	0,54	0,66	0,33	0,13
19/ ATTI3	Openness to others	0,41	**0,70**	0,27	0,18

(*continued*)

Table 3. (*continued*)

Item (To me/To my team/To my organization <X> is important.) "org"-prefix relates to model 2a		*Value-oriented*		*Action/solution-oriented*	
		Teamwork	Attitude	Organisation	Customer Centrism
20/ orgBINS1	Self-organization	0,33	0,46	0,39	0,31
21/ ORGA1	Maintain a steady pace of work	0,10	0,11	**0,67**	
22	Ability to collaborate	0,57	0,56	0,19	0,24
23	Sharing knowledge and results	0,63	0,37	0,30	0,17
24	Asking questions in case of insufficient knowledge	0,61	0,35	0,36	0,20
25/ ORGA2	Finishing the current task before taking the next one			**0,79**	
26/ ORGA3	Transparency in decision-making and actions	0,47	0,15	**0,70**	0,14

in the middle of the table, underlining the thorough approach Miler and Gaida took to capture mindsets in breadth.

Taking the covariance matrix as a basis for further analysis of principal components, scree plot, eigenvalues (>1) and proportional variance explained by each component (>0.1), overall variance explained (>.7) by the principal components found suggest up to four components underlying the data – for the individual level.

This divergence is supported when doing factor congruence checking between the canonical PCA results and the factor analysis. Congruence is still given for two factors, after adding a third, results become inconsistent. However, cluster analysis convinced to continue analysis with more rather than less components, as these were semantically convincing and would still fall into one of the two categories "value-oriented" and "action/solution-oriented", splitting each up in half effectively. The identified components along with the items and their loadings are listed in Table 3.

Examination of the PCA loadings of all items onto the four components followed. Cut-off values were defined, so that items with low discriminant properties were discarded. For this, items should load at least with .6 onto their strongest loading component. Also, the second highest loading should not exceed two thirds of the highest loading to ensure discriminant validity. Including them in order of descending loadings, only three items were then selected to be included in each component. The components were then inspected and named according to their semantic focus: 'teamwork', 'attitude, 'organization', and 'customer-centrism'.

The same procedure was applied onto the team-based and organizational level. Qualitative differences were noted: The 'attitude' and 'organization' component were identical on the team-level when applying the same ruleset. Hence, one 'value-oriented' and one 'action/solution-oriented' component were already identified, giving more weight to the factor analysis results.

A 'teamwork' component found on team-level, however, included partly different items ('self-organization', 'openness', 'responsibility'), while a 'customer-oriented' component was not identified. Instead, a two-item-component including 'being motivated' and 'continuous delivery' was generated. This mindset model was not included in further analyses as explained below.

Assessing the organizational mindset in a similar fashion failed more profoundly. Eigenvalue decomposition suggested only three factors, factor analysis only one. Analysis of PCA loadings in conjunction with cluster analysis and semantic considerations led to the formulation of the following three components. First the "customer-focus" component includes items on customer focus and customer partnering, but also items regarding openness and positive attitude like the first component on the individual level. The second component could be described as self-organized, equality-based information sharing or in short, the organization as a "balanced information system". And the third component included aspects like continuous delivery, direct communication and trust-based communication which was summarised as "strong delivery network".

In the following, one step back will be taken and the seeming divergence of conceptualizations of mindsets along the organizational levels will at first be deliberately ignored. Reliability measures as well as in-depth analysis with different ramifications for missing values hint at the potential that the conceptual divergence might be caused by worse data quality for the organizational level. The approach to continue with a unified mindset conceptualization will help see how far the individual mindset concept can be applied successfully in the application of PLS regression models. However, at last, calculations with alternate mindset conceptualization will be shown and finally discussed.

5 Structural Equation Modelling

In the following, SEM approaches were used to calculate path coefficients between the components measured on the three different organizational levels. The aim is to obtain a clearer view on the interconnectedness between the different constructs. If there was anything like an individual agile mindset, one should be able to discover links to the same mindset construct on other organizational levels – especially in an organization that has undergone self-acclaimed agile transitioning for months.

The use of PLS SEM is debated a lot in literature. Scholars seem to have settled on the point that it is more useful than covariance-based SEM in exploratory, rather than confirmatory contexts. Also, it suits the collected data better, because the covariances of closely related constructs are high, and almost none of the collected data pass normal distribution tests. Lastly, PLS SEM is the preferred approach considering the amount of data available, especially since in low n-cases path coefficients are likely underestimated [12].

Construct definitions, correlations, and composite reliabilities for all constructs on all levels are listed in the appendix. HTMT tables are not included in the appendix for brevity, but results will be discussed in the text. The same applies to VIF for items and constructs. Path coefficients are displayed in the figures corresponding to each model. Effect sizes are omitted there but are in general small to moderate.

The first model all four identified components of the individual mindset construct onto corresponding constructs on the team level, following the identical item selection. Figure 1 shows the inner and outer model, path coefficients, confidence intervals and R^2 values (See Fig. 1).

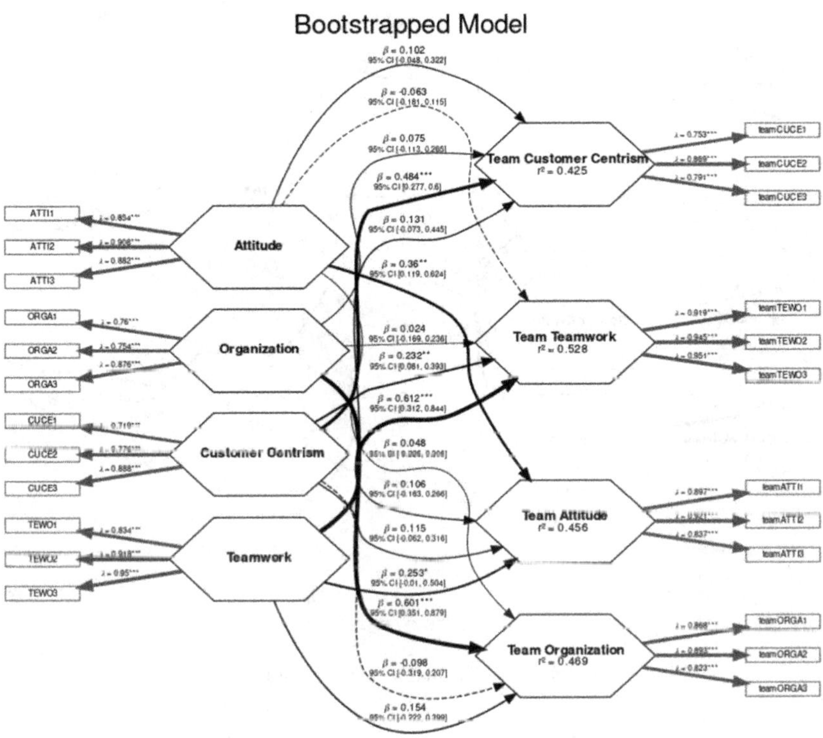

Fig. 1. Model 1 – Individual and team mindset

Clearly, the team mindset is inferable from the individual mindset components. The presence of each construct predicts the presence of the corresponding mindset component with moderately high path coefficients. This result is in support of the path taken, not to model a unique team mindset.

Going forward with the organizational mindset, this approach does not hold any longer – as could be expected from the more divergent analytical analysis. While the two teamwork constructs are still linked significantly, a link from teamwork to customer centrism seems erratic. This is explainable, as VIF values of individual indicators of the organizational mindset were unacceptably high, and correspondingly, HTMT matrix items for organizational mindset components did not meet acceptance criteria. This also rendered the attempt impossible to relate the team-level mindset to the organizational-level mindset in a model not shown here (see Fig. 2).

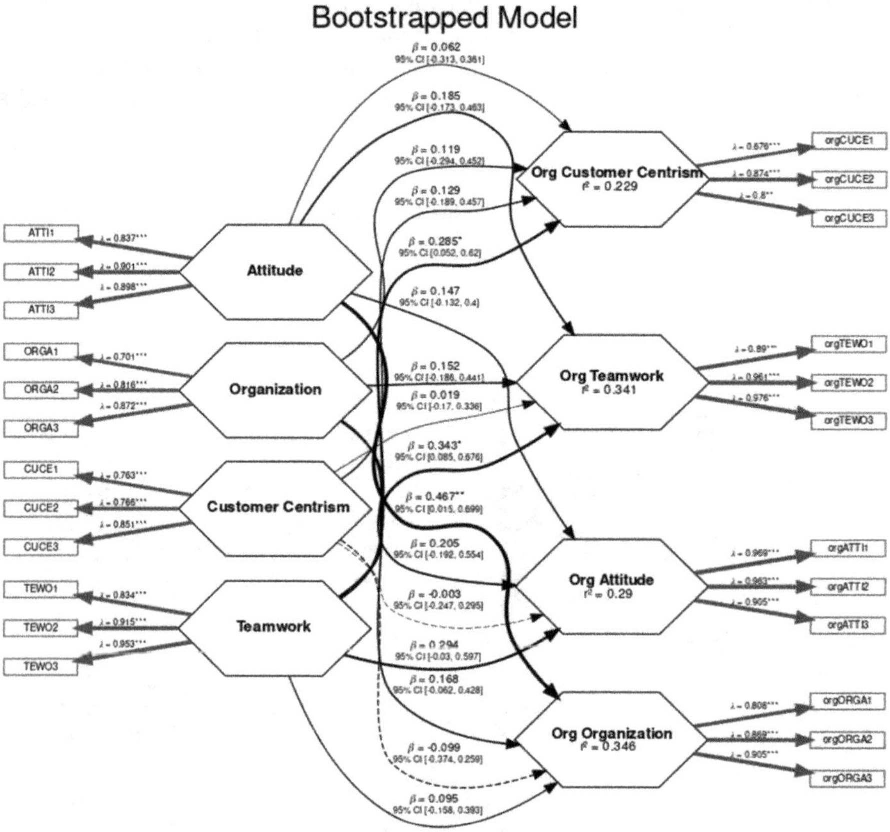

Fig. 2. Model 2 – Individual and organizational mindset

To see whether the 3-component model is a better choice, a corresponding model was built and fitted (see Fig. 3). As can be seen, neither does this model have any greater significant predictive power. However, the high number of missing values that had to be inferred and the already low number of observations may be the main reason for the lack of systematic significant findings. The found significant paths are likely due to high construct correlations and high conceptual overlaps, as indicated again by unacceptably high HTMT matrix entries.

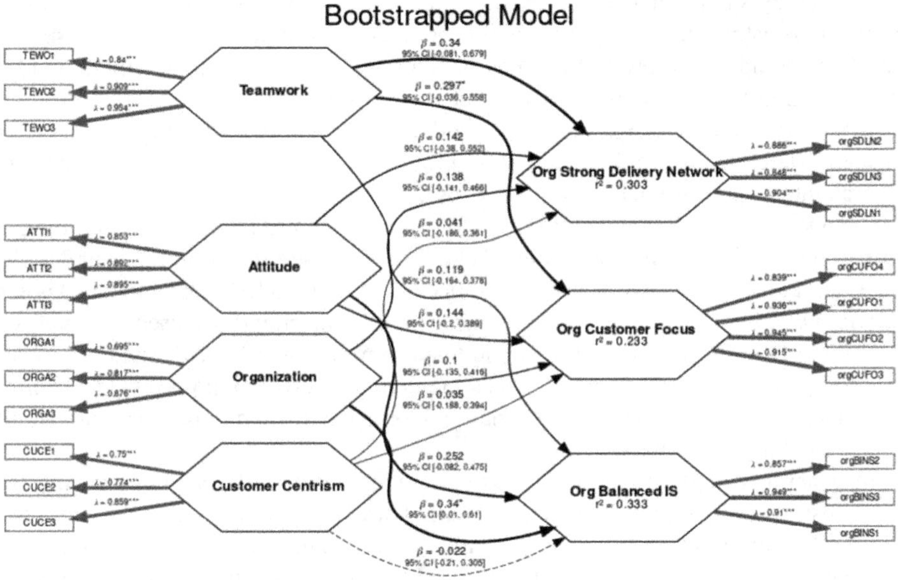

Fig. 3. Model 2a – Individual and organizational 3-component mindset

6 Limitations, Discussion, and Future Research

While this research report is worth being written now, as it clearly delivers some interesting insights, this research avenue has hardly reached its end. 79 responses do not satisfy the condition for successful application of EFA (15 observations per variables required), especially not when considering missing values in up to 11 cases in organizational-level contexts. Too little variance within the data and resulting problems of multi-collinearity demand for future additions to the extant dataset.

The survey was accompanied by some additional qualitative research. One of the most interesting findings from there showed that item naming is a problem: some participants in interviews corresponding to the online survey mentioned that they did not understand what was meant by "my organization" in the specific context. This information fits well into the picture that the afore-mentioned missing values had to be handled for organization level items, while for individual and team-based items almost no missing values are found. So, the understanding of the "organizational" items is flawed. This leads to open pathways for future development of constructs capturing the agile mindset of maybe "companies" rather than "organizations".

Controlling for age, role and education, singular significant but minor effects onto the organization- and team-level construct were discovered. The fact that exactly these concepts showed significant effects is taken as an argument in favour of the existence of mindsets, such that agile masters' mindsets would reasonably be different from regular team members' mindset. Very noteworthy: despite the company being in agile transition, the "next transition stage gate" control variable was not significant.

The survey's individual items asked: "what is important to you/the team/the organization", thus asking about importance in general and not importance towards a specific goal. For the individual level, there may be a discrepancy between what the individual thinks is important and how they act. Yet this difference may be conceptually non-existent for the other two levels as one individual may not able to discriminatingly judge between awareness, aspiration and action [13]. This opens a very interesting path for research into organizational transparency for agile organizations.

Despite these limitations, bottom line of this research to this point is: Building on prior studies, agile mindset definitions have been analysed regarding their underlying structures and this attempt was in general successful. Two underlying factors and potentially four components were identified that allowed for meaningful subsequent analyses. This supports hypothesis H1. Also, agile team mindsets seem to exist and are by and large compatible with individual mindset structures, supporting hypothesis H2. No direct support was found for hypothesis H3, claiming organizational mindsets are comparable to individual agile mindsets.

In general, if agile mindsets indeed exist, they seem to differ with growing levels of abstraction from the individual. While it was possible to stretch the individual mindset definition to a team's perceived mindset due to the closely related nature of its components, such attempts were not successful for the organizational mindset anymore.

This insight received additional support from examination of the means of the responses per item. Consistently for all items, the ratings on the individual level are higher than those on the team level, which in turn are rated higher than those on the organizational level. In other words: the more graspable the reference point for the mindset is, the higher the ratings; the more distal and fuzzy, the lower it will be. It would be interesting to see in follow-up studies how for instance a customer's mindset, or the mindsets of some completely unrelated stakeholder groups are rated.

The research described within this article helps in the definition of the entire concept of agility, namely by assessing agile mindsets. With its help, practitioners can explain the behavioural patterns of agile individuals, teams, and organizations better, which is important as agile modes of working are harder to plan and need to be managed in indirect ways [14]. Furthermore, researchers can investigate into antecedents and outcomes of agile mindsets on different organizational levels by linking the mindset concept to environmental factors and performance indicators. Also longitudinal studies in change initiatives – often called for – become feasible.

Appendix

(See Tables 4, 5 and 6).

Table 4. Reliability measures.

Construct	alpha	rhoC	AVE	rhoA
Model 1				
Teamwork	0.884	0.929	0.813	0.903
Attitude	0.859	0.911	0.773	0.920
Organization	0.719	0.841	0.639	0.756
Customer Centrism	0.712	0.838	0.633	0.716
Team Teamwork	0.932	0.957	0.881	0.933
Team Attitude	0.863	0.916	0.785	0.876
Team Organization	0.827	0.896	0.743	0.839
Team Customer Centrism	0.728	0.847	0.649	0.734
Model 2				
Org Teamwork	0.937	0.960	0.889	0.937
Org Attitude	0.941	0.962	0.895	0.942
Org Organization	0.830	0.897	0.743	0.869
Org Customer Centrism	0.691	0.836	0.639	0.692
Model 2a				
Org Customer Focus	0.932	0.951	0.829	0.952
Org Balanced IS	0.891	0.933	0.822	0.896
Org Strong Delivery Network	0.855	0.911	0.774	0.869

Table 5. Construct Correlations – Model 1.

	Individual				Team-level			
	Teamwork	Attitude	Team Org	Customer Centrism	Teamwork	Attitude	Team Org	Customer Centrism
Individual								
Teamwork	1,00	0,59	0,53	0,48	0,70	0,58	0,46	0,46
Attitude	0,59	1,00	0,43	0,39	0,40	0,60	0,36	0,40
Organization	0,53	0,43	1,00	0,36	0,41	0,43	0,67	0,36
Customer Centrism	0,48	0,39	0,36	1,00	0,51	0,41	0,21	0,61
Team-level								
Teamwork	0,70	0,40	0,41	0,51	1,00	0,76	0,56	0,68
Attitude	0,58	0,60	0,43	0,41	0,76	1,00	0,59	0,70
Organization	0,46	0,36	0,67	0,21	0,56	0,59	1,00	0,47
Customer Centrism	0,46	0,40	0,36	0,61	0,68	0,70	0,47	1,00

Table 6. Construct Correlations – Model 2.

	Individual				Team-level			
	Teamwork	Attitude	Team Org	Customer Centrism	Teamwork	Attitude	Team Org	Customer Centrism
Individual								
Teamwork	1,00	0,60	0,53	0,46	0,54	0,49	0,42	0,42
Attitude	0,60	1,00	0,42	0,39	0,46	0,41	0,55	0,34
Organization	0,53	0,42	1,00	0,34	0,42	0,42	0,38	0,32
Customer Centrism	0,46	0,39	0,34	1,00	0,30	0,26	0,18	0,31
Organizational								
Teamwork	0,54	0,46	0,42	0,30	1,00	0,90	0,71	0,79
Attitude	0,49	0,41	0,42	0,26	0,90	1,00	0,74	0,83
Organization	0,42	0,55	0,38	0,18	0,71	0,74	1,00	0,66
Customer Centrism	0,42	0,34	0,32	0,31	0,79	0,83	0,66	1,00

References

1. Denning, S.: The ten stages of the Agile transformation journey. Strategy Leadersh. **47**(1), 3–10 (2019)
2. Li, H., Wu, Y., Cao, D., Wang, Y.: Organizational mindfulness towards digital transformation as a prerequisite of information processing capability to achieve market agility. J. Bus. Res. **122**, 700–712 (2021)
3. Miler, J., Gaida, P.: On the agile mindset of an effective team an industrial opinion survey. In: Proceedings of the 2019 Federated Conference on Computer Science and Information Systems. IEEE (2019)
4. Park, Y., Pavlou, P.A., Saraf, N.: Configurations for achieving organizational ambidexterity with digitization. Inf. Syst. Res. **31**(4), 1376–1397 (2020)
5. Söderlund, J., Hobbs, B., Ahola, T.: Project-based and temporary organizing: reconnecting and rediscovering. Int. J. Project Manag. **32**, 1085–1090 (2014)
6. Lundin, R.A., Söderholm, A.: A theory of the temporary organization. Scand. J. Manag. **11**(4), 437–455 (1995)
7. Weick, K.E., Roberts, K.H.: Collective mind in organizations: heedful interrelating on flight decks. Adm. Sci. Q. **38**(3), 357 (1993)
8. Nuhn, H.F.R., Heidenreich, S., Wald, A.: Performance outcomes of turnover intentions in temporary organizations: a dyadic study on the effects at the individual. Team Organ. Level **16**, 255–271 (2019)
9. Dikert, K., Paasivaara, M., Lassenius, C.: Challenges and success factors for large-scale agile transformations: a systematic literature review. J. Syst. Softw. **119**, 87–108 (2016)
10. Santistevan, D., Josserand, E.: Meta-teams: getting global work done in MNEs. J. Manag. **45**(2), 510–539 (2018)

11. Gregory, P., Taylor, K.: Defining agile culture: a collaborative and practitioner-led approach. In: 2019 IEEE/ACM 12th International Workshop on Cooperative and Human Aspects of Software Engineering (CHASE). IEEE (2019)
12. Jannoo, Z., Auchoybur, N., Lazim, M.: The effect of nonnormality on CB-SEM and PLS-SEM path estimates. Int. J. Comput. Sci. Eng. **8**, 29–35 (2014)
13. Pfeffer, J., Sutton, R.I., et al.: The Knowing-Doing Gap: How Smart Companies Turn Knowledge into Action. Harvard Business Press (2000)
14. Tuczek, H.C., Flore, A., Nuhn, H.F.R., Schaffitzel, N.: A systemic approach to agile management and self-organization for a sustainable transformation of organizations. In: Ding, R., Wagner, R., Bodea, C.-N. (eds.) Research on Project, Programme and Portfolio Management. LNMIE, pp. 29–47. Springer, Cham (2022). https://doi.org/10.1007/978-3-030-86248-0_3

Analysing Cost Overrun in Building a Construction Project Using Building Information Modeling

Lukas Beladi Sihombing[1(✉)] and Ari Tiandaru Baskoro[2]

[1] Universitas Pembangunan Jaya, Tangerang Selatan, Banten 15413, Indonesia
lukas_sihombing@yahoo.com

[2] Pelita Harapan University, Tangerang, Banten 15811, Indonesia

Abstract. The growth of construction projects and technology in Indonesia is very rapid in the implementation process which has resulted in several innovations. One of the innovation models in construction modeling is building information modeling (BIM). Meanwhile, in implementing a construction project, there is often cost overrun and delays. This paper aims to analyze the causes of cost overrun in building construction projects in the construction phase using BIM. A structural equation model (SEM) method with a smart partial least square tool was used to analyze what causes and influences cost overrun that can be tackled by BIM. The factors that influence cost overrun are the high frequency of change in the construction phase, schedule delays, inadequate control and monitoring, poor finance control in the field, inadequate supervisor placement and qualifications, shipping delays, inadequate equipment, a lack of material, and poor management and supervision at the site. This paper contributes to project managers and project teams to know what factors influence cost overrun in building a construction project.

Keywords: Cost Overrun · Building Information Modeling · Structural Equation Model (SEM)

1 Introduction

According to Santoso [1], a building construction project is a physical form resulting from the conversion of the design carried out by the planner. In the implementation process, there are many construction projects that experience cost overruns and delays [2]. Cost overrun is an obstacle that often occurs in the implementation of construction projects.

Therefore, in managing the process of implementing a project, project management is needed for control efforts to ensure that each work stage can be carried out effectively to reduce various project risks that may occur, including the possibility of cost overrun and execution time delays [3].

According to the Director General of Construction Development at the PUPR Ministry, Syarif Burhanuddin, in 2019, "National construction sector policies must of course

S. Bushuyev et al. (Eds.): IPMA 2021, LNNS 704, pp. 180–197, 2023.
https://doi.org/10.1007/978-3-031-34629-3_16

encourage the use of technology that provides solutions to the problems faced today, so that infrastructure development can be done faster, easier, and certainly better" [4]. In the construction industry, several innovation models have been developed, including innovations in modern construction implementation methods, innovations in project management tools, and building information technology using building information modeling (BIM).

Building information modeling (BIM) can also provide many benefits in construction projects and its implementation can improve project quality because it is useful in helping construction actors to be successful in building projects that have low or high risk [5]. In another statement, Latiffi et al. [5] also conveyed that there are five main benefits of using BIM. The advantages of using BIM can be seen in the design, budget, communication, documentation, and scheduling.

Therefore, the formulation of this research is: 1) What is the role of building information modeling (BIM) in controlling the causes of cost overrun in the implementation of building construction projects as part of cost control efforts? 2) What are the important variables that cause cost overrun in the implementation of building construction projects that can be controlled by using BIM in this study?

2 Literature Review

One of the project risks that may occur in project implementation related to costs is the risk of cost overrun. According to Andrić et al. [6], cost overrun is the difference between the actual costs which represent the costs determined at the time of the project completion and the estimated costs determined at the time of the formal decision to build, or as in the following Eq. 1:

$$\text{Cost overrun} = (\text{actual costs} - \text{appraisal costs})/\text{Appraisal costs x } 100\% \quad (1)$$

where the actual costs are estimated after the project completion, and the appraisal costs are determined at the time of the formal decision to build.

The basic understanding of the use of BIM is the collaboration of various stakeholders through good planning results on a project using a digital design application [7], so as to minimize uncertainty, improve safety aspects, overcome problems, and assess potential impacts [8], and to be able to design and manage project implementation virtually in all stages of the building project implementation [9].

Overall, the references related to the factors and variables of this study can be seen in Table 1.

3 Methodology

After seeing the risk of cost overrun as one of the problems that occur related to project cost management in building construction projects and realizing how using BIM can reduce this risk, this research began by identifying the factors that cause cost overrun in the planning and implementation stages of building projects based on a literature study and relevant research.

From these factors, it was necessary to examine the significant factors that cause the risk of cost overrun in relation to project cost management. Then after this was determined, we looked for the relationship between the use of BIM and risk reduction based on a literature study and relevant research. It was validated with expert interviews. The conclusion of this study is expected to provide recommendations on the implementation of BIM in building projects in reducing the risk factors of cost overrun, so that it can be applied in future projects.

3.1 Evaluation of the Modeling Structure

An evaluation of the outer model in SEM - PLS was done by making a path diagram (Fig. 1), which reflects the relationship between exogenous (measurement model) and endogenous (structural model/inner model) factors as well as the relationship between exogenous and endogenous factors on their respective variables/indicators (measurement model/outer model). The factors/constructs in the modeling structure in this study used variables/indicators with a confirmatory reflective relationship pattern towards the relationships between each factor. Reflective variables/indicators are generally seen as problems that can surface from what we observe [10].

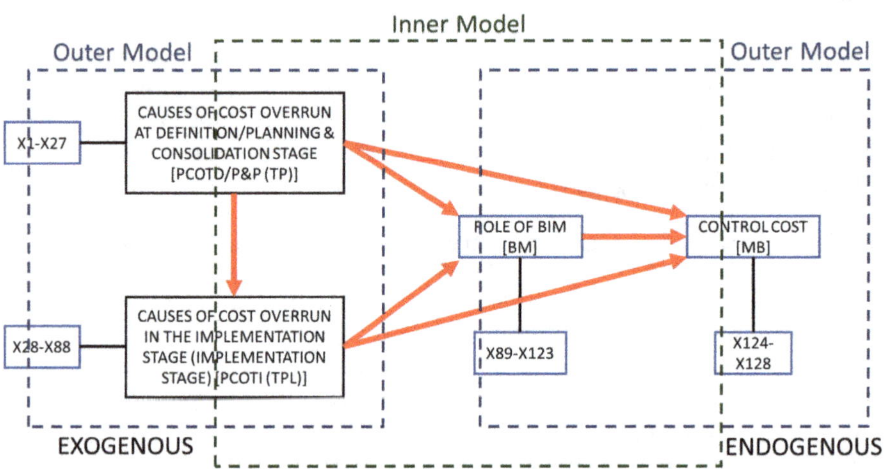

Fig. 1. Modeling Structure

Clarification:

- PCOTD/P&P (TP):Causes of Cost Overrun in the Planning Stage (Definition/Planning and Consolidation Stage);
- PCOTI (TPL): Causes of Cost Overrun in the Implementation Stage (Implementation Stage);
- BM:Role of BIM;
- MB:Control Costs.

Table 1. Research Factors and Variables

Factor	Variable	References	Notation
PCOTD/P&P (TP)	Poor design and design delay	[11, 12]	X1
	Design mistake	[11]	X2
	Difficult design	[11]	X3
	Slow in drawing preparation and approval	[11, 12]	X4
	Changes in the material type and specifications	[12]	X5
	Incomplete design during tender	[12]	X6
	Frequent design changes	[12, 13]	X7
	Error in determining the types of work to be subcontracted	[13]	X8
	Mistake in determining the quantity of work to be subcontracted	[13]	X9
	Mistake in predicting field conditions and future events	[3, 13, 14]	X10
	Unclear working drawings and specifications	[13]	X11
	Inaccurate/unrealistic estimation of subcontractor work costs	[3, 11, 13]	X12
	Inadequate time and land arrangements for the work of subcontractors who will work	[13]	X13
	Mistake in subcontractor selection	[12, 13]	X14
	Incomplete data and information on subcontractor performance	[13]	X15
	Incomplete project data and information	[3, 12, 14]	X16
	Lack of calculation of anticipated contingency costs	[3, 12, 14]	X18
	Incorrect cost estimate	[11, 12, 14]	X19
	Bureaucracy in the tender implementation stage	[11]	X20
	Incomplete subcontract clauses	[13]	X21
	Unrealistic contract duration and terms imposed	[11]	X24
	Poor contract management	[11, 12]	X25
	Contract-related claims, such as time extensions with claim fees	[11]	X26

(*continued*)

Table 1. (*continued*)

Factor	Variable	References	Notation
	There is no clause in the subcontract that explains about the work added or reduced (change in orders)	[13]	X27
PCOTI (TPL)	Cash flow and financial difficulties faced by contractors	[11]	X28
	Poor financial control in the field	[3, 11, 12]	X29
	Financing, guarantee, and payment system	[11]	X33
	Late delivery/a lack of materials at the time of implementation	[11, 12, 14]	X36
	A lack of materials	[11, 12]	X37
	An excess of materials	[11]	X38
	A worker shortage	[11, 12, 14]	X39
	Worker productivity	[11, 12]	X42
	The placement of supervisors that do not match the qualifications	[12, 13]	X44
	Late delivery of equipment	[11, 12]	X47
	An insufficient amount of equipment	[11]	X49
	Poor communication and coordination between the main contractor and the subcontractors	[3, 12, 13]	X51
	A lack of firmness of the main contractor in giving sanctions for violations committed by the subcontractors	[13]	X52
	Very poor coordination in meetings	[13]	X53
	The relationship between management and the workforce	[11]	X54
	Poor field management and supervision	[11]	X55
	Poor relationships between the owner, planner, and contractor	[12, 14]	X56
	A lack of coordination between the construction management, planner, and contractor	[12, 14]	X58
	There are differences/disputes that occur in the project	[14]	X59
	Slow in decision making	[11]	X60
	Inadequate control and monitoring	[3, 11, 12]	X61

(*continued*)

Table 1. (*continued*)

Factor	Variable	References	Notation
	Poor project management	[11]	X62
	Slow flow of information between parties	[11]	X64
	A lack of communication and coordination between the parties	[3, 11]	X65
	Owner intervention	[11]	X66
	There is a delay in the activities carried out previously (predecessor)	[13]	X67
	A high frequency of changes in orders during the execution of work	[2, 3, 9, 14]	X68
	A high rate of work repetition due to the poor quality of work	[2, 11, 12, 14]	X69
	The number of projects being handled at the same time	[11, 12, 14]	X70
	Inadequate planning and scheduling	[11]	X71
	Errors during construction	[11]	X72
	Inaccurate time estimation	[11]	X73
	Inaccurate analysis of field conditions	[11]	X74
	Added work	[11]	X75
	Late schedule	[11]	X76
	An inaccurate volume of work at the time the work is carried out	[11]	X77
	Delays in checking and approving completed work	[3, 11, 12]	X78
	Outdated or inappropriate construction methods/technology	[11, 13]	X79
	A lack of field productivity from subcontractors	[13]	X80
	A lack of subcontractor knowledge regarding project characteristics	[13]	X81
	Unexpected location conditions	[11]	X88
PB	Minimize revisions	[8]	X89
	Facilitate coordination between subcontractors	[8]	X90
	More efficient mockup costs	[8]	X91
	Errors can be found at the beginning	[8]	X92

(*continued*)

Table 1. (*continued*)

Factor	Variable	References	Notation
	Improve the ease of information exchange and document control, including the exchange of information for object modeling, meeting minutes, project drawings, etc.	[5, 8, 9, 15, 16]	X93
	Interoperability, i.e., the ability to share data across applications	[15]	X94
	Eliminate/reduce unbudgeted changes to the project	[15]	X95
	Accuracy rate for the cost estimation up to 3%	[15]	X96
	Can reduce the time in making a cost estimation of up to 80%	[15]	X97
	Detect discrepancies/non-conformances (especially in execution) early through model visualization, providing time and cost savings	[5, 9, 15, 16]	X98
	Reduce the project completion time	[15]	X99
	Reduced repair work due to improved quality control and design coordination	[15, 16]	X100
	Structural analysis capabilities built into the equipment, which is key for manufacturers	[15, 16]	X101
	Better concepts and feasibility	[9]	X102
	Effective site analysis to understand environmental and resource-related issues	[9, 15]	X103
	An effective design review leads to a sustainable design	[9, 16]	X104
	An increase in energy efficiency	[9]	X105
	Allows for faster and more accurate cost estimations	[9, 16]	X106
	Facilitates better communication and faster design decision making	[5]	X107
	Allows project managers and contractors to view stages of construction work, equipment, and materials as well as track progress against set logistics and schedules	[5]	X108
	It is possible to make work accelerations, calculations, and measurements because it can be viewed directly from a 3-dimensional (3D) project model	[5, 8]	X109

(*continued*)

Table 1. (*continued*)

Factor	Variable	References	Notation
	Reduces the demand for utility and demolition costs	[5]	X110
	An evaluation of the construction of complex building systems to improve resource planning and alternative stages	[9]	X111
	Effective project resource storage and procurement management	[9]	X112
	Efficient fabrication of various off-site building components using design models as a basis	[9]	X113
	BIM enables better site utilization	[9]	X114
	Reduces excess resources on site and improves health and safety	[9]	X115
	Can demonstrate the construction process, including access and egress, traffic flow, site materials, and machinery/equipment	[5]	X116
	Provides better tracking for cost control and cash flow	[5]	X117
	Can track work in real time with faster resource flow and better field management	[5]	X118
	Improved facility management data accessibility	[16]	X119
	Automated security check platform	[16]	X120
	Integration between key stakeholders	[16]	X121
	A risk assessment of facility design components for prevention through design	[16]	X122
	Clarification of the scope of work	[16]	X123
MB	Project management plan	[17]	X124
	Project documentation	[17]	X125
	Project funding requirements	[17]	X126
	Work performance data	[17]	X127
	Asset management process	[17]	X128

Source: Expert validation results

3.2 Research Respondents

The respondents in this research were the parties from the assignee (contractors) who had knowledge and experience in the use of BIM in building construction projects with the design and build type of contract. The respondents who answered the questions in this study understand the BIM implementation process in building construction projects. The respondents in this study are described in the Table 2.

Table 2. Profile of the Research Respondents

Background	Characteristic	Total
Education level	High school/equivalent	1
	3-year diploma	1
	Bachelor's Degree	37
	Master's Degree	6
Work experience	2–5 years	15
	6–10 years	15
	11–15 years	8
	>15 years	7
Experience in using BIM	1 year	13
	2 years	10
	3 years	11
	4 years	4
	5 years	2
	>5 years	5

Source: Processed by the researcher

The other respondents came from experts selected with the criteria that have knowledge and experience in the use of BIM for at least three years in building construction projects.

4 Results

To determine the discriminant validity, an indicator is declared valid if it has the highest loading factor for the intended construct compared to the loading factor for other constructs. Therefore, these three items must be removed from the model. Thus, the modeling after the removal of the three items can be seen in Fig. 2.

4.1 Modeling Hypothesis Testing

This section examines the role of BIM in reducing the factors and variables causing cost overrun in an effort to control costs in building construction projects. In Fig. 2, the role of BIM can be seen as a connecting or intermediate factor (variable) which has a mediating role between exogenous and endogenous constructs. The exogenous constructs based on Fig. 2 above are the factors (variables) "The Cause of Cost Overrun in the Planning

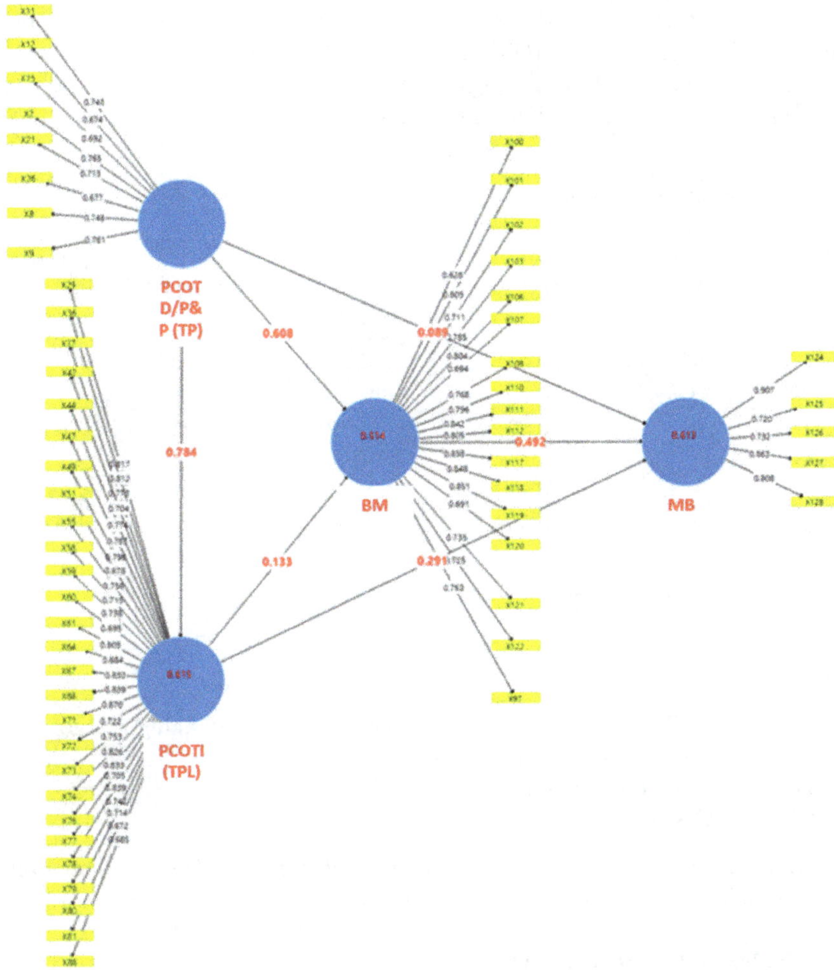

Fig. 2. Modeling Structure

Stage (Definition/Planning and Consolidation Stage)/PCOTD/P&P (TP)" and the factor (variable) "The Cause of Cost Overrun in the Implementation Stage/PCOTI (TPL)". Meanwhile, the endogenous construct based on Fig. 2 above is the factor (variable) "Controlling Costs/MB".

Testing the First Modeling Structure

The first modeling structure test examines the magnitude of the influence of the exogenous factors (variables) "Causes of Cost Overrun in the Planning Stage (Definition/Planning and Consolidation Stage)/PCOTD/P&P (TP)" and the factors (variables) "The Causes of Cost Overrun in the Implementation Stage/PCOTI (TPL)" towards the endogenous factor (variable) "Controlling Costs/MB". This relationship can be seen in Appendix 4. Based on the opinion of Baron and Kenny, in the first modeling structure, the effects of exogenous factors (variables) on endogenous factors (variables) must be significant at T-statistics > 1.96 (Ghozali & Latan, 2015) [8].

The results of the hypothesis testing on the modeling which were carried out through bootstrapping procedures using SmartPLS are shown in the Table 3.

Table 3. The Results of the First Modeling Structure Hypothesis Test

Relationship Pattern	Original Sample (O)	Mean Sample (M)	Standard Deviation (STDEV)	T-Statistics (|O/STDEV|)	P-Value
PCOTD /P&P (TP) → MB	0.393	0.421	0.152	2.586	0.010
PCOTD /P&P (TP) → PCOTI (TPL)	0.792	0.801	0.050	15.825	0.000
PCOTI (TPL) → MB	0.363	0.355	0.168	2.160	0.031

Source: Processed by the researcher

Clarification:

- PCOTD/P&P (TP): Causes of Cost Overrun in the Planning Stage (Definition/Planning and Consolidation Stage);
- PCOTI (TPL): Implementation Stage;
- MB: Control Costs.

The test results based on Table 3 reveal that T-statistics > 1.96 for each relationship, which indicates that the relationship is significant.

Testing the Second Modeling Structure

The second modeling structure test is intended to simultaneously test the magnitude of the influence of the exogenous factors (variables) "Causes of Cost Overrun in the Planning Stage (Definition/Planning and Consolidation Stage)/PCOTD/P&P (TP)", the factors (variables) "Causes of Cost Overrun in the Implementation Stage/PCOTI (TPL)" and the mediating factor (variable) "Role of BIM/PB" towards the endogenous factor (variable) "Controlling Costs/MB". This relationship can be seen in Fig. 3. Based on the opinion of Baron and Kenny, in the second modeling structure, the effects of exogenous factors (variables) on endogenous factors (variables) are not significant with the T-statistics < 1.96, while the effects of the mediating factors (variables) on the endogenous factors (variables) must be significant with the T-statistics > 1.96 [10].

The results obtained from the hypothesis testing on the modeling carried out through the bootstrapping procedures using SmartPLS are shown in the Table 4.

Clarification:

- PCOTD /P&P (TP): Causes of Cost Overrun in the Planning Stage (Definition/Planning and Consolidation Stage);

Table 4. Results of the Second Modeling Structure Hypothesis Test

Relationship Pattern	Original Sample (O)	Mean Sample (M)	Standard Deviation (STDEV)	T-Statistics (|O/STDEV|)	P-Value
PCOTD /P&P (TP) → MB	0.089	0.095	0.169	0.528	0.598
PCOTD /P&P (TP) → PCOTI (TPL)	0.784	0.789	0.056	14.038	0.000
PCOTD /P&P (TP) → PB	0.608	0.623	0.147	4.142	0.000
PCOTI (TPL) → MB	0.291	0.296	0.159	1.829	0.068
PCOTI (TPL) → PB	0.133	0.134	0.190	0.703	0.482
PB → MB	0.492	0.489	0.186	2.641	0.009

Source: Processed by the researcher

- PCOTI (TPL): Implementation Stage;
- PB: Role of BIM;
- MB: Control Costs.

The test results based on Table 4 show that the relationships between the exogenous factors (variables) "Causes of Cost Overrun in the Planning Stage (Definition/Planning and Consolidation Stage)/PCOTD/P&P (TP)" and the exogenous factors (variables) "Causes of Cost Overrun in the Implementation Stage/PCOTI (TPL)" towards the endogenous factor (variable) "Controlling Costs/MB" has a T-statistics value < 1.96, which shows an insignificant influence on the relationship. While the relationship between the mediating factor (variable) "Role of BIM/PB" towards the endogenous factor (variable) "Controlling Costs/MB" has a T-statistics value > 1.96, which depicts a significant relationship. This demonstrates that the role of BIM has a significant influence between exogenous and endogenous factors (variables).

The results obtained from model testing were then grouped into five significant variables at each stage based on the highest rating of the loading factor value, so that five important variables caused cost overrun at the planning and implementation stages of building construction projects that can be controlled by the use of BIM such as seen in the Table 5.

5 Discussion

From Table 5 above, each of the five significant variables in the Planning Stage (Definition/Planning and Consolidation Stage) and Implementation Stage can be seen.

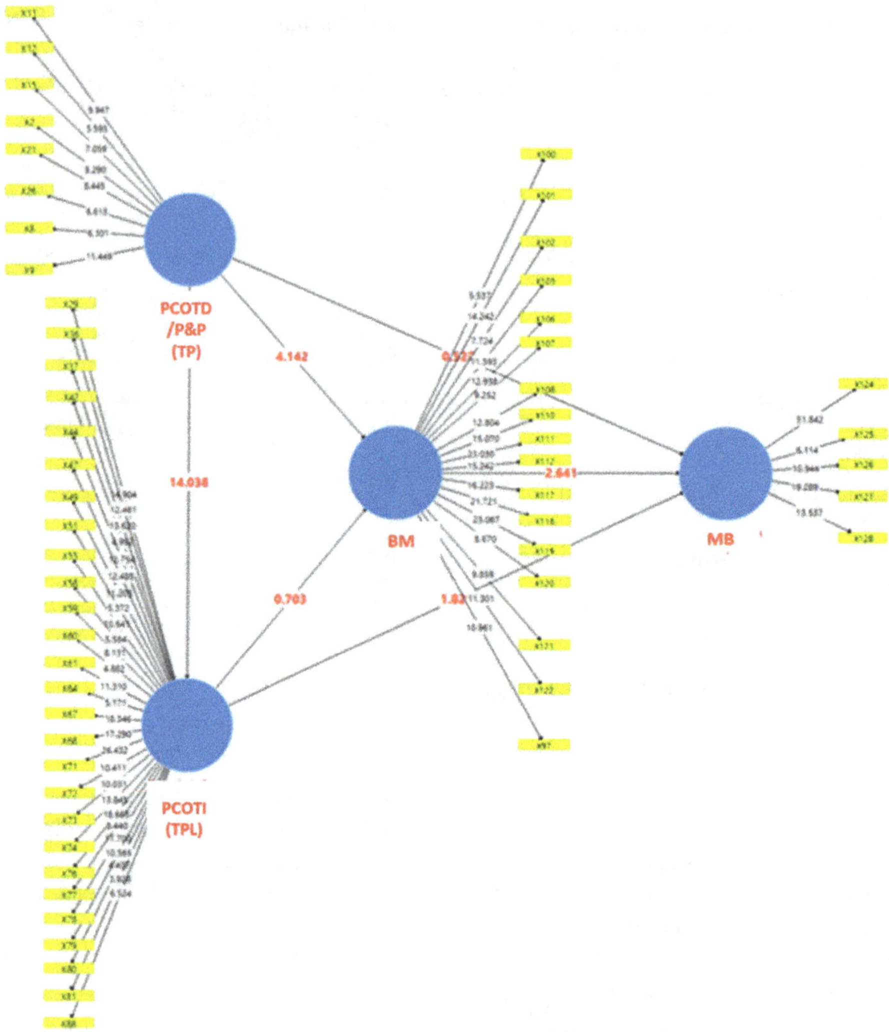

Fig. 3. Second Modeling Structure Test

5.1 Planning Stage (Definition/Planning and Consolidation Stage)

Of the important variables that cause cost overrun in the "Planning Stage (Definition/Planning and Consolidation Stage)", they can be grouped into variables that are closely related to factors connected to the design process and variables that are closely related to factors connected to the contract process.

The variables which have strong connections to the design process include:

1. Design errors (X2).
2. Mistakes in determining the quantity of work to be subcontracted (X9).
3. Errors in determining the type of work to be subcontracted (X8).

Table 5. Important Factors and Variables that Cause Cost Overrun that Can Be Controlled Using BIM

Ranking	Variable	Notation	Loading Factor
Planning Stage (Definition/Planning and Consolidation Stage)			
1	Errors in the design	X2	0.765
2	Mistakes in determining the quantity of work to be subcontracted	X9	0.761
3	Errors in determining the types of work to be subcontracted	X8	0.748
4	Unclear working drawings and specifications	X11	0.743
5	Incomplete subcontract clauses	X21	0.713
Implementation Stage			
1	Inadequate planning and scheduling	X71	0.870
2	Delays in checking and approving completed work	X78	0.839
3	A high frequency of changes in orders during the execution of work	X68	0.839
4	Late schedule	X76	0.833
5	There is a delay in the activities carried out previously (predecessor)	X67	0.833

Source: Processed by the researcher

4. Unclear working drawings and specifications (X11).
5. While the variables that are closely related to a contract include:
6. Mistakes in determining the type of work to be subcontracted (X8).
7. Incomplete subcontract clauses (X21).

Regarding the variables causing cost overrun which are closely related to factors connected to the design process based on a literature study, it can be clarified as follows:

- Errors in design, mistakes in determining the quantity of work, and errors in establishing the types of work can be caused by planners and project management working unprofessionally, or as a result of frequent requests for design replacement by the owner [18]. Meanwhile, in other studies, it is stated that failure in the construction of a project can be caused by an error in the design process. This can lead to a misunderstanding of the design concept between the relevant parties which can affect the quality, cost overrun, and project delays [19].
- According to Kulkarni, 3D functions in the use of BIM have uses in terms of making parametric design models and space programming tools, namely the use of the spatial dimensions of width, length, and depth to represent an object, which allows 3D visualization and tracing, non-conformance detection and coordination, and item scheduling, so as to control the occurrence of errors in the design [15].
- While errors in determining the quantity of work in the implementation of building construction can be controlled by applying the use of BIM with the 5D function. The

5D function in using BIM can perform the design integration process with estimating, scheduling, and setting costs, including the creation of quantity invoices and reducing the level of productivity and labor costs [15].
- In terms of errors in determining the quantity of work, Remi [12] in his research stated that mitigation efforts related to errors in determining the quantity of work include, among others, controlling the consistency and choosing a professional estimator.

Meanwhile, the variables that cause cost overrun that are closely attached to contract-related factors based on a literature study and expert interviews can result from and be controlled by the following:

- Related to the problem of the quantity of work to be subcontracted, as quoted by Achirwan [13], it is necessary to control the sub-contractors because 80%–90% of the project implementation budget is in the management of the sub-contractors [13]. For this reason, in controlling this problem it is necessary to determine the type of contract that will be applied to the implementation of a project.

5.2 Implementation Stage

Regarding the variables causing cost overrun in the Implementation Stage based on the literature study, it can be explained as follows:

1. Inadequate planning and scheduling (X71). Inadequate planning and scheduling can cause delays in project completion, in which delays in project completion can also result in cost overrun. Efforts to control this require consistency to control costs, work schedules, materials, and labor [12]. In an effort to control the occurrence of inadequate planning and scheduling related with cost overrun, according to Kulkarni [15], the 4D BIM function has the ability to connect parts or series of 3D functions with the project implementation/completion schedule, including the quantity, resources, as well as modular prefabrication to assist project tracking and phasing. In addition to collaboration in terms of information/data exchange, 4D simulation serves as a communication tool to uncover potential bottlenecks in project implementation [15].
2. Delays in checking and approving completed work (X78). This variable can cause delays in the project implementation process to the next stage and project handover when a project has been completed. This situation can certainly disrupt the previously planned schedule. This can happen, among others, due to the poor relationship between the parties involved. For this reason, efforts are needed to build good relationships by coordinating, communicating, and exchanging information between parties, especially between owners, contractors, and consultants [12].
3. The high frequency of changes in orders during the execution of work (X68). Requests for job changes (changes in orders) often occur in the process of implementing construction projects. Requests for job changes can cause delays and cost overrun as a result of additional work time. According to Widhiawati [20], there are four groups of factors that cause changes in orders, including project owners, planning consultants, contractors, and project external factors. The results of this study indicate that the most influential elements on the occurrence of requests for job changes are factors originating from the owner in the form of requests for design changes [20].

4. Delayed schedules (X76). Late schedules in the implementation of building construction projects can cause cost overrun. According to Wirabakti [18], there are ten factors that cause delays in the implementation of building construction projects, including delays in material delivery, limited availability of materials to the market, less available manpower, a high intensity of rainfall, a lack of worker presence in project implementation, workers who lack discipline, a lack of workforce expertise, poor communication between contractors and owners, poor communication between workers and management, and errors in the design process by planners. In controlling this problem, collaboration and coordination between stakeholders is needed. According to Messner [7], the basic assumption of using BIM is the collaboration of different stakeholders at a phase of the life cycle of a project to include, describe, update, or modify information in BIM to support and reflect the roles of stakeholders [5].
5. There is a delay in the activities carried out previously (predecessor) (X67). This of course can affect the planned schedule of activities that will be carried out at a later stage. With the delay in activities carried out previously, it can also affect the schedule of all activities that have been set related to the work that has experienced the delay. To prevent this problem, a good planning and scheduling process is needed. According to Glavinich [21], project scheduling is needed to build communication between the parties in carrying out the plan of the construction process, determining the production goals, monitoring and measuring the progress of work, and managing any changes that may occur [21].

6 Conclusion

From the several stages of research carried out, several conclusions were obtained based on several problems in the research, including:

1. The implementation of Building Information Modelling (BIM) in doing building construction projects consists of several basic principles, including:
   - Determining the dimensions of using BIM in various building construction processes based on their functions, which are differentiated as follows:
     - 3D: serves the purpose of visualization and analysis through building 3D modelling.
     - 4D: functions as a time/schedule simulation through a virtual construction approach that is made and performs simulations for each stage in the construction process.
     - 5D: operates for the benefit of cost estimation/simulation.
   - Doing mapping about the planned project execution process.
   - Making a mapping outline of the BIM implementation plan.
   - Constructing a detailed map of the BIM implementation plan.
   - Verifying each goal achieved at key decision points in the implementation process.

2. The role of Building Information Modeling (BIM) in controlling the occurrence of cost overrun in the implementation of building construction projects as part of cost control efforts can be seen from the results of hypothesis testing using SmartPLS through the modeling tests conducted.
3. From the test results, there were 35 important variables that cause cost overrun in the implementation of building construction projects that can be controlled by the use of BIM, which were sorted by rank based on the highest loading factor value. These variables were divided into two stages in the implementation of building construction projects, namely 8 important variables in the planning stage (definition/implementation and consolidation stage) and 27 significant variables in the implementation stage.
4. The description of the results of the discussion of the fourth problem shows that the main concept of implementing BIM in building construction projects is a collaboration based on building modeling data, which can be accessed by stakeholders involved in the implementation of building construction projects. Problems in each project implementation process usually often arise from the beginning of the implementation process. In an effort to prevent any obstacles from surfacing, BIM needs to be carried out from the beginning of the project implementation stage, starting from the planning stage, implementation stage, until the work handover stage by making an implementation plan for the use of BIM in all resources involved in all stages of the implementation process.

From the description above, it can be concluded that in order to reduce the factors and variables that cause cost overrun, which is an obstacle that still often occurs in the implementation of construction projects, the BIM execution plan must be done from the early stages of the construction project implementation process, so that the benefits can be achieved and the advantages of using BIM can have a more significant role.

References

1. Santosa, B.: Manajemen Proyek - Konsep & Implementasi. Graha Ilmu, Yogyakarta (2009)
2. Wisudanto, A.W.: Factor Contributing Delay in Building Constructing Projects at Kediri [In Indonesian]. Master Thesis, Institut Teknologi Sepuluh Nopember (ITS) Surabaya (2012). https://chrome-extension://efaidnbmnnnibpcajpcglclefindmkaj/https://repository.its.ac.id/63037/1/3109203006-Master%20Thesis.pdf
3. Dapu, Y., Dundu, A., Walangitan, R.: Faktor – Faktor Yang Menyebabkan Cost Overrun Pada Proyek Konstruksi. Jurnal Sipil Statik (4), 641–647 (2016)
4. Tri, I.: Teknologi Turut Mendukung Industri Jasa Konstruksi Lebih Efisien (2019). https://pu.go.id/berita/view/17195/teknologi-turut-mendukung-industri-jasa-konstruksi-lebih-efisien
5. Latiffi, A., Mohd S., Kasim, N., Fathi, M.: Building information modeling (BIM) application in Malaysian construction industry. Int. J. Constr. Eng. Manag. **1**(6) (2013)
6. Andrić, J., Mahamadu, A., Wang, J., Zou, P., Zhong, R.: The cost performance and causes of overruns in infrastructure development projects in Asia. J. Civil Eng. Manag. **25**(3), 203–214 (2019)
7. Messner, J., et al.: BIM Project Execution Planning Guide - Version 2.2. Computer Integration Construction Research Program, Penn State - University Park, PA, USA (2019). https://psu.pb.unizin.org/%0Abimprojectexecutionplanningv2x2/

8. Berlian, C., Adhi, R., Hidayat, A., Nugroho, H.: Perbandingan Efisiensi Waktu, Biaya, Dan Sumber Daya Manusia Antara Metode Building Information Modeling (BIM) Dan Konvensional (Studi Kasus: Perencanaan Gedung 20 Lantai). Jurnal Karya Teknik Sipil **5**(2), 220–229 (2016)
9. Ullah, K., Lill, I., Witt, E.: An overview of BIM adoption in the construction industry: benefits and barriers. In: Emerald Reach Proceedings Series, pp. 291–303 (2019)
10. Ghozali, H., Latan H.: Partial Least Square - Konsep, Teknik dan Aplikasi Menggunakan Program SmartPLS 3.0. 2 ed. Badan Penerbit – Undip, Semarang (2015)
11. Memon, A., Rahman I., Azis, A.: Preliminary study on causative factors leading to construction cost overrun. Int. J. Sustain. Constr. Eng. Technol. **2**(1) (2011)
12. Remi, F: Kajian Faktor Penyebab Cost Overrun Pada Proyek Konstruksi Gedung. Jurnal Teknik Mesin 06 (2017)
13. Achirwan, Y., dan Abidin, I.: Pola Hubungan Antara Kinerja Biaya Proyek Dan Dampak Penyimpangan Biaya Proyek Dengan Pendekatan Indikator Cost Overrun Pada Pengelolaan Sub Kontraktor. Jurnal Konstruksia **4**(2), 61–75 (2013)
14. Sahusilawane, T., Bisri, M., Rachmansyah, A.: Analisis Faktor-Faktor Penyebab Terjadinya Pembengkakan Biaya (Cost Overrun) Pada Proyek Konstruksi Gedung Di Kota Ambon. Jurnal Rekayasa Sipil **5**(2), 118–128 (2011)
15. Kulkarni, S., Mhetar, G.: Cost control technique using building information modeling (BIM) for a residential building. Int. J. Eng. Res. Technol. (2017)
16. Tahir, M., et al.: Improving cost and time control in construction using building information model (BIM). A review. Pertanika J. Sci. Technol. **26**, 1 (2018)
17. ICB - IPMA Competence Baseline, Version 4.0. International Project Management Association, The Netherlands (2015)
18. Wirabakti, D., Abdullah, R., Maddeppungeng A.: Studi Faktor-Faktor Penyebab Keterlambatan Proyek Konstruksi Bangunan Gedung. Jurnal Konstruksia **6**(1) (2014)
19. Fuadie, D.: Analisis Faktor Penyebab Dan Dampak Kesalahan Desain Pada Proyek-Proyek Di Pt. Pertamina Mor V, Institut Teknologi Sepuluh Nopember Surabaya (2017)
20. Widhiawati, I.A.., Wiranata, A.A., Wirawan I. P.Y.: Faktor-Faktor Penyebab Change Order Pada Proyek Konstruksi Gedung. Jurnal Ilmiah Teknik Sipil **20**(1) (2016)
21. Glavinich, T.: Construction Planning & Scheduling. Associated General Contractors of America (2004)

Project Manager's Challenges in Sustainable Project Management: Can Digital Environment Empower the Action?

Vladimir Obradović, Marija Todorović, and Danijela Toljaga-Nikolić(✉)

Faculty of Organizational Sciences, University of Belgrade, Belgrade, Serbia
danijela.toljaga.nikolic@fon.bg.ac.rs

Abstract. This paper examines specific challenges for project managers when introducing sustainable project management since they need to take on new responsibilities and develop new competencies. The paper also examines whether the digital environment creates advantages or disadvantages for project managers facing these new challenges. Project managers were always responsible for delivering project results, and now the focus is shifting more towards the sustainability of project deliverables and their impact on the organization and society. The paper examines what competencies are required and significant when sustainably managing a project through literature review and empirical research. The key findings show that project managers need to think beyond the boundaries and develop open communication with stakeholders on the long-term and short-term economic, social, and environmental project effects. Stakeholder management, open communication, and transparency about what has been done and what will be achieved are important for managing sustainable project results. The paper particularly discusses research results in the digital environment and the benefits that digital technologies bring to project managers and sustainability.

Keywords: Project Manager · Competencies · Sustainable Project Management · Digital Environment

1 Introduction

Today, we are witnessing that industries and enterprises of all sizes reinvent their businesses influenced by the usage of advanced digital technologies, innovative business models and processes. They are running their businesses in the digital environment, offering smart products, smart services, and new value propositions while reaching out to new clients. This is the fourth industrial revolution that redefines the rules of the global economy and creates new markets and business opportunities at the same time [1]. Therefore, managers are encouraged to view digital innovation as an integral part of their strategy [2]. It will accelerate the development that needs to take on new responsibilities and develop new competencies of all people involved, hence the project stuff as well.

S. Bushuyev et al. (Eds.): IPMA 2021, LNNS 704, pp. 198–210, 2023.
https://doi.org/10.1007/978-3-031-34629-3_17

The authors stated in [3] that the global digital business environment has many challenges. All companies that are going digital will face these challenges and experience the necessity of changing their products, business models, business processes, required skills and knowledge. Sustainable work in a digital environment requires a lot of collaborative competencies but, at the same time, executing projects within the digital environment benefits everyone involved. According to the authors in [4], project management in a digital environment - virtual project management, is vital to the company's growth and helps teams work in the most efficient and human-sensitive way by using information technology and digital tools. Therefore, virtual project teams are challenged due to the increasing globalisation of companies and many new technologies that can improve project team collaboration.

A large number of organizations today integrate sustainability strategies into their business strategies and missions, in order to survive and achieve competitive advantage, which makes the concept of sustainability a global business trend. By adopting these business strategies and conducting activities accordingly, organizations meet the needs of stakeholders today while at the same time improving and preserving social and natural resources for the future [5]. Organizations should develop their business models, cultures, and practices around sustainability concepts rather than treat sustainability as an "add-on" to their businesses [6].

Strategic changes in organizations are most often introduced through the implementation of projects, and the organization's decision to operate sustainably must be incorporated at all management levels. In this way, the alignment and sustainability of the business goals of organizations with the global goals of sustainable development are ensured. According to the authors in [7], linking the concepts of sustainability and project management contributes to fundamental changes in the way of thinking, operations, cooperation, and partnerships at different levels of business and organization. It is important to emphasize that the implementation of sustainable project management requires flexibility and the readiness to introduce and accept changes at the project level.

The digital environment offers many opportunities for managing projects and requires a lot of flexibility from a project manager and project team members, incorporating and learning how to use new technologies and digital collaborative tools. The authors in [8] consider experience and competencies to be the most influential critical success factors in sustainable project management. In order to provide sustainable project management, the authors in [9] argued that a project manager should be flexible in dealing with preventive risk reduction, consider stakeholder risks, and actively communicates with them. The authors concluded that a project manager in sustainable project management is required primarily to have competencies related to open communication, stakeholder management, and transparency about what has been done and what will be achieved.

2 Literature Review

Companies should develop a collaborative digital work environment and rethink about the roles, expertise, capabilities, culture, talent, skill-set, and leadership of their employees [10]. The integration of sustainability into project management challenges project managers, who need to take on new responsibilities and develop new competencies. Competence is defined as „applying knowledge, skills and abilities to achieve the desired results" [11]. In order to manage projects in a sustainable manner, project managers must think beyond the boundaries within which they have moved and develop open communication with stakeholders and the environment about the project's long-term and short-term social and environmental effects [12–14].

According to the authors in [15], the key competencies of a successful project manager are achievement orientation, initiative, information seeking, focus on client needs, influence, direction, leadership, analytical thinking, conceptual thinking, self-control, and flexibility. When it comes to flexibility, the authors [16] point out that in sustainable project management, a project may have other goals that are not fully included in the project's initial goals, so at some point in the project, certain changes will be necessary. That is why the flexibility of a project manager is important, especially when it operates in a digital environment with many changes and adjustments. The authors in [17] concluded that by introducing sustainable project management, the list of project stakeholders becomes more comprehensive, complicating the project manager's responsibility. Besides the need to be competent in terms of having the knowledge, skills, and capabilities required for project management, sustainability in project management requires project managers to be ethical, righteous, and fair in managing projects.

The authors in [18–20] argued that managers have to decide about the necessary changes for a successful digital transformation, meaning what existing assets can be leveraged, which capabilities can be used in new ways, and what new competencies are needed. Companies start the digital transformation process by considering new business models, redesigning business processes, and developing new ways of communication among stakeholders. This new environment brings very specific demands to a project manager, especially when sustainability aspects of project management are considered. The following competencies of the project manager and project team members are considered significant to sustainable project management [11, 16, 21]:

- Flexibility for change,
- Ethics in work, fair and responsible behaviour,
- Responsibility for sustainable development in the organization,
- Communication about sustainable results that need to be achieved through project processes,
- Communication on the short-term and long-term social and environmental effects of the project,
- Communication with an expanded list of stakeholders due to the long-term effects of sustainable project management,
- Awareness of available and accessible sustainable solutions that can be included in the project,
- Risk analysis related to social, environmental, and economic aspects of sustainability in the project.

The International Project Management Association IPMA has published the Individual Competence Baseline (ICB4) that defines competencies that enable project managers to operate projects through three main areas of competencies: people, practice, and perspective [11]. Competencies that are presented in ICB4 are considered significant to sustainable project management and listed (see Table 1).

Table 1. IPMA ICB4 competencies that are significant to sustainable project management.

Competencies for sustainable project management	IPMA ICB4 competencies
Flexibility for change	Strategy, Governance, structures and processes, Personal integrity and reliability, Relationships and engagement, Personal communication, Leadership, Conflict and crisis, Resourcefulness, Change and transformation
Ethics in work	Governance, structures and processes, Personal integrity and reliability, Compliance, standards and regulations, Culture and values
Fair and responsible behaviour	Personal integrity and reliability, Power and interest, Culture and values, Self-reflection and self-management, Teamwork
Responsibility for sustainable development in the organization	Strategy, Compliance, standards and regulations, Organization and information, Personal integrity and reliability, Plan and control, Procurement
Communication about sustainable results that need to be achieved through project processes	Personal communication, Requirements and objectives, Results orientation, Scope, Project design, Stakeholders
Communication on short-term and long-term social and environmental effects of the project	Personal communication, Time, Stakeholders
Communication with an expanded list of stakeholders due to the long-term effects of sustainable project management	Strategy, Personal communication, Stakeholders
Awareness of available and accessible sustainable solutions that can be included in the project	Project design, Quality, Procurement, Resources, Finance, Negotiation
Risk analysis related to social, environmental and economic aspects of sustainability in the project	Conflict and crisis, Risk and opportunities

To analyze how managers apply knowledge, skills, and abilities to support and promote sustainability during the project life cycle, the author [21] suggests examining to what extent the project manager pays attention to the achievement of financial goals, improvement of environmental protection, contribution to social development, fostering innovation, and supporting creativity and creative solutions. Important elements of sustainability that also need to be analyzed are related to the following social aspects: ethics, human rights, equality, health, and safety at work [22–25].

In addition to the many advantages that virtual communication and the digital environment can bring to project management practice, project teams that operate in the virtual environment face obstacles regarding information exchange efficiency because they rely mostly on information technology to communicate [4]. Also, these teams lose the non-verbal aspects of communications that make up 65% to 93% of a message's meaning due to the lower opportunity for face-to-face communication [26]. Another challenge when working as a project manager in a digital environment is the need to develop new skills through continuous education related to software, content management systems, digital marketing, databases, etc.

When introducing sustainability into project management practice, the authors [27] stated that organizations are concerned about sustainability in project management, but there is a gap between the perception of importance and the actual use in practice. The modern challenges of a sustainability concept could be overcome if more young people are educated to implement sustainable development projects, where education for sustainable development thus becomes lifelong learning [28].

3 Research Methodology

The authors conducted empirical research in the field of sustainable project management. Based on the literature review, in one part of the research, the authors focused on researching what competencies of the project manager and project team members are significant for sustainable project management as a research question. For sending the online questionnaire, a list based on the authors' professional, academic, and business contacts was used. The list was comprised of project managers and professionals in project management who have a lot of experience managing projects in different industries. The questionnaire was sent to 352 contacts in May 2021. The total number of acquired responses was 207, providing a response rate of 58.8%. To write this paper, the authors used the part of the survey results related to the competencies of the project manager and team members for sustainable project management. The list of the measured survey items is given in the appendix. All the items are measured on a 5-point Likert scale: Relating to tag A, the range of items a1–a7 is from 1 - 'not at all to 5 -'completely'; and the range of items b1–b9 for tag B is from 1 - 'not significant' to 5 - 'completely significant'.

The main characteristics of the acquired samples are summarized in Figs. 1, 2, 3, 4 and 5. Among the 207 respondents, 60% were male and 40% were female (see Fig. 1). Respondents stated that 53% of projects were implemented in organizations belonging to the private sector, 40% in organizations belonging to the public sector, and 7% in the civil sector (see Fig. 2).

If you look at the industries to which the organizations belong, they are mainly from IT (38 in total), followed by education (24 in total), energy (16 in total), etc. (see Fig. 3).

Regarding the professional experience of the respondents who participated in the research, most of the respondents have worked on more than ten projects (66%), followed by those who worked on 3–10 projects (30%), and only 4% have worked on less than three

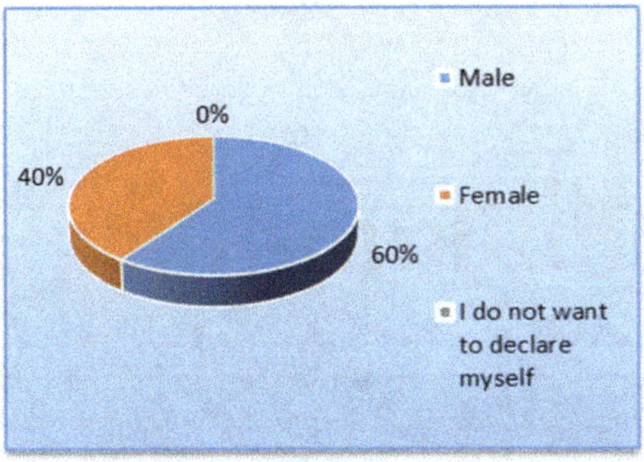

Fig. 1. Gender of the respondents

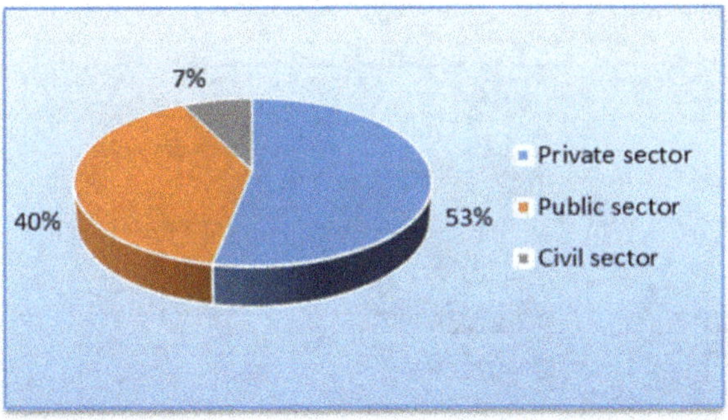

Fig. 2. Ownership of the organization in which the project was implemented

projects (see Fig. 4). In addition, 71% of respondents stated that they held the project manager position on the project that was the subject of the questionnaire, while 20% participated as a member of the project team (see Fig. 5). Accordingly, it can be concluded that the respondents had relevant professional experience in project management.

4 Research Results and Discussion

4.1 Elements of Sustainability Aspects Supported and Promoted Throughout the Project Life Cycle by the Project Manager

The variables in this study are measured on a 5-point Likert scale (Appendix A). To analyze the results and the difference among the following elements of the sustainability aspects, the authors preferred nonparametric tests such as Friedman (Frd) [29–31]. Regarding the elements of sustainability aspects that the project manager has supported and promoted throughout the project life cycle, there is a statistically significant difference among these elements (Friedman, $p < 0.001$). The most supported and promoted element, according to the value of the Friedman Mean Rank, is *Achievement of the financial goals, followed by Health and safety at work, Support creativity and creative*

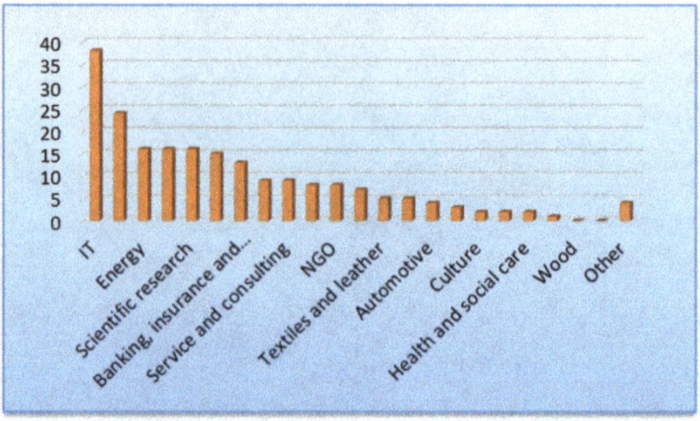

Fig. 3. Industry of the organization in which the project was implemented

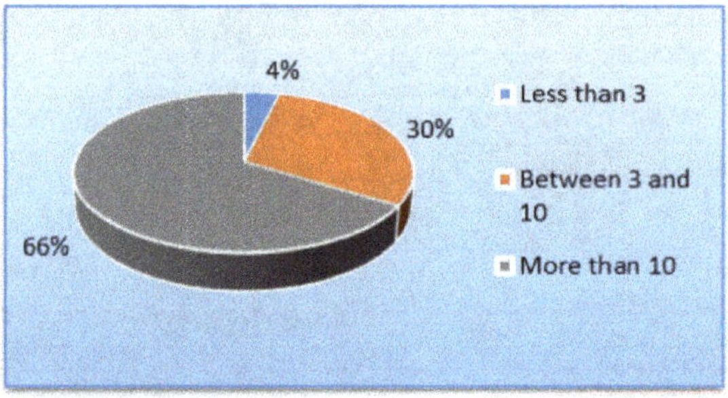

Fig. 4. Professional experience of the respondents

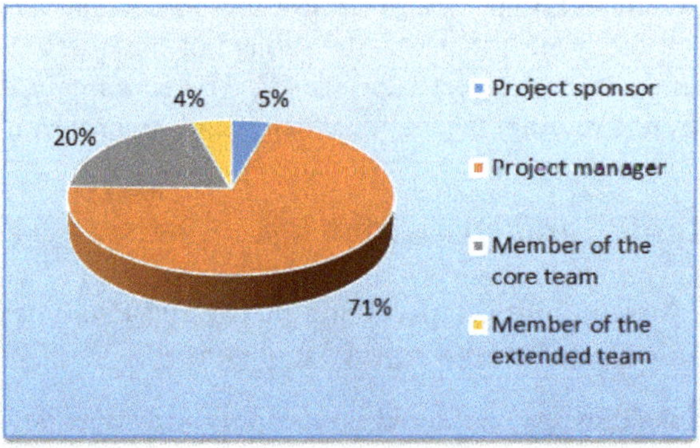

Fig. 5. Project role of the respondents

solution, Ethics, human rights and equality, Fostering innovation, Contribution to social development, and Improvement of environmental protection.

In support of the findings of the present research (see Table 2), the research results are consistent with the conclusions in [21], namely that the economic and social aspects

Table 2. Elements of sustainability aspects that were supported and promoted throughout the project life cycle by the project manager.

Elements of the sustainability aspects	Median	Friedman Mean Rank	Friedman Significance
Achievement of the financial goals	5.0	4.62	<0.001
Health and safety at work	5.0	4.34	
Support creativity and creative solution	5.0	4.30	
Ethics, human rights and equality	4.0	4.19	
Fostering innovation	4.0	4.11	
Contribution to the social development	4.0	3.63	
Improvement of the environmental protection	3.0	2.81	

of sustainability are more highlighted than the other ones and that the company should pay more attention to environmental protection. Since most organizations in the survey were from the profit sector and IT industry, it can be concluded that the economic aspects are most important for those organizations. When it comes to managing projects in the digital environment, one of the most important benefits is cost savings due to various digital collaborative tools, which positively influence the financial result. The social aspect's ranking is higher than the environmental aspects, which corresponds with the research results in [21]. As the authors in [22] stated, implementing new technologies as part of the Industry 4.0 concept will improve the process impact on the environment by reducing resource usage and waste, while the authors in [32] recognized project management as one factor for achieving Industry 4.0 and environmental sustainability harmony. Therefore, project managers should pay more attention to the environmental aspects of sustainability and incorporate those benefits, e.g. reduction of gas emissions, reduction of water consumption, waste reduction, etc. [33], while for the lack of knowledge, the solution could be additional education for all individuals engaged in managing projects.

4.2 Competencies of the Project Manager and Project Team Members for Sustainable Project Management

To analyze the research results and the difference among the competencies of the project manager and project team members for sustainable project management, the authors used the nonparametric Friedman (Frd) test [29–31]. There is a statistically significant difference among these competencies (Friedman, $p < 0.001$). The competence that was recognized as the most important for sustainable project management, according to the value of the Friedman Mean Rank, is *Fair and responsible behaviour, followed by Flexibility for change, Ethics in work, Communication about sustainable results that need to be achieved through project processes, Awareness of available and accessible sustainable solutions that can be included in the project, Responsibility for sustainable development in the organization, Risk analysis related to social, environmental and economic aspects of sustainability in the project, Communication with an expanded list of stakeholders due to the long-term effects of sustainable project management, and Communication on short-term and long-term social and environmental effects of the project.*

Table 3. Competencies of the project manager and project team members for sustainable project management

Elements of the sustainability aspects	Median	Friedman Mean Rank	Friedman Significance
Fair and responsible behaviour	5.0	6.30	<0.001
Flexibility for change	5.0	5.85	
Ethics in work	5.0	5.83	
Communication about sustainable results that need to be achieved through project processes	4.0	4.84	
Awareness of available and accessible sustainable solutions that can be included in the project	4.0	4.66	
Responsibility for sustainable development in the organization	4.0	4.62	
Risk analysis related to social, environmental and economic aspects of sustainability in the project	4.0	4.49	
Communication with an expanded list of stakeholders due to the long-term effects of sustainable project management	4.0	4.23	
Communication on short-term and long-term social and environmental effects of the project	4.0	4.17	

The research findings presented in Table 3 show that *Fair and responsible behaviour* is the highest-ranked competence, which is consistent with the conclusions in [9], where the authors emphasize how important it is that project managers set a personal example and behaviour to lead. This research result is also consistent with the conclusion in [17], where the authors stated that sustainability in project management requires project managers to be ethical, righteous, and fair in managing projects. The authors in [9] argued that open communication with stakeholders is an extremely important competence for project managers and the team in sustainable project management. It can be concluded that open communication with stakeholders can be empowered by managing projects in a digital environment and using many available collaborative tools.

According to the authors in [34], the resistance to change is one of the barriers that need to be overcome in a situation where the company wants to gain the benefits of sustainably oriented Industry 4.0, which supports the research result presented in Table 3, where the competence *flexibility for change* is highly ranked. The competence *Responsibility for sustainable development in the organization* is also highly ranked and consistent with the conclusion in [35], where the authors link sustainability to the project manager's values in the context of ethics, honesty, equality, and responsibility. One of the challenges is communication, especially on the project's short-term and long-term social and environmental effects. In Table 2. *Improvement of environmental protection* is one of the challenges, so it can be concluded that the respondents to the research did not recognize the importance of considering the environmental aspects of sustainability.

5 Conclusion

Project management is recognized as one factor for achieving Industry 4.0 and environmental sustainability harmony [32]. The research results have shown that the economic and social aspects of sustainability were more highlighted, supported, and promoted by the project manager throughout the project life cycle, while the environmental aspects were less recognized as important. The solution might be additional education in the field of sustainability aspects. Managing projects in the digital environment brings cost savings, which positively influences the financial result, and the implementation of new technologies as part of the Industry 4.0 concept will improve the process impact on the environment by reducing resource usage and waste [22].

The level of the project manager's responsibilities and the set of required competencies change when introducing the social, environmental, and economic dimensions of sustainability into project management. In this situation, the digital environment can empower the actions taken by a project manager and project team members when managing projects since it requires a lot of flexibility in planning, communication with an expanded list of stakeholders due to the long-term effects of sustainable project management, the incorporation of new technologies, and digital collaborative tools that bring many savings.

When it comes to the competencies required of a project manager for sustainable project management, the results of this paper have shown that fair and responsible behaviour and flexibility for change are of the highest rank. These competencies can be leveraged by operating in a digital environment because it offers the opportunity to keep stakeholders informed and gain consensus and transparency throughout the project with available digital tools as well as the ability to make changes quickly.

Appendix A

(See Table 4).

Table 4. Survey items

Tag	Survey items	
A	Extent to which the project manager has supported and promoted the following elements of sustainability aspects throughout the project life cycle:	
	a1	Ethics, human rights, and equality
	a2	Health and safety at work
	a3	Achievement of the financial goals
	a4	Improvement of the environmental protection
	a5	Contribution to the social development
	a6	Fostering innovation
	a7	Support creativity and creative solution
B	Extent to which the respondent considers the competencies of the project manager and project team members to be important for sustainable project management	
	b1	Flexibility for changes
	b2	Ethics in work
	b3	Fair and responsible behaviour
	b4	Responsibility for a sustainable development in the organization
	b5	Communication about sustainable results that need to be achieved through project processes
	b6	Communication on short-term and long-term social and environmental effects of the project
	b7	Communication with an expanded list of stakeholders due to the long-term effects of sustainable project management
	b8	Awareness of available and accessible sustainable solutions that can be included in the project
	b9	Risk analysis related to the social, environmental, and economic aspects of sustainability in the project

References

1. European Commission: Digital Transformation Scoreboard 2017: Evidence of Positive Outcomes and Current Opportunities for EU Businesses. European Commission, Brussels (2017)
2. Kaufman, I., Horton, C.: Digital transformation: leveraging digital technology with core values to achieve sustainable business goals. Eur. Financ. Rev. 63–67 (2015)
3. Blanchet, M., Rinn, T., Von Thaden, G., De Thieulloy, G.: Think Act Industry 4.0: The new industrial revolution, How Europe will succeed. Roland Berger Strategy Consultants Gmbh, Munich (2014)

4. Nauman, S., Iqbal, S.: Challenges of virtual project management in developing countries. In: Conference: Engineering Management Conference, Proceedings. IEEE International, vol. 2, pp. 579–583 (2005)
5. Chofreh, A.G., Goni, F., Shaharoun, A.M., Ismail, S.: A review on sustainability transformation roadmaps using project management methodology. Adv. Sci. Lett. **21**(2), 133–136 (2015)
6. Stubbs, W., Cocklin, C.: Conceptualizing a "sustainability business model." Organ. Environ. **2**, 103–127 (2008)
7. Schipper, R.P., Silvius, A.J.G.: The sustainable project management canvas. J. Mod. Project Manag. **4**(3), 50–59 (2017)
8. Banihashemi, S., Hosseini, M.R., Golizadeh, H., Sankaran, S.: Critical success factors (CSFs) for integration of sustainability into construction project management practices in developing countries. Int. J. Project Manag. **35**(6), 1103–1119 (2017)
9. Goedknegt, D., Silvius, A.J.G.: The implementation of sustainability principles in project management. In: Proceedings of the 26th IPMA World Congress, Crete, Greece, pp. 875–882 (2012)
10. Kane, G.C., Palmer, D., Phillips, A.N., Kiron, D., Buckley, N.: Strategy, not technology, drives digital transformation. MIT Sloan Manag. Rev. 1–25 (2015)
11. IPMA ICB4. Individual Competence Baseline for Project Management. International Project Management Association, Zurich, Switzerland (2015)
12. Khalfan, M.M.A.: Managing sustainability within construction projects. JEAPM **8**(1), 41–60 (2006)
13. Silvius, A.J.G., Schipper, R.: Sustainability in project management: a literature review and impact analysis. Soc. Bus. **4**(1), 63–96 (2014)
14. Taylor, T.: Sustainability Interventions-For Managers of Projects and Programmes. Centre for Education in the Built Environment. The Higher Education Academy, Salford (2010)
15. Cheng, M.-I., Dainty, A.R.J., Moore, D.R.: What makes a good project manager? Hum. Resour. Manag. J. **15**(1), 25–37 (2005)
16. Maltzman, R., Shirley, D.: Green Project Management. CRC Press, New York (2010)
17. Toljaga-Nikolić, D., Todorović, M., Dobrota, M., Obradović, T., Obradović, V.: Project management and sustainability: playing trick or treat with the planet. Sustainability **12**(20), 8619 (2020)
18. Hess, T., Matt, C., Benlian, A., Wiesbock, F.: Options for formulating a digital transformation strategy. MIS Q. Exec. **15**(2), 123–139 (2016)
19. Ross, J.W., Sebastian, I.M., Beath, C.M.: How to develop a great digital strategy. MIT Sloan Manag. Rev. **58**(2), 7–9 (2017)
20. Westerman, G., Bonnet, D.: Revamping your business through digital transformation. MIT Sloan Manag. Rev. **56**(3), 10–13 (2015)
21. Szabo, L.: Sustainability, creativity and innovation in project management - model development for assessing organizational performance through projects. Vezetéstudomány **10** (2016)
22. Abubakr, M., Abbas, A.T., Tomaz, I., Soliman, M.S., Luqman, M., Hegab, H.: Sustainable and smart manufacturing: an integrated approach. Sustainability **12**(6), 2280 (2020)
23. Garbie, I.: Sustainability in Manufacturing Enterprises Concepts, Analyses and Assessments for Industry 4.0. Springer, Cham (2016). https://doi.org/10.1007/978-3-319-29306-6
24. Garbie, I.: Sustainability in small and medium-sized manufacturing enterprises: an empirical study. J. Eng. Res. (TJER) **13**(1), 42–57 (2016)
25. Leksono, E.B., Suparno, S., Vanany, I.: Integration of a balanced scorecard, DEMATEL, and ANP for measuring the performance of a sustainable healthcare supply chain. Sustainability **11**(13), 3626 (2019)

26. Harris, T.E.: Applied Organizational Communication: Perspectives, Principles, and Pragmatics. Lawrence Erlbaum Associates, Hillsdale (1993)
27. Martens, M.L., de Carvalho, M.M.: The challenge of introducing sustainability into project management function: multiple-case studies. J. Cleaner Prod. **117**, 29–40 (2016). https://doi.org/10.1016/j.jclepro.2015.12.039
28. Ostojić, B., Berić, I., Pavlović, K., Poćić, M.: Management education and sustainable development projects. Eur. Project Manag. J. **10**(1), 69–77 (2020)
29. Friedman, M.A.: The use of ranks to avoid the assumption of normality implicit in the analysis of variance. J. Am. Stat. Assoc. **32**, 675–701 (1937)
30. Friedman, M.A.: Correction. J. Am. Stat. Assoc. **34**, 109–109 (1939)
31. Friedman, M.A.: Comparison of alternative tests of significance for the problem of m rankings. Ann. Math. Stat. **11**, 86–92 (1940)
32. Jabbour, A.B.L.S., Jabbour, C.J.C., Foropon, C., Filho, M.G.: When titans meet - can industry 4.0 revolutionize the environmentally-sustainable manufacturing wave? The role of critical success factors. Technol. Forecast. Soc. Change **132**, 18–25 (2018)
33. Green Project Management. The P5 Standard for Sustainability in Project Management, Version 2.0. GPM Global, USA (2019)
34. Raj, A., Dwivedi, G., Sharma, A., Jabbour, A.B.L.S., Rajak, S.: Barriers to the adoption of industry 4.0 technologies in the manufacturing sector: an inter-country comparative perspective. Int. J. Prod. Econ. **224**, 107546 (2019)
35. Silvius, A.J.G., Schipper, R., Planko, J., van der Brink, J., Köhler, A.: Sustainability in Project Management. Gower Publishing (2012)

Hybridization of Project Management Methodologies. Complementary or Contradictory?

Sergey Neizvestny(✉)

Financial University under the Government of the Russian Federation, Moscow, Russia
sineizvestny@fa.ru

Abstract. The subject of this article is the development of hybrid methodologies for managing projects, programs and project portfolios. Significant reductions in the life cycles of programs, which contain projects of different lines of activity and different management methodologies, for example, waterfalls with a rigid life cycle and Agile with a flexible methodology life cycle, require the use of convergence tools to form "hybrid" methodologies. The goal is to develop a convergent approach to building hybrid project management methodologies in terms of decision-making processes in project management based on various platforms. Results of the studies were obtained using a convergent approach to building methodologies for managing projects and programs. A substantial model of the resulting hybrid methodology for managing projects and programs is presented. Conclusions: Hybrid methodologies for managing projects and programs are becoming more popular in this category of programs. The main reason is the presence in the program of components with different life cycles, and, consequently, methodologies that require integration and harmonization. The methodology is tested by the practice of using mechanisms of hybrid multilevel management of projects and programs. The application of the hybrid project management methodology allowed the authors to execute different projects in essence within the stipulated timeframe with the given budget and quality of construction projects and projects of creation of information and communication management systems. Further areas of research are related to the elaboration of harmonization, integration, convergence and actualization mechanisms.

Keywords: Convergence Model · Hybrid Management Methodologies · Program · Basic Project Management Methodology Interaction Model · Program Manager

1 Introduction

The analysis of the tendencies of the development of projects and programs has determined that the key factors for their success are taking into account the specifics of managing each class of projects included in the program, reliability of implementation, efficiency in value creation and ecological harmonization in the light of turbulent environment.

S. Bushuyev et al. (Eds.): IPMA 2021, LNNS 704, pp. 211–229, 2023.
https://doi.org/10.1007/978-3-031-34629-3_18

At the same time, improving the level of implementation of programs is a strategic direction for Ukraine's development. The conducted analysis of the status and problems of the existing s of regions, cities, settlements and consumers of Ukraine has determined the relevance and practical importance of research on the implementation of complex projects in turbulence based on the construction of a hybrid methodology for project and program management within a converged balanced approach [1]. The authors have developed a holistic model for solving problems of implementation of projects and programs on the basis of dual (management with simultaneous learning, rational models) management within hybrid methodologies that combine different principles and life cycles of the management model. Experimental studies of the proposed hybrid approaches, models and methods of managing programs have confirmed their adequacy and effectiveness. Today, there is relatively little work in the known literature in which research would focus on various aspects of such an important direction of development as the effective management of projects and programs to successfully implement them through a hybrid management methodology with parallel training of models to reduce uncertainty in program implementation.

Also, insufficient attention was paid to project management of the implementation of programs on the basis of prudent management [2, 3]. Of particular relevance to the uncertainty, challenges and challenges of the program's external and internal environments is the acquisition of such large-scale and long-lasting programs. This class includes state targeted programs for the development of territorial systems, as a major component of Ukraine's development. Currently, the configuration of development programs is based, as a rule, on the use of stationary models that do not take into account the variety of projects and management methods, and the dynamics of the external environment over the period of program implementation. On the other hand, in certain segments of the critical path of the program not only the values of the parameters of the environment, but also their methodologies and the priority in making management decisions change.

Current trends in the global development of project oriented organizations that implement program are accompanied by the development of their competence and organizational maturity [1]. The activities of project-oriented organizations are generally mobile and based on generally accepted rules for implementing programs and programs. This concerns the involvement of productive forces caused by cost minimization and environmental compliance. The basis for balanced development in the implementation of projects is an innovative upgrade aimed at increasing the quality of products and services, enhancing their competitiveness through the implementation of best international practice [5]. The key factors are quality enhancements and the ability to create value and generate value [6] programs, regardless of their type, operate in accordance with their organizational management strategies, which are based on a focus on the development and satisfaction of stakeholders. In a competitive environment, such strategies of programs are in many cases designed to support or create competitive advantages over other organizations or the competitive value of program products [7, 8]. In the course of their activities, organizations must adapt the strategies of programs to the influences of the external environment and internal dynamics. At the same time, it is necessary to find ways of effective investment of management resources in the application of appropriate methodologies, for timely achievement of the planned strategic indicators of programs and programs [9].

In the practical business, enterprises are often faced with the problem of selecting the best methodologies necessary for the effective management of a portfolio of programs and projects. As a rule, in specific unique conditions for the implementation of programs and projects, it is difficult for the enterprise's methodologists to choose a methodology that is "suitable" for the goals and scope of the functional scope of the projects from the methodologies presented on the global or national "market" [1–3]. If, however, a systematic approach is made to the problem of choosing a methodology, then it is obvious that a "boxed" product in this situation cannot be ideally suited, since this product was developed for specific application conditions that cannot coincide with the conditions at this enterprise. This problem poses a challenge for the enterprise's methodologists either to adapt the "boxed" methodology, or to develop their own, for this portfolio, under the conditions in which the enterprise operates. An optimal project management methodology will guide the project manager through a controlled, manageable and visible set of activities to achieve project results [4, 5].

However, the development of a new high-quality methodology is a very costly undertaking, requiring highly qualified methodological resources, diverse professional experience, time, etc. In professional cases, when solving this problem, methodological managers follow the path of adapting existing methodologies. Adaptation of methodologies can take place using convergence and/or hybridization.

By the convergence of project management methodologies we mean systematically verified merging, the unification of methodologies, subject to the conditions of consistency of elements of the combined methodologies. The convergence of methodologies involves the evolutionary convergence/transformation of some elements of different methodologies into new or modified elements. This rapprochement occurs mainly under the influence of external and internal conditions of project activities. The main purpose of applying the methodology convergence mechanisms is the synergy of the elements of converged methodologies in the development of project management methodologies for specific conditions [6].

By hybridization of methodologies we mean the combination of elements of several different methodologies into one. This methodology can be called hybrid. In a hybrid methodology, elements of the original methodologies do not undergo transformations.

Ontologically, the convergence of project management methodologies arises under the influence of external and internal conditions in a situation of sufficiently large-scale, multifunctional projects and/or programs [7]. Performing complex, multifunctional project programs, the enterprise management is consciously or not logically forced to follow Ashby's law: to effectively manage the variety of projects included in the program, it is necessary to have a set of methodologies larger than the variety of projects.

The purpose of the article is to develop a convergent approach to building hybrid methodologies for managing projects and programs from the perspective of decision making processes within boundaries based on different methodological platforms and standards.

2 Research Methodology

This diversity is determined by several factors, for example, the feasibility of saving (not scattering) resources: a functionally specific project, in terms of economic feasibility, quality requirements, time, other parameters of the project, dictates the choice of a specific methodology that is optimal for this type of project. If at the enterprise the methodologists systematically approach the application of different methodologies, "specialized" for specific projects of the enterprise, then evolutionarily starting from the scale of the life cycle of the portfolio of projects, within the life cycle of the enterprise, under the influence of both internal and external conditions for the implementation of projects, the different methodologies applied some general characteristics appear [1, 7, 8]. Over time, a set of methodologies shows signs of "convergence", a single "new" methodology appears in the enterprise that works effectively within the framework of this enterprise. This is a direct result of the convergence of methodologies.

Convergence of project management methodologies is an objective consequence of the impact of project implementation conditions. This is possible precisely in conditions of a high level of maturity of management processes - not lower than the third level in the IPMA-Delta system. In other words, in conditions of high maturity of the project management, convergence inevitably occurs, without the decisive active role of methodologists, but under the influence, first of all, of the corporate culture and systematic management processes [7].

In conditions of a low level of maturity, the convergent manifestations in the development of methodologies are insignificant and can occur on a small scale or even be absent.

The situation is different with the use of hybridization of methodologies. In conditions of low maturity of project management, in the presence of professional methodologists in the organization, hybridization of project management methods can be very effective. An experienced project management methodologist can take two or more project management methodologies that are not directly suitable for implementing the projects of an enterprise portfolio, create an optimal hybrid version of a single methodology by selecting from each of the initial methodologies the elements that are necessary in the specifics of a particular project portfolio of a given enterprise.

In this process of creating a hybrid methodology, the developer, using the Occam principle, cuts off "unnecessary" elements (functionally unclaimed or elements that duplicate functions) of methodologies. Thus, the hybridization of methodologies is the result of the active influence of methodologists on the initial "input" set of methodologies.

In conditions of a high level of project management maturity can be used and convergence, and hybridization of methodologies. In this combined process, convergence can actually proceed as a synergistic objective effect of the transformation of methodologies with a relatively insignificant ("passive") impact of the methodologists, and hybridization, on the contrary, is realized with the active influence of a methodologist acting like a sculptor, cutting off everything "superfluous" from the milestone.

In general, from the point of view of time parameters and changes in management structures in the development of project management methodologies, convergence technologies can be conditionally classified as evolutionary, and hybridization technologies as revolutionary. It should be noted that convergence and hybridization can be carried out simultaneously, giving a synergistic effect.

Both convergence and hybridization imply the creation of a methodology for complex multifunctional projects consisting not only of a relatively small set of basic elements, but of an expanded set that meets the wide variety of project characteristics, criteria for its success, and features of the organizational structure in which it is implemented [8, 9]. That is, both of these technologies are effective for "heavy" multi-element methodologies [3].

Converged technologies for the development of project management methodologies are relatively well developed (see, for example, [6, 7]) here we only partially affect them; we will dwell a little more on some aspects of hybridization technologies for project management methodologies.

Programs should not only be aimed at building (development) of a complex or large-scale system, they should take into account the development of concepts, changes in organizational structures and teams of managers [9]. In program designed to develop a new type of product (service), the use of new technologies [10, 11] and the concept of overall success management are crucial. Strategic management of the program identifies elements of strategy that are critical to achieving the mission of the program, and creates a program structure for combining key elements [12, 13]. Managing program strategy is primarily about establishing a mission and managing its mental space [14].

In the practice of managing programs, three approaches to strategy implementation are applied. The first type of approach is to establish certain principles and/or implement an appropriate structure for the day-to-day operations, basic and minor decisions to enhance the professionalism and culture of the organization's members. This technique is developed to create a well-managed organization. The second type approach is aimed at the development of new techniques and mechanisms for their implementation, which the organization can apply in work with both technological processes (equipment and technology) and management (business processes and business models). The third type approach defines the mechanism of cyclical creation of new values, such as products and services, which provide organizations with success, which in turn defines the existing values of the organizations themselves. For each approach, a specific activity profile scheme should be defined [15, 16].

Value creation is seen as a core concept of program that is phased out. The program begins by identifying the kind of values that you want to create, then moving on to value creation actions, and determining when to manage a value adjusting activity. Most important in this work is the process of defining the vision, mission and strategy, which aims to prepare the program by identifying problems and formulating key areas of the strategy. This process consists of:

- contextual analysis of the situation for the correct presentation of the complete picture (status) of the program, including the environment;
- preparation of alternative scenarios.

In order for the management of the program to be dynamic, with adequate response and to provide compensation for changes in the environment, the program manager must forecast trends in the environment in the short and long term. It is necessary to model numerous scenarios of the situation development. To achieve this, the program team must begin with a thorough analysis of the current state and modelling of the desired state in the future.

Leaders of the organization should formulate the main purpose of the program in the form of a mission and justify in it the reason for the existence of the program itself. This sets the stage for stakeholders to share the concept of the management team, increase their involvement in the implementation of the program, and increase awareness of the issues through active collaborative learning and motivation. In the next stage, the manager sets out a clear goal for which management activities should be directed. This objective is formulated in the form of a concept or task of finding ways to implement program, in other words, identifying ways of achieving a mission to decide which program management methodologies to use.

At this stage, it is desirable, whenever it's possible to describe the steps required to carry out the mission of the program in the form of specific objectives program management tools investigate, narrow, and sift through numerous scenarios and related management methodologies for the implementation of the program, selecting a specific scenario that becomes the basis for further management and evaluation of the program architecture. Consider contextual analysis as a method of comprehending and presenting a holistic picture of program. This analysis is used to interpret the mission and strategy mainly when the interacting multiple values of the program are expressed in abstract terms. In context analysis, the rules that shape the system or the appropriate methodology for generating a general context must be presented visually, and the requirements of the client or specialist from the planner should be maximally embodied.

Obviously, the need has ripened in the development of intelligent project management methodologies that generate control methods in environments of natural, artificial and combined intelligences.

The difference in hybridization of the development of project management methodologies is that in these methodologies fundamentally different objects, and subjects, and management processes [12]. To do this, just look at an example of the structure of the areas of knowledge of the AI control subject (Fig. 1) and the structure of the hybrid control process (Fig. 2).

If in traditional project management methodologies we do not think about many aspects, for example, in the block "Subjects of management and behavioural competence" that influence decision making, in hybrid methodologies "Natural-artificial intelligence" these aspects require careful development. Hybrid methodologies "Natural-artificial intelligence" needs to expand the elements that are the structure of the functional behavioural properties of AI (see the example in Fig. 3).

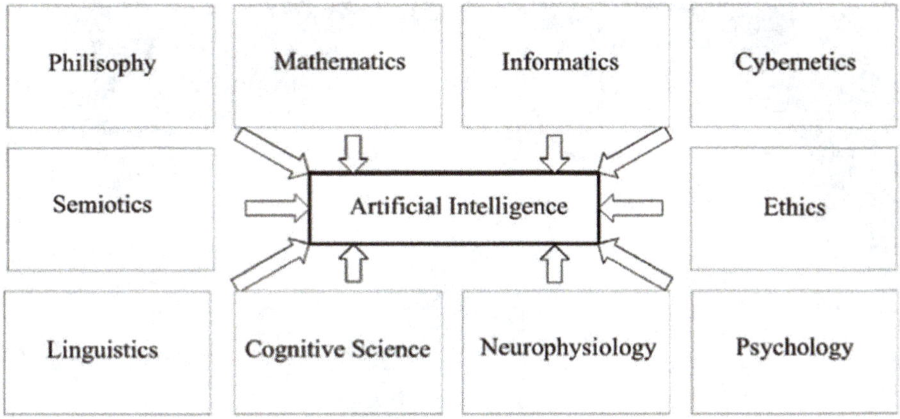

Fig. 1. An example of the structure of the areas of knowledge of the subject of AI control

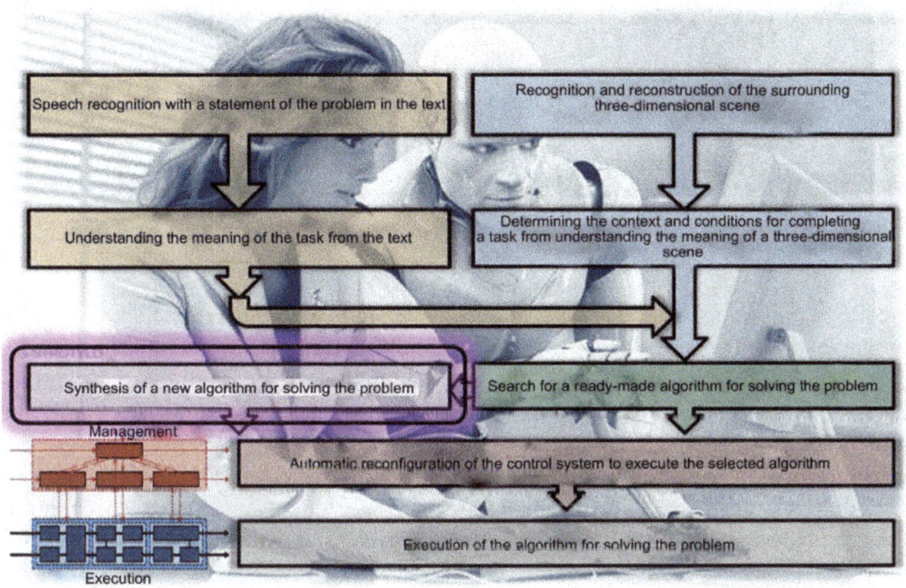

Fig. 2. The structure of the part of the hybrid process control "Natural-artificial" intelligence.

Hybridization gives a special impetus to the development of decision support systems (DSS). The practical application of modern DSS shows their lack of effectiveness, which is determined both by the limited number of analysed "entity-attribute" relationships, and by limiting the consideration of the influence of external and internal conditions on management processes. This makes it impossible to give final decision-making to artificial intelligence [13]. The problems of using modern DSS arise in both social and legal aspects. Thus, in the social aspect, the problem of utilizing the released human resources is sharply aggravated, in the legal aspect, conflicts arise associated with the

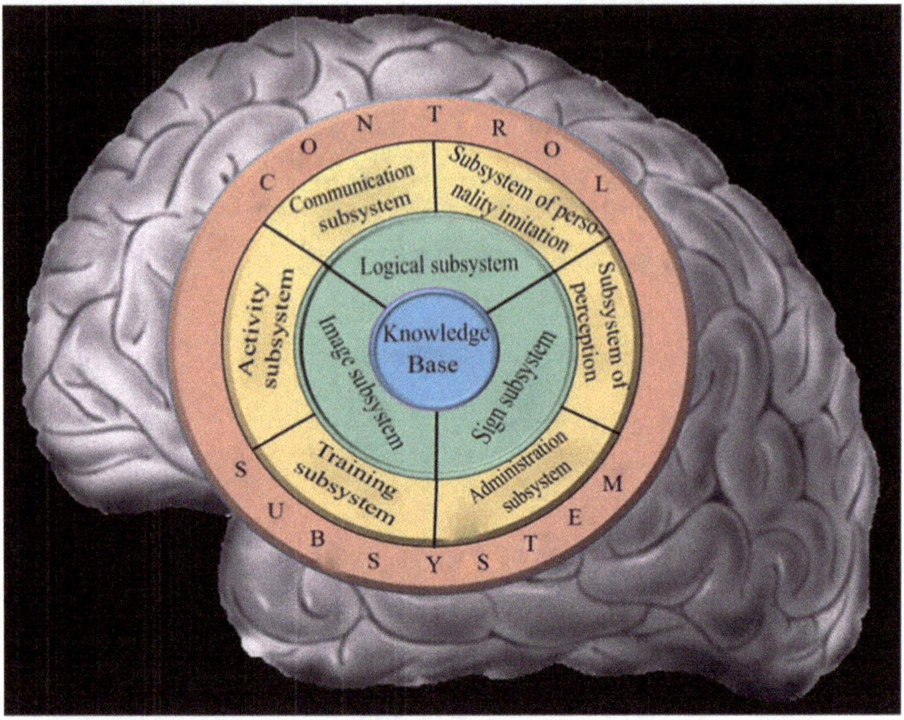

Fig. 3. An example of the functional architecture of the core of the AI control system.

so-called "guillotine" problem for the cyber manager - responsibility for the negative consequences of decisions made by artificial intelligence [14].

In the future, a situation is possible when in hybrid methodologies of project management, artificial intelligence will completely supersede modern (traditional) methodologies (Fig. 4).

Hybridization of traditional and cybernetic (on the basis of AI) project management methodologies allows creating a highly efficient "conveyor" of solutions; a real economic effect can be brought only in case of a complete business transformation on the principle of "AI as a process".

Fig. 4. For project management artificial intelligence methodologies, traditional methodologies may be of only retrospective interest.

3 Study Results

The question is not whether it is possible to combine the "traditional" project management methodologies and the "new, flexible" methodologies? Here, the problem is rather the terminology and the replacement of simple entities. In modern business, it is widely believed that there are "old" and "new" project management methodologies. If we consider the actual nature of the applied "old" methodologies, then it consists in the use of well-planned, developed packages of documents for the entire life cycle of the project.

Documents ranging from concept, technical, assignment, passport, project charter to an integrated project management plan with all the regulations and system-developed management processes (for example, in accordance with PMBOK PMI ICB IPMA). This project management option is used in those enterprises where there is a professional pool of managers, methodologists, specialists in project management and where there is no "rush" in project implementation, thanks to a well-developed business strategy and high-quality planning. In such organizations, managers typically manage the situation. In those cases when the level of maturity of business management is not high, where there is insufficient professionalism of people managing projects, when there are problems with planning resources, time, etc., when there is a rush, an emergency, managers go to the so-called " manual "management. In conditions of lack of time, as a rule, arising due to the lack of professional long-term business planning (when there is no well-developed planning strategy, project implementation, there is no clear common understanding by the participants of the goal, tasks), the project team works practically without a detailed high-quality technical task, managers begin apply technologies (methodologies) of "flexible" management. In fact, flexible management methodologies are used in cases where the non-manager manages the situation, but rather, when the goals and objectives of the project are adjusted, change during its implementation, when methodological tools and tools change on the go.

Appropriate knowledge is applied to solve complex program management problems, diagnostics are made, and the relationship between the whole and parts is improved by maintaining the right balance between the overall management of the whole and the autonomy of the parts, often different in nature of implementation and life cycles. This is necessary because the essence of managing program is to generate decisions while reducing managerial uncertainty, the programs are considered as a whole, are reciprocal, and must be coordinated to ensure the smooth progress of the program. Therefore, bureaucratic obstacles between the implementation of different components of program must be eliminated in order for the programs to interact in a harmonious way and to produce results defined by the mission. For this reason, program management establishes specific rules for the interaction between component and projects, maintains and ensures the inherent autonomy of programs.

In order to build relationships between the whole and parts and to align them with the content of the mission and the strategy of the program, the project management should analyze and determine the principles and organization of the relationship, the behaviour of the whole and its components. Both the rules for managing the program and the principles and methodologies of management for reaching a compromise and determining the direction of action in the organizations implementing the programs should be established and, if necessary, revised and improved. Program management is designed to interpret mission-defined, mission-relevant components and values at the strategic level to identify synergies across the subject areas, objectives, goals and tools, implementation of integration management, which focuses on prioritizing implementation of components in all program implementation processes.

In the dual control system, more actively accumulated information about the secondary order is transmitted, the characteristics of the control are reflected. At this entry of the object, the "first-hand" infusion will be given in, and the reaction of the object will be analysed by the cherubic annex. I do not have to manage victorious for managing an object, but at the same time it's the same for brining. The principle of activation of the control system in the process of function of the system is locked and in the higher classes of systems, for example: adaptive control and control with an identifier. An adaptive system (a system that needs to stick on its own, a self-adjusting system) is a system that automatically changes to the algorithm of its functionality and (in) its structure by saving and I will be able to keep the optimal. The structure of the hybrid system and the management of projects are shown on the Fig. 5.

On the given scheme of images, the multilevel projects are managed. The project is being stored up with the components of different nature (business, purchase, installation of ownership, creation of automatic control systems, development of software technologies and technologies). The door to encourage interchange of parts, terms and conditions, the need for strategic infrastructure, project management, and analysis there are many instances of delayed installation, use, redistribution, and thorough refinement of the rules for managing the infrastructure program, as well as the principle and methodology of managing the way to reach a compromise and a separate agreement.

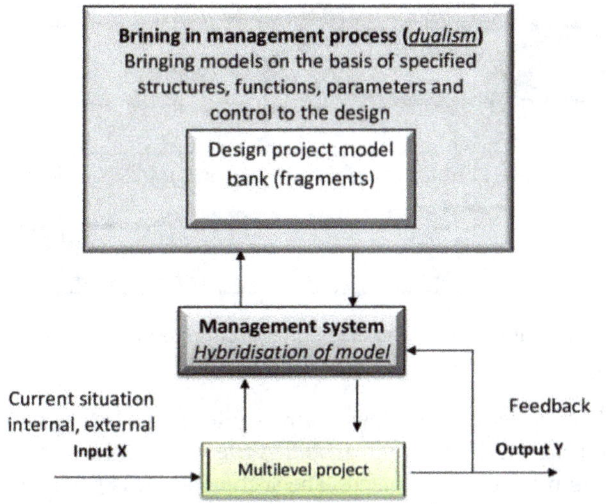

Fig. 5. Adaptive system model of hybrid dual project management

The conceptual model proposed in the study is based on the organization's life cycle, problems and challenges related to specific situations during the project lifecycles, early warning indicators, and proactive impact model generation during program implementation [17].

For the purpose of understanding the terms that are not well known, here are the basic definitions of the study.

- Definition 1. By the hybrid methodology, we define a methodology that combines different management principles and methodology approaches. For example: Agile with flexible lifecycle, Waterfall lifecycle PMBOK, Value-based P2M and Prince 2 with Gateway project phase management model.
- Definition 2. Proactive management is based on forecasting and decision making based on a vision of the future and challenges or perturbations.
- Definition 3. Threat is a net risk that is damaging and has a critical or a negative impact on the status and success of a project or program.
- Definition 4. Bifurcation is a critical state of the system in which two scenarios of its development are possible. It's a rise or fall. A bifurcation point is a change in the system's default mode.
- Definition 5. Value determines the benefits a stakeholder receives. In projects and programs, value must be balanced across all stakeholders.
- Definition 6. Turbulence is a state of the system or its environment that generates the uncertainties (vortices, tsunamis, or other forms) in organizational development processes.

In this case, the development of the system is significantly inhibited or accelerated. The conceptual diagram of program management based on the hybrid methodology is shown in Fig. 6.

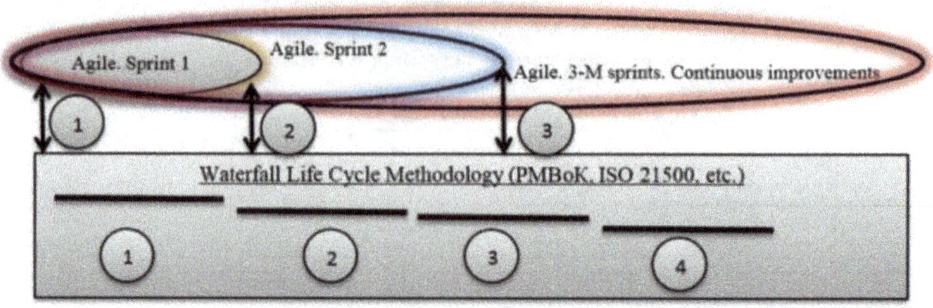

Fig. 6. Conceptual scheme for managing programs based on hybrid methodology

This example provides an overview of how Agile and PMBOK methodologies work together. Two sprints are highlighted in the Agile methodology. At the first sprint a minimum working product was created. The second sprint is where a working product is created with the necessary features that support the functioning of this component of the project. The third and subsequent sprints are dedicated to the system of continuous improvement (Kaizen). The waterfall life cycle methodology outlines 4 phases of the projects. Hybrid methodology models 1–3 are harmonized, integrating, converging and updating based on relationships 1–3.

These operations are shown in Fig. 2 when developing a hybrid methodology for managing programs based on genomic representations of components.

Managing the success of program can only be guided by value vectors for different stakeholder groups, use trends to accelerate development, convergence of knowledge, timely change in management paradigms, understanding the philosophy of life cycles, [18]. Each of these elements is part of the success of the program, and the formula for success is determined by the synergy of their interaction [1]. The return from a proactive to a reactive management method is associated with the transformation of a challenge into a problem. The term proactivity is called the period of transformation of a call into a problem. The proactive term assessment approach is used in software development projects based on Agile methodology. For example, in a definition such as proactive protection of program means a set of technologies used in information and communication systems. The main purpose is to identify potentially dangerous software and eliminate its impact or the code itself within the operating software systems.

Unlike other technologies, they warn and try to block dangerous activity, rather than detecting a known malware on the system. It uses a model of active bifurcation points that are linked to life-cycle events and change the priorities of the components to implement one of the strategies – anti-crisis, stabilization and sustainable development. Stakeholder management philosophy assumes that their claims are legitimate and therefore subject to consideration. It should be noted that those interpretations of the stakeholder concept that limit the legitimacy of their claims should be rejected for economic reasons as they ignore the fact that the organization's resources are limited. Unlimited satisfaction is not possible because it will cause a shortage of the exchange product. For their part, the stakeholders in many cases do not intend to relinquish their claims without resistance. In practice, this leads to political processes in the relations between groups, in particular, to the struggle for the distribution of material and other goods. Let's consider a generalized model for implementing programs in which companies from several countries are involved.

The model is represented as:

$$K = < \check{E}, O >,$$

where Ě – the environment of the program; O – model of program implementation activities.

The environment of the organization $\check{E}$ is also considered as a fractal consisting of self-similar models of the cultural, political and economic environment of the countries where the organizations that are involved in the implementation of the program are operating.

$$\check{E} = \{E1, E2, Ei ..., En\},$$

where *n* number of countries where the organization does business. Accordingly, a project-oriented organization contains a number of separate units in these countries and is displayed by a fractal model $\check{D}$ in the organization of self-similar in structure and functions of organizations.

$$\check{D} = \{D1, D2, Dj ..., Dm\},$$

where *m* number of organizational units.

On the basis of the divisions of the organization $\check{D}$, a model of its activity *O* is formed on the basis of a conceptual scheme of value creation and migration in the cultural, political and economic environment. At the same time, organizational units in different countries have selfsimilar structures. This makes it possible to apply fractal models to the formation of control mechanisms. Model of the environment of the organization:

$$\forall E_i \in \check{E} \ \exists M^o = < \check{R}_i, \hat{G}_i, C_i >,$$

where 0 *Mi* – the environment model of the organization in the *i*-th country; *Ri* – model of market and country influence on organization activity; *Gi* – model of interaction with partners in the i-th country; *Ci* – model of interaction with competitors in the i-th country.

Let the model of activity in each country be formed as:

$$\forall D_j \in \check{D} \ \exists O^o = < P_j, \hat{R}_j, \hat{U}_j >,$$

where *Oj* – model of the organization's activity in the *j*-th country; *Pj* – model of product creation in the *j*-th country of the organization; *Rj* – production model in the *j*-th country; *U j* – management model in the *j*-th country.

The interaction model is determined by the matrices Ψ, Ψ*p*, Ψ*c* expressed in the form of multipliers that interact through dampers or risk acceptors of market influence on the organization, partners, and competitors.

Here Ψ is a matrix of the impact of the environment on organizations; Ψ*p* – matrix of influence of environment on partners; Ψ*c* is the matrix of influence of environment on competitors. The elements of the matrices *Ψp [1, j], Ψп [1, j], Ψк [1, j]* determine the influence of the environment on the activity of the unit of the j organization. The elements of the matrices *Ψp [2, j], Ψп [2, j], Ψк [2, j]* determine the influence of the activity of the unit j of the organization on the environment. The model of influence, which determines the immune mechanisms, includes coefficients distributed over five risk zones of influence of critical factors on the state of the organization and future destruction. The model is based on trends monitored by the project team on a scale of estimates Ω.

Suppose that for each element of the matrices Ψ *Ψp [1, j], Ψп [1, j], Ψк [1, j]* defined *Fp (Ωj), Fп (Ωj), Fк (Ωj)* mechanisms of control of the hybrid methodology – market dampers and competitors and risk absorbers passed on to the partners of the program are known. Then, the organization's management mechanism implements the following steps for each unit: Assessment of the state of the environment for each unit of the organization Ω*j*.

For each unit of the organization, the value of risk absorption is assessed through their transfer to contract partners and the company as a whole.

$$\forall j = 1, J : F^{\text{мк}} = \sum F\text{п}(\Omega j),$$

where j – the number of countries in which the organizational units are located, FМК the total amount of potential losses transferred to partners. For each unit of the organization, the values of accepted threats, risks of market behaviour and possible damping of their negative impact on the organization as a whole are evaluated.

$$\forall j = 1, J: F^{\text{мк}} = \sum F\text{p}(\Omega j),$$

where $F^{\text{МК}}$ the total amount of potential losses from the negative impact of the market in the *j*-th country. For the organization the values of the accepted threats, risks and their partial transfer to the panthers are evaluated.

$$\Phi = F_{\text{р}}^{\text{мк}} + F_{\text{к}}^{\text{мк}} - F_{\text{п}}^{\text{мк}},$$

where *F* – the total amount of losses from the negative impact of the environment of the program.

The values of losses that form the maximum negative impacts of the market or competitors are highlighted.

$$\forall j = 1, J, \Psi\text{e} = max(Fp(\Omega j), F\text{к}(\Omega j)).$$

For each maximum loss value Ψ, an anti-crisis project is formed, which assesses the *Ce* costs and the BCR cost benefit indicator. If the BCR factor is greater than 1, then this project is placed in the pool of projects of crisis management.

If there are not yet considered *Fp (Ωj), Fκ (Ωj),* and then iteration occurs and goes to p.3. After the program pool is formed, it is selected into the current project portfolio. The selection of projects is subject to limitations

$$\sum^{e} \mathrm{Ce} \leq Bp,$$

where *Bp* is the budget of the program for the current period.

In order to build an effective hybrid project management methodology, the feasibility of using genomic representations of the portfolio management, program and project management methodologies is substantiated [2, 3].

We will analyze the use of genomic representations of management methodologies, taking into account existing problems of financial organizations, turbulence of internal and external environment. The problems can be divided into four groups: definition of purpose, goal achievement, organizational competence and organizational behavior (Fig. 7).

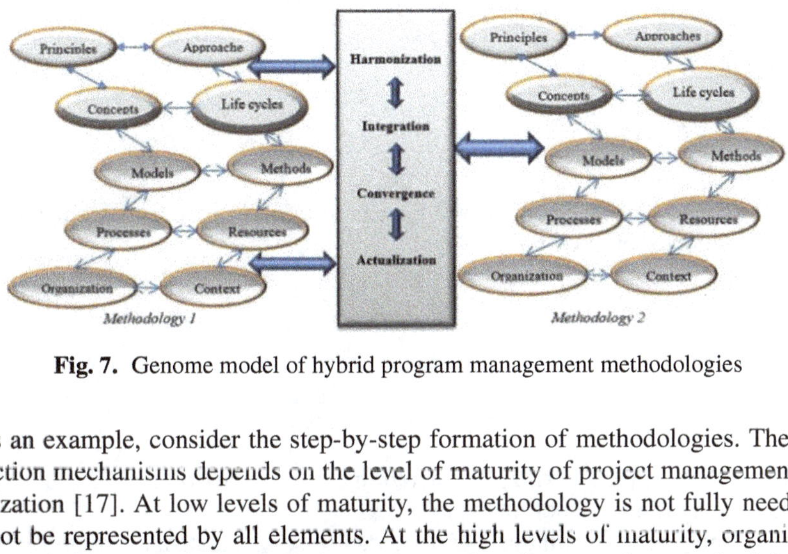

Fig. 7. Genome model of hybrid program management methodologies

As an example, consider the step-by-step formation of methodologies. The use of interaction mechanisms depends on the level of maturity of project management in the organization [17]. At low levels of maturity, the methodology is not fully needed and may not be represented by all elements. At the high levels of maturity, organizations implementing large-scale, integrated programs and projects require a methodology that has the full range of elements that contain all the variety of elements needed. That is, for the most complex and hypothetically diverse portfolio of programs and projects, the methodology approaches a "hypothetically complete system".

The knowledge carrier of the methodology formed is based on this model. The formalized model of the methodology genome is written as follows

$$\mathrm{M} = <\mathrm{P}, \mathrm{K}, \mathrm{A}, \mathrm{L}, \mathrm{N}, \mathrm{D}, \mathrm{V}, \mathrm{O}>.$$

Let there be a set of principles that define the methodology:

$$\mathrm{P} = \{\mathrm{p1}, \mathrm{p2}, \ldots, \mathrm{pn}\}.$$

This set must have the properties of completeness and consistency. As part of the implementation of the principles and alternative concepts, many approaches are known to be used in determining the methodology:

$$A = \{a1, a2, \ldots, am\}.$$

Alternative concepts can be formed based on these principles:

$$K = \{k1, k2, \ldots ki\}.$$

The methodology should be acceptable for multiple project life cycles:

$$L = \{l1, l2, \ldots, lr\}.$$

Models and methods for managing projects, programs, and portfolios are superimposed on combining models. These models, methods and mechanisms are implemented based on the process components of the methodology:

$$N = \{\Pi 1, \Pi 2, \ldots, \Pi q\}.$$

The documents that accompany the management process system are defined as a set of:

$$D = \{d1, d2, \ldots, dz\}.$$

Each methodology is linked to the corporate culture of management (culture is expressed through a set of cultural values):

$$V = \{v1, v2, \ldots, vc\}.$$

This methodology should be linked to the organizational environment and context:

$$O = \{o1, o2, \ldots, og\}.$$

where the indices n, m, s, r, q, z, c, g determine the number of elements of each set, which describes the genome of methodologies.

In this model, we will distinguish two mechanisms: systematization within each level and harmonization between the levels of the model. The harmonization mechanism provides a systematic elaboration of models at each level. The elements of each level are systematized based on the adjacency matrix. The integration mechanism ensures the formation of a complete model of the hybrid methodology. This integration depends on the specifics of each methodology that is included in the hybrid. The convergence mechanism ensures the interpenetration of best practices (solutions) within the framework of applied methodologies. The update mechanism prepares a hybrid methodology for implementation.

Each of these mechanisms has a complex structure, functions inputs and outputs. Such mechanisms will be defined in detail in the following publications of the authors.

For example, relationships at the first level between principles and approaches are systematized by the adjacency matrix. At the same time, the matrix elements determine the system compatibility of individual principles and approaches. A value of 0 indicates that elements do not match or do not interact with each other, and 1 indicates that they are fully consistent and have the maximum interaction. Similarly, matrices are formed at other levels of the model. The task of the systematization mechanism is to determine the system compatibility of the selected elements of methodologies. In this case, the term "genetic code of the project (program)" will mean its system model, which includes an initial idea of the "vision" of the project or program product, an integrated process for the development of a specific functional activity, built for the entire life cycle of the project, tools for its interaction with the external environment. At the same time, the genome of project management methodologies, project and program portfolio management have the same spiral structure for different content.

We introduce the concept of algebra of project management methodologies:

$$A = <\check{S}, \Omega>,$$

where: $\check{S}$ – carrier of knowledge of methodologies, placed in the genome; Ω – a signature that defines a set of operations on methodology elements.

The signature contains operations for projecting individual elements of the methodology and their groups, combining, intersecting, and supplementing methodologies, and so on. The purpose of the genetic code of projects, programs and portfolios is navigation in the implementation of functions and development processes. At the same time, the Navigator, moving through the phases and steps of the life cycle, uses information from the genetic code – its interconnected structures in the management processes. Most often, the process of forming the genetic code of a project occurs spontaneously, based on the intuition and practice of project management of the organization. Moving further in the application of project management analogies in the field of genomic models, we will distinguish two additional genomes of the hybrid model - the genome of competencies in methodology 1 and the genome of competencies in methodology 2. These genomes intersect at the point where project competencies are implemented at the expense of project manager's competence. This interaction forms the "chromosome" of the project [17].

Can we say that "flexible" methodologies use hybrid technologies? Yes, it is possible when during the course of the project one methodology (a fragment of a methodology) is replaced/supplemented by another methodology or methodological improvisation takes place [5]. However, in many cases of applying "flexible" methodologies, hybridization does not resemble systematically thought-out and ahead of time planned processes of synergistic combination of certain methodologies, but sporadic actions in a spontaneously established "zoo" of methodologies. Correspondingly, in "flexible" project management methodologies, the convergence of methodologies can be present only fragmentarily and more often generally not.

If we consider the Agile tool and related tools (KANBAN, Scram, etc.) as an example of a "flexible" project management methodology, then a retrospective analysis of project management shows that there is nothing fundamentally new in terms of knowledge gain in management. So, for example, back in 1924, at the Central Institute of Labour, under the guidance of A.K. Gastev, methodologies for lean manufacturing, labour ergonomics, and flexible ways of planning and implementing work in the absence of time and other resources were developed [16–19].

Other examples of this can be found in systematic scientific research and reviews (see, for example, [20, 21]), which are generally prone to conclusions regarding "flexible" methodologies.

In modern business, Agile is actively used in extreme conditions, in particular, in extreme programming projects [22, 23]. However, from a systemic point of view, the creation of an extreme situation is the result of an insufficiently mature management process [24, 25].

Despite its methodological fragmentation, it is also the fact that "flexible" methodologies are characterized by additional material, human, and sometimes financial costs due to "throwing from side to side" for reasons of reaction to internal and external changes (reaction to situations). These methodologies are partially applied at high-maturity enterprises, usually as a hybrid complement to the main methodologies [7, 20]. Hybrid methodologies built on the basis of traditional ones using AI and elements of "flexible" methodologies will reduce the risks of dispersing resources, improve productivity and quality of managerial decision-making.

4 Conclusion

The study of hybrid methodologies for managing programs is determined by the practice of implementing program components that differ in their essence.

The use of a hybrid project management methodology allowed the authors to complete essentially different projects within a specified time frame with a specified budget and quality of construction projects and projects for creating information and communication systems for management.

Further research areas are related to detailed elaboration of mechanisms for harmonization, integration, convergence and actualization.

References

1. Bushuyev, S., Verenych, O.: Organizational maturity and project: program and portfolio success, developing organizational maturity for effective project management. In: Silvius, G., Karayaz, G. (eds.), pp. 104–127. IGI Global (2018)
2. ISO 21500:2012. Guidance on project management. Project Committee ISO/PC 236, ISO (2012)
3. The Standard for Program Management – Fourth Edition. Project Management Institute (2017)
4. Azarov, N.Ya., Yaroshenko, F.A., Bushuev, S.D.: Innovative Mechanisms for Managing Development Programs. Summit Book, Kyiv (2011)
5. Lefley, F., Sarkis, J.: Applying the FAP model to the evaluation of strategic information technology projects. Int. J. Enterp. Inf. Syst. **1**, 69–90 (2005)
6. Turner, J.: A Guide to Project-Oriented Management. Dom Grebennikova, Moscow (2007)

7. Archibald, R.D.: Management of High-Tech Programs and Projects. DMK Press, Moscow (2004)
8. Burkov, V.N., Novikov, D.A.: How to Manage Projects. Sinteg, Moscow (1997)
9. Bushuyev, S., Verenych, O.: Competencies as an instrument for the blended mental space management of a project. In: The 11th International Conference on Project Management (ProMac2017), Munich, Germany, pp. 850–859 (2017)
10. Verenych, O., Bushuyev, S.: Interaction researching mental spaces of movable context, stakeholder and project manager. Organ. Technol. Manag. Constr.: Int. J. **10**(1), 1684–1695 (2018)
11. Phillips, J.J., Bothell, T.W., Snead, G.L.: The Project Management Scorecards. Elseiver, Amsterdam (2003)
12. Verenych, O., Bushuieva, V.: The blended mental space as one of the factors for sustainability and success in the project implementation process. In: International Research Conference, Dortmund, Germany, pp. 118–125 (2017)
13. Molokanova, V.M.: Investigation of synergetic levels of design-oriented management. Project management and production development. Coll. Sciences. Ave. Lugansk, SNU them. Dalia **3**, 30–37 (2011)
14. Verenych, O., Dorosh, M.: Blended Mental Space: the methodology for creation and approaches for its management. Comput. Sci. Inf. Technol. (CSIT) **2**, 30–37 (2017). https://doi.org/10.1109/STC-CSIT.2017.8099423
15. Verenich, O.V.: Management of projects and programs as a key element for the development of social and economic systems. Management of the development of complex systems. Kyiv, KNUBA **25**, 23–31 (2016)
16. Yaroshenko, Y.: Organization development in bifurcation. Manag. Dev. Complex Syst. **15**, 93–97 (2013)
17. Bushuyev, S.D., Bushueva, N.S.: Proactive management of organizational development programs. Project Program Manag. **4**(12), 270–282 (2007)
18. Bushueva, N.S.: Models and methods of proactive management of organizational development program. Kyiv, Science. Light (2007)
19. Bushuyev, S., Kozyr, B., Zapryvoda, A.: Nonlinear strategic management of infrastructure programs. Innov. Technol. Sci. Solutions Ind. **4**(10), 14–23 (2019)
20. Bushuyev, S.D., Neizvestnyj, S.I.: Sixth sense in development management. intuition as a channel of communication and a means of decision making. Manag. Dev. Complex Syst. **7**, 8–15 (2009)
21. Neizvestny, S.I., Chernykh, N.V.: The role of altered states of consciousness in decision-making in management. Project Program Manag. 1, 56–65; 2, 114–123 (2012)
22. Collyer, S., Warren, C., Hemsley, B., Stevens, C.: Aim, fire, aim – project planning styles in dynamic environments. Proj. Manag. J. **41**(4), 108–121 (2010)
23. Nelson, K.M., Ghods, M., Nelson, H.J.: Measuring the effectiveness of a structured methodology: a comparative analysis. In: Proceedings of the Thirty-First Hawaii International Conference on System Sciences, Kohala Coast, HI, pp. 492–499 (1998)
24. Cockburn, A.: Selecting a project's methodology. IEEE Softw. **17**(4), 64–71 (2000)
25. Office of Government Commerce. Managing Successful Projects with PRINCE 2. The Stationary Office, Norwich (2009)

Project Management Methodology in the Practice of Evidence-Based Development, Evidence-Based Policy

Olga Agatova(✉)

Russian Academy of Education, Pogodinskaya. 8, 119121 Moscow, Russia
olga_agatova@mail.ru

Abstract. It is analyzed: how the principles of evidence-based management, evidence-based policy based on big data analysis are integrated with project management methodology (ISO 21500: 2021). The results of changes in the management practices of state projects and programs are presented. The tasks of the competence transition from big-data to smart-data for data-management and evidence-based policy are formulated. The practice of implementing the methodology of evidence-based management, evidence-based policy is associated with the implementation of the concept of "Open Data", the National Data Management System in Russia, the development of the technological infrastructure of big data and digital data analytics services. Our research under the RFBR grant No. 19-29-14016 "Methodology for Big Data Analysis and Its Integration into Professional Development Programs for Executives" revealed how data analysis and evidence-based management principles are used in project management activities for state development programs. At the strategic management level: strategic performance indicators, infrastructure and data sources are defined. At the level of project management activities, sources of alternative data analysis are determined based on feedback from citizens (public policy). At the level of project life cycle management, it is determined which data sets will indicate the dynamics of value development and the effectiveness of products in the portfolio of projects of state development programs. At every level of evidence-based project management, this requires data competencies from leaders.

Keywords: Project management methodology · evidence-based development · evidence-based policy · impact investing · big data analysis methodology

1 Introduction

The practice of implementing a management methodology based on data analysis is associated with the implementation of the Open Data Concept, the National Data Management System in Russia, the development of the technological infrastructure of big data and digital data analytics services.

The practice of implementing project management is associated with the need to ensure development through differentiation in state programs of project and process

S. Bushuyev et al. (Eds.): IPMA 2021, LNNS 704, pp. 230–241, 2023.
https://doi.org/10.1007/978-3-031-34629-3_19

components (Decree of the Government of the Russian Federation "On the transfer of the state program" Development of education "to project management") [1].

The new Regulation on the management system of state programs of the Russian Federation [2], which establishes the rules for the development, implementation, monitoring and evaluation of the effectiveness of state programs of the Russian Federation (comes into force in 2022), updates new monitoring models congruent with the set of planned results, as well as actualizes demand on the methodology of data analysis (methodology for calculating indicators) and digital data analytics services based on artificial intelligence technologies. The regulation provides for the allocation in the structure of state programs - PROJECTS, determined, formed and implemented in accordance with the Decree of the Government of the Russian Federation of October 31, 2018. No. 1288 "On the organization of project activities in the Government of the Russian Federation" (Regulation on project activities).

Thus, the State Program is a tool for strategic development management, a tool for budgeting development directions, a tool for consolidating communities participating in development projects.

The analysis of data as conditions and results of development is organized in the state automated information system "Management", which collects and processes data, analyzes the implementation and evaluates the effectiveness of the implementation of state programs and their structural elements (PROJECTS).

In the analytical information system "Portal of state programs" (www.programs.gov.ru) [3], publicly available information is posted on the results of data analytics and on the assessment of the effectiveness of state programs.

Probably, the new management practice actualizes the need for the development of data-competencies among educational leaders. The project competencies of managers have been evaluated and certified for a long time [4], and competitive practices [5] make it possible to identify the most effective organization of Project offices responsible for the implementation of state projects and state programs and analysis of their effectiveness. But data competencies are not included in certification and competency assessment systems. It is these competencies that change the competence structure of management activities.

In many countries of the world - top 100 according to the Networked Readiness Index, digital environment standards and data analysis standards are developed. Such standards make it possible to assess the technological maturity and maturity of the management system in the implementation of government projects and government programs.

In 2021, Rosstandart introduced the Big Data terminology standard [6] GOST "Information Technologies. Big data. Overview and vocabulary", which is identical to the international standard "Information technology - Big data - Overview and vocabulary". After terminology standards, technology standards are enacted. But already now in the terminological standard the sequence of actions in working with big data is fixed: "data analytics - receiving, collecting, checking and processing data, evaluating them, visualizing and interpreting", "data variability - the format and structure of data, semantics and quality of the data array, variability of values" [6].

If technologically working with big data is standardized in international and national standards, and standardized technologies imply the universality of working with big data, then with regard to data analytics in the field of education - methodologically working with data is more complicated.

This complexity is primarily due to the fact that in education as a field of human development, not only the Performance Management Systems methodology is applied, but also the Human Capital Management Systems methodology, which characterizes the optics of humanitarian systems, digital humanities, big-data development analytics person.

1.1 Problematization, Research Issues

The development of the practice of evidence-based management is associated with the need to justify project decisions on the development of the state, society, and economic sectors. It is also connected with the need for reasonable project budgets and an assessment of their investment efficiency, the impact of investment projects.

In the conditions of a developed system of generation of project management standards [5], investment standards [4], there is clearly an insufficiently developed methodology and technologies for big data analysis, as well as the integration of data analysis methodology with project management methodology and impact investment methodology.

In fact, such an integrated methodological framework is the basis for standardizing activities in the field of evidence management, evidence policy, both at the state federal level and at the territorial regional level, in the sectoral and organizational-institutional aspect.

In this regard, a number of research questions arise:

1. How are the project management standard and big data analysis technologies and the principles of evidence-based data management methodologically integrated?
2. How does the module of competencies of a project management specialist, a project manager develop in the conditions of the need for proof of project decisions based on data analytics?
3. How do the management decision styles of project managers change in practice in the conditions of working with big data?

1.2 Methodology and Methods of Research

The methodology of the study was the concept of project management [6]; the concept of the national data management system [5]; the methodology of big data analysis [3]; the concept of management decision-making styles [8].

Research methods:

a) the method of structural and functional analysis of project management activities of state programs/development programs;
b) the method of content analysis of public reports of the governing bodies;
c) method of research of styles of management decisions;
d) a method of theoretical and methodological analysis of project management standards and generated standards for big data analysis for evidence-based management.

2 Analysis of the Development of the Concept and Practice of Evidence-Based Management and Evidence-Based Policy

In the conditions of a developed system for the formation of project management standards [7, 8], investment standards [9], the methodology and technologies of big data analysis for the management of state projects and state programs are not sufficiently institutionalized. In fact, the methodology is the basis for the standardization of activities in the field of evidence-based management, project management based on data.

In the field of education, data analytics standards are technologically linked to the standards of the digital educational environment, which determines the architecture of data on human development, the conditions of his education, and are ideologically linked to the principles of human development.

In this regard, a number of research questions arise:

1. How are the project management standard and big data analysis technologies methodologically integrated in management activities in the context of the implementation of the principles of evidence-based management based on data?
2. How does the manager's competence assessment model change in the context of working with big data and the tasks of implementing evidence-based decisions based on data analytics?
3. Do the styles of management decisions among managers change in practice in the context of working with big data?
4. How are data architecture and assessment tools changing, monitoring education development, what is the digital data infrastructure?

The research methodology was made up of the methodology for the analysis of big data in education [10]; the concept of project management [11]; the concept of a national data management system [12]; big data analysis methodology [13]; the concept of styles of forming managerial decisions [14].

The concept of evidence-based governance and evidence-based politics is studied in science and also institutionalized in the practice of governance. In the countries of the world, special institutions are being created as elements of the ecosystem of evidence-based management - management based on data analysis.

These are technological infrastructures and data analysis services, and institutions for the professional training of competent personnel, institutions of expertise and evidence-based policies. The USA has the Chief Evaluation Office, as well as The U.S. Commission on Evidence-Based Policymaking (CEP). CEP organizes the implementation of national EBPM projects.

In Japan, the Council for Science, Technology and Innovation (CSTI) is responsible for collecting and analyzing data and coordinating the e-CSTI platform as a technology infrastructure for evidence-based government project management. The evidence-based policymaking principles were approved by the Cabinet of Ministers.

A technological data infrastructure is being developed in Russia and the Concept of the National Data Management System has been approved. Evidence-based policymaking principles are integrated into the activities of governing bodies.

In the UK, there is Administrative Data Research UK (ADR UK), whose analytical recommendations form the basis of projects. ADK UK accredits data analysts as evidence-based policy professionals.

The Corporation for Social Research is established in Canada, which conducts data analysis and evaluates the effectiveness of projects based on data analysis.

In the study [10], we analyzed the development of data management in the countries of the world. We compared the E-Government Development Index - E-Government Development Index, EGDI - and the Open-Data/Open Government Index.

2.1 Factors Determining the Development of Evidence-Based Management Practices in Project Management

A study of evidence-based management standards, national databases of analytics and statistics revealed the following:

1. Trends in the implementation of management policies based on data and the ideology of open data are associated with ensuring the organizational coherence of institutions that provide educational data through legal regulations and laws of countries, as well as with the development and implementation of integral digital technologies that allow integrating different types of educational data through their digital programs. Processing for building a system of connections and forecasting opportunities.
2. Countries implementing a policy of human development through projects of state programs for the development of education organize the collection and analysis of data based on two Data-Analysis methodologies: activities; b) institutional methodology, in which the nature of the data is related to reports on the conditions of educational activities, reports on the implementation of educational development programs. Subsequently, these two variables in educational data are compared in the search for correlations for evidence-based analytics as the basis for making effective decisions.
3. Countries that implement the policy of openness of educational data and openness of data analytics services on digital educational data platforms, as a rule, implement an ethical policy by explaining to users the ethical protocols of working with educational data.
4. According to the management structure, the institutions responsible for ensuring the functioning and development of digital databases (repositories) of educational analytics and statistics data are directly subordinate to the government of the country or a sectoral ministry - the Ministry of Education.
5. Countries with a developed technological infrastructure of analytics and statistics of educational data and highly functional repositories of educational data integrate for processing and interpretation educational data of the institutional, national and international levels of assessment of educational results related to school systems for assessing educational results, national certification exams and the results of international quality studies education. This allows for timely alignment and harmonization of education monitoring systems and assessment tools.

2.2 Results of the Study of Management Decision Styles in the Implementation of Development Projects and Programs, Analysis of the Principles of Evidence in Public Reports

With regard to the sectoral aspect - education - in the study, we analyzed:

1. What data in education do heads of the federal, regional level in the field of education work with in managing state programs for the development of education and projects in the structure of state programs?
2. What data do the heads of educational organizations work with when implementing development programs for educational organizations and projects in the program structure?
3. What data do teachers work with?

The results of the study of pedagogical work with data, methods and technologies of data analysis in various information and analytical systems and educational platforms are presented by us in publications [11]. Both the existing infrastructure of education and human development data and the projected infrastructure and architecture of educational data reflecting new indicators of human development in post-industrial society are analyzed.

With regard to data analytics at the managerial level, the infrastructure and services for data analysis are developing in the field of education, new objects of managerial analysis are emerging, and the monitoring system is developing.

The data architecture of the state program for the development of education is constructed in the integration of the areas of analysis:

- anthropological,
- sectoral, territorial,
- inter-program (in connection with other state development programs).

This requires the heads of project offices of ministries/departments/departments of education to have an integral vision, cross-sectoral understanding of development tasks, strategizing ways to achieve goals, taking into account data analysis.

It should be noted that the methods of calculating the indicators of educational development programs are changing, focusing the calculation formulas on a person and the development opportunities of each. For example, the coverage of additional education for children and its availability were previously calculated according to the data on the occupancy rate of programs of additional education for children of different directions, at present - according to the data on the participation of each child in programs of additional education, which represent opportunities for the development of the human potential of each.

Thus, the ideology of human orientation, territorial socio-economic integration of education development programs into the socio-economic development of regions and the country, impact investment - changes the technologies for designing data architecture, data analysis, and, consequently, changes the requirements for the competencies of educational leaders, their managerial thinking. And activities (see the description of the structure of the set of open data [12], information on the achievement of planned values of indicators (indicators) for government programs [13].

Changing methodological approaches to the development and implementation of state programs (education, including), change the methodology of monitoring in education. The role of monitoring changes from ascertaining (fixing data) to risk-oriented and motivating [14].

The new methodology changes the structure of projects that make up the education development program [15–17], and the assessment of their effectiveness in connection with the new design of the education development data architecture [18].

In this regard, many regions are beginning to change the models of training and certification of management personnel in the field of education [19, 20].

One of the backbone competencies in management is the competence of analyzing data on the development of education. For example, the attestation certificate of the director of a Moscow school includes indicators reflecting the competence of the director to work with data analytics and organize education development projects based on data analysis: a) data on the effectiveness of resource management, b) data on the contribution to quality education, c) data on development teaching staff, d) data on the conditions for providing education, e) data of public assessment of activities [21].

It is not only the competence of management that changes, the style of making management decisions in general changes: from situational to conceptual.

The task of the current stage is to ensure a competence-based transition from big data to smart big data: from the accumulation of an array of educational data - to the smart use of data for evidence-based management of development projects.

The analysis of public reports in the context of the application of data for evidence-based management and evidence-based policy was carried out by us on the basis of research questions:

1. How is the evidentiary discourse of public reports changing ?,
2. What types of data are used in public reports?
3. What digital services and data analysis methods - smart big data - are used in Public Reports?

Using the method of discourse analysis, the authors studied the application of the concept of "evidence-based management", "evidence-based politics" in the discourse of public reports. In the analysis, the species range of concepts was differentiated: frame (statistical) and scenario (dynamic).

As a result of the analysis, the following conclusions were formulated:

1. The discourse of public reports is changing towards the development of evidence-based education management based on educational data analytics.
2. The concepts of "evidence-based management", "evidence-based policy" are used in Public Reports, mainly not as a "scenario", but as a "frame" - a descriptive data frame, stating data in reports, but not using data to predict regional development scenarios, municipal education systems.
3. Public reports reflect those educational data that are regulated by the requirements for the mandatory part of public reports, as well as those data, the possibility the use

of which is associated with the availability for the region, the municipality of a digital platform, analytics service and statistics of educational data.

The style of making management decisions in the context of data analytics was studied during the use of digital platforms and data analytics services by program participants.

The following methods were applied: a) a structured interview on the use of data sources and digital data analysis services to form management decisions; b) Alan Row's methodology (cognitive-situational model) for defining management decision-making styles. The classification of decision-making styles was taken as a basis, based on the criteria: 1) the cognitive complexity of the tasks being solved, 2) an orientation towards the way to achieve the goal. The criterion "cognitive complexity of the tasks to be solved" characterizes the complexity of the analyzed control object, the system of data sources about the controlled object, the complex of applied data analysis methods. The criterion "focus on the way to achieve the goal" characterizes the priorities of the leader, namely: focus on solving a problem or on people.

Figure 1 shows the criterion characteristics of the styles of management decisions with a highlighted analytical style, which we considered as the dominant of evidence-based management of educational systems.

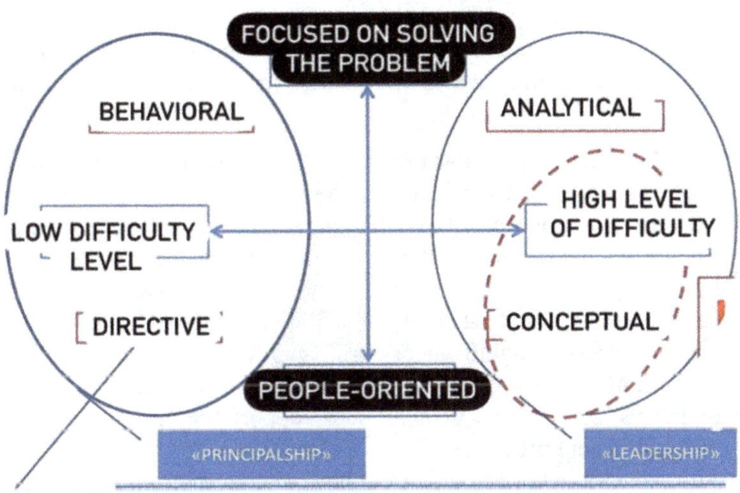

Fig. 1. Essence of Management decision making styles (Alan Rowe)

Since Alan Rowe's methodology for studying the style of managerial decisions is based on two coordinates: cognitive complexity (preference for structure or uncertainty) and preferential orientation (toward people or tasks) [14], it allows us to identify two fundamentally opposite styles of forming managerial decisions - "directive-behavioral" and "conceptual-analytical" (Fig. 1). The styles of managerial decision-making are changing: from directive-behavioral to conceptual-analytical - in the context of the development of data analysis for the formation of managerial decisions and evidence-based development of education.

Measurements according to the Alan Rowe method were carried out within the framework of the implementation of the program of additional professional education "Data analytics in public administration" twice: at the beginning of the development of the program and after its completion.

The program was originally designed in the structure of educational modules that develop the ability to use the technological data infrastructure, choose digital platforms and data analytics services that are appropriate for management tasks, the ability to identify and apply appropriate data analysis methods to various objects of management analytics, and the ability to form management decisions. This made it possible to study the dynamics of management decision-making styles built into the program.

The markers of management decision-making styles built into the methodology made it possible not only to see how immersion in evidence-based management affects the reflection of managers, but also to observe individual external circumstances of the choice of style, such as the level of managerial responsibility, work experience in the position, age, general managerial experience, professional installations.

3 Conclusions and Recommendations

Based on research on management practices based on data, a competency model has been identified.

The structure of data-competence, characterizing the subject managing the educational system, as implementing the strategies of evidence-based development of such systems:

- reflected professional attitude towards the management of education development projects based on data in the conditions of "open-data", "open-governance";
- the formation of basic concepts of data-based management methodology; knowledge of regulatory and legal regulators of assessment and monitoring in education;
- knowledge and application of information systems for assessing the quality of education; a comprehensive understanding of analytical services and systems of federal, regional and institutional significance;
- readiness and ability to form, at the imputed level of official responsibility, teams of "data-analysts" as subjects of analytical support for project management and education development programs;
- willingness and ability to develop a culture of data analysis and evidence-based management in a professional environment and organize the necessary measures to consolidate this culture in ethical standards/regulations of education management;
- the willingness and ability to respond in a mobile way to evidence based on the study of public opinion and evidence supported by scientific research in education.

For the formation of the above competencies, specialized programs of higher [24] additional professional education are required, as well as tools for assessing and certifying qualifications, data analytics standards for evidence-based project management and evidence-based policy.

In the management of public education development programs, evidence-based management methodology is integrated with project management methodology at all levels.

At the strategic management level: strategic performance indicators, infrastructure and data sources are defined. At the level of project management activities, sources of alternative data analysis are determined based on feedback from citizens (public policy). At the level of project life cycle management, it is determined which data sets will indicate the dynamics of value development and the effectiveness of the project system in state development programs. At every level of evidence-based project management, this requires data competencies from leaders.

Currently, the scope of application of data-competencies in education management is expanding in connection with the integration of industry analytics with the National Data Management System [25], conceptually and technologically integrating all levels of analytics.

The methodology of evidence-based management is integrated with the methodology of project management at all levels. At the level of strategic management: strategic performance indicators, infrastructure and data sources are determined. At the level of project management activities, sources of alternative data analysis based on feedback from citizens (public policy) are determined. At the level of project lifecycle management, it is determined: which data sets will indicate the dynamics of value development and the effectiveness of the products of the project portfolio of state development programs. At every level of evidence-based project management, this requires data-competencies from managers.

Data-competencies are the basis in the structure of project management competence. The methodology of evidence-based management is integrated with the methodology of project management at the methodological level - project lifecycle management-defining: how the analysis of data sets in the project portfolio determines the configuration of changes at the strategic level and the level of project management activities through the evidence-based public policy.

Acknowledgments. The research was prepared with the grant "Methodology for the analysis of bulk data in education and its integration into training programs for teachers and heads of educational institutions in the logic "Pedagogy based on data," "Management of education based on data" No. 19-29-14016 from Russian Foundation for Basic Research in the competition for the best projects of interdisciplinary fundamental research "Fundamental scientific support of the general education digitalization processes".

References

1. Decree of the Government of the Russian Federation of December 26, 2017 No. 1642 On the transfer of the state program "Development of Education" to project management from 2018. http://government.ru/docs/30832/. Accessed 22 Jan 2021
2. Decree of the Government of the Russian Federation of May 26, 2021. No. 786 "Regulations on the management system of state programs of the Russian Federation". http://government.ru/docs/all/134475/. Accessed 22 Jan 2021
3. Portal of state programs of the Russian Federation. https://programs.gov.ru/. Accessed 22 Jan 2021
4. Levels of certification/Center for assessment and development of project management. https://www.isopm.ru/sertifikatsiya/urovni/. Accessed 22 Jan 2021

5. Project Olympus/Annual competition for professional project management. http://olimp.ac.gov.ru. Accessed 22 Jan 2021
6. GOST R ISO/IEC 20546-2021 "Information technology. Big data. Review and Dictionary". Federal Agency for Technical Regulation and Metrology. http://protect.gost.ru/document.aspx?control=7&id=240981. Accessed 22 Jan 2021
7. Project, program and portfolio management - Context and concepts. ISO 21500: 2021. https://www.iso.org/ru/standard/75704.html. Accessed 22 Jan 2021
8. GOST R 54869-2011. National standard of the Russian Federation. Project management. Requirements for project management
9. Impact Reporting & Investment Standards. Montreal, Quebec, Canada (2011)
10. Fiofanova, O.A.: Big data analysis in education: methodology and technologies. Delo, Moscow (2020)
11. Fiofanova, O.A.: The problem of integrating digital data analytics services: the teacher's competence in working with educational data. Moscow University Bulletin, Series 20, Pedagogical Educ. **9**, 38–49 (2020)
12. Description of the structure of the open data set/Versions of the structure of the open data set. https://programs.gov.ru/opendata/7710349494-consolidatedreport2020/. Accessed 22 Jan 2021
13. Consolidated reports: information on the achievement of planned values of indicators (indicators) under state programs. https://programs.gov.ru/Portal/opendata. Accessed 22 Jan 2021
14. Motivating monitoring: The Ministry of Education will increase the responsibility of the regions for the implementation of national projects. https://edu.ru/news/nacionalnyy-proekt-obrazovanie/motiviruyushchiy-monitoring-minprosveshcheniya-pov/. Accessed 22 Jan 2021
15. State program of the city of Moscow "Development of education in the city of Moscow ("Capital education"). https://docs.cntd.ru/document/537907408. Accessed 22 Jan 2021
16. State program of the Sverdlovsk region "Development of the education system and implementation of youth policy in the Sverdlovsk region until 2025" (as amended on April 15, 2021). https://docs.cntd.ru/document/561648866. Accessed 22 Jan 2021
17. Public declaration of the state program of the Khanty-Mansiysk Autonomous Okrug-Yugra "Development of Education" (Yugra 2020–2030). https://depobr-molod.admhmao.ru/gosudarstvennye-programmy/gosudarstvennaya-programma-khanty-mansiyskogo-avtonomnogo-okruga-yugry-razvitie-obrazovaniya-vstupae/2251775/prezentatsiya-gosudarstvennoy-programmy-razvitie-obrazovaniya-vstupaet-v-silu-s-2019-goda. Accessed 22 Jan 2021
18. The procedure for the development, implementation and evaluation of the effectiveness of state programs of the Perm Territory was approved by the Resolution of the Government of the Perm Territory dated November 21, 2019 No. 835-p "On approval of the Procedure for the development, implementation and evaluation of the effectiveness of state programs of the Perm Territory"
19. The new corporate university will solve the problems of metropolitan education/Vesti obrazovaniya. https://vogazeta.ru/articles/2021/7/2/city_education/17515-novyy_korporativnyy_universitet_budet_reshat_zadachi_stolichnogo_obrazovaniya. Accessed 22 Jan 2021
20. Certification of managers and candidates for compliance with the position of the head of a state educational organization. https://www.mos.ru/donm/function/activities/attestatciia-rukovoditelei-i-kandidatov-na-sootvetstvie-dolzhnosti-rukovoditelia-gosudarstvennoi/. Accessed 22 Jan 2021
21. Education of Moscow: mechanisms, resources, results. A guide to open resources DONM. Moscow (2020)
22. Fiofanova, O.A.: Smart Big Data in public reports/Educational policy. **4**(84), 70–77 (2020). https://edpolicy.ru/smart-big-data. Accessed 22 Jan 2021

23. Rowe, J., Boulgarides, D.: Decision styles - a perspective. Leadersh. Organ. Dev. J. **4**(4), 3–9 (1983)
24. The project of the magistracy of the IOM RANEPA was supported by the Vladimir Potanin Foundation. https://www.ranepa.ru/sobytiya/novosti/proekt-magistratury-iom-rankhigs-podderzhan-fondom-vladimira-potanina/. Accessed 22 Jan 2021
25. National Data Management System. Analytical Center under the Government of the Russian Federation (2020). https://ac.gov.ru/projects/project/nacionalnaa-sistema-upravlenia-dannymi-nsud-41. Accessed 22 Jan 2021

Assessment of the Impact of Sustainability on the Competencies of Project Managers

Ekaterina Khalimon(✉), Ariana Kalganova, Elena Ryabtseva, and Daniil Rakhmankulov

State University of Management, 99, Ryazansky Prospect, Moscow, Russia
guu.konf@yandex.ru

Abstract. Project management and sustainable development are ones of the most important concepts that have gained great popularity in the scientific community in recent years. The need to integrate sustainability into project management is increasing every year changing the requirements for project managers' competencies. Despite of the growing scientific interest in this topic, there is a small number of research papers that consider the impact of the concept of sustainability on the competences of a project manager. Thus, the purpose of this study is to identify the most important knowledge, skills and abilities and personal qualities needed by project managers for successful sustainable project management. The research is based on the analysis of the sources of scientific literature, international, and national standards in the field of project management. The results showed that over the past 5 years the project manager profession has undergone significant changes, and in order to achieve sustainability project managers need to develop new professional competencies based on knowledge, skills and abilities, and personal qualities.

Keywords: Competencies · Sustainability · Project Manager · Knowledge · Skills · Personal Qualities

1 Introduction

Considering sustainability requires changing the mindset of project managers, obtaining new knowledge, mastering competencies and developing personal and professional qualities necessary for successful project management. Corporations around the world are actively including the concept of sustainable development: they are increasingly thinking about the impact of their activities on the world and are striving to take responsibility for the results of their projects, including aspects of the sustain-ability of this result.

The introduction of the concept of sustainable development has changed not only the general approach to project management but also the profession of a project manager itself. To achieve sustainable goals, project managers are required to develop new, diverse skills related to sustainable project management.

S. Bushuyev et al. (Eds.): IPMA 2021, LNNS 704, pp. 242–255, 2023.
https://doi.org/10.1007/978-3-031-34629-3_20

Researchers confirm that the project manager takes a central and influential position in sustainable project management. It is project managers who are essential agents of change and catalysts in ensuring the sustainability of the project and the success of project management depends on the development of their competencies.

Current trends and ongoing changes have caused great interest in the scientific community in these topics. In the period from 2015 to 2019, 73% of scientific articles were published at the junction of sustainable development and project management [22].

2 Methodology

In the research there was adopted a qualitative research method was a way to make a more efficient study of the topic of sustainable project management, to better understand its influence on the profession of project managers and serve as a basis for future research. To answer the research question of whether the requirements for the professional competencies of specialists in the field of project management have changed in the conditions of sustainability or not, in-depth research of literature sources and international and national standards in the field of project management was conducted. When choosing literature to use, it was very important to make sure of the author's knowledge and experience in this field, also the rating of journals played an important role in choosing literature. Materials, such as articles, conference papers, and some books on the subject of sustainability and professional competencies of specialists in the field of project management, have been carefully analyzed. Some notes have been made during analysis for the future comparison and collation of the thoughts on the theme of sustainability and project management competencies. Several phrases have been explained for a better understanding of the final decision to combine either of the chosen statements.

As a result of the analysis of literature and standards in the field of project management, it was possible to form a list of knowledge, skills, abilities and personality qualities that specialists in the field of project management should possess. Also, based on the analysis of scientific literature, hypotheses have been put forward regarding the impact of sustainability on the requirements for the professional competence of project managers. The results presented in Table 2 of the Literature Analysis section laid the foundation for the formation of an expert survey for further research on the topic.

In the second stage, scientific sources were reviewed in order to determine which methodologies the authors used in their research and which one suits the best for the research. As a result, the survey of experts was chosen.

The survey of experts refers to expert methods of development and decision-making since the main participants in this method are experts. Experts can be: heads of enterprises of different industries and spheres of activity; middle managers (heads of the marketing department, production department, planning and financial management, etc.); specialists (economists, engineers, technologists, marketers, etc.). In our case, the experts were from the Russian Federation, selected specialists in the field of project management from different areas who have more than five years of experience in project management, as well as have specialized education and/or certification in project management (international or national).

The application of the survey of experts has significant advantages: the method of expert evaluation through the processing of individual expert assessments allows obtaining a general opinion based on independent assessments with a sufficient degree of validity and reliability. Also, thanks to independent surveys, the method allows avoiding social pressure on experts due to their different statuses, bandwagon effect, group thinking that leads to mistakes and getting a consensus of experts.

To form an expert group, the authors have established strict selection criteria: experts must be functioning project managers, be certified specialists and have professional experience in the field of project management, not less than five years. These requirements allowed us to identify and consider changes in the competencies of project managers under the influence of sustainable development. Further stages of using the survey of experts are presented in Table 1.

Table 1. The stages of using the survey of experts.

Stage	Characteristic
Formation of the group of participants	- Search for potential experts from the professional community of the Project Management according to the requirements put forward - Verification of candidates for compliance with the requirements As a result, a list of experts has been formed who are ready to take part in the survey using the Survey method. It included representatives of 6 professional areas: IT, education, real estate, construction, public administration, and ecology, with a total number of 91 respondents
Preparation of the survey	A questionnaire is being formed, which contains 23 questions. Questions are closed and have the maximum number of answers to them. The questionnaire was prepared in the Russian language and has been entered into the docs.google system and links have been sent to experts to complete the survey. The deadline for providing feedback was up to 7 calendar days
Conducting the survey	Experts click on the links to access the survey and mark the answers to the questions. Based on the results of the answers to all the questions, they complete the testing and send the results
Processing of results	The survey results are processed by the authors of this study and checked for answers to all questions. Discussions, collective meetings, explanations or comments are not provided. Experts got acquainted with the results

The presence of specialists with different experiences and distinct points of view allowed to get a meaningful assessment and reflected the actual national situation and, in general, increased the reliability of scientific research.

3 Literature Review

In the first phase of the exploratory research the analysis of scientific literature was conducted through the world's largest scientific publishing house Elsevier.com and the international platform Researchgate.net. The search for suitable scientific literature sources was conducted in the leading journals in the field of project management, using such keywords as sustainability, sustainable development, sustainable project management, and project manager competencies.

A systematic review of the literature revealed that there is a small number of re-search papers that address the key competencies of project managers in the field of sustainable project management. As a result of the review of the research topic, 32 scientific sources were selected. Based on a thorough analysis of the literature, the competencies that are put forward to project managers under the influence of sus-tainable development were identified (Table 2).

Table 2. Literature analysis.

N	Name of the article	Key competencies (knowledge, skills and personal abilities)	The methodology used in the source
1	Managing project sustainability in the extractive industries: Towards a reciprocity framework for community engagement [3]	Effective and efficient in terms of sustainability; take into account the needs of stakeholders; knowledge of the concept of capacity building	Capability approach developed by philosophers Marta Nussbaum and Amartya Sen [18, 19, 23, 24]
2	Exploring a variety of factors that stimulate project managers to address sustainability issues [27]	Pragmatic, internally motivated and task-oriented. A supportive attitude toward stakeholders, knowledge of standards and tools, the ability to deal with uncertainty, and success-oriented. Certain competencies in the field of sustainable development and normative values. The behavior of the project manager in relation to sustainability is a personal trait	Q-methodology is based on the factors provided by the Theory of Planned Behavior (TPA) Based on the factor analysis of 49 Q-varieties
3	25 years of 'sustainable projects'. What we know and what the literature says [22]	System thinking competencies, proactive competencies, normative competencies, strategic competencies, and interpersonal competencies. Internally motivated to work on a sustainable project and achieve sustainable results. Certain competencies and skills in the field of sustainability, are not traditionally taught in the process of training project managers	An extensive systematic review of the literature from 770 publications of that period from 1993 to 2017
4	Project management knowledge and skills for green construction: Overcoming challenges [11]	Communication management knowledge, transparency in communication, greater coordination with various stakeholders and dealing with unexpected circumstances	Literature review, surveys and interviews with project managers

(continued)

Table 2. (*continued*)

N	Name of the article	Key competencies (knowledge, skills and personal abilities)	The methodology used in the source
5	Project Management and Sustainability: Playing Trick or Treat with the Planet [28]	Responsibility and result-oriented, ethically correct and taking fair decisions, and excellent communication skills with stakeholders. Flexibility, leadership	An empirical survey in project-oriented organizations of both public and private sectors
6	New Approach for Managing Sustainability in Projects [15]	Personal competencies: 1. The ability to introspect and self-management; 2. Personal honesty; 3. Interpersonal communication skills; 4. The ability to have interpersonal relationships, participate in project activities and encourage the participation of others; 5. Leadership qualities; 6. Teamwork opportunities; 7. Conflict and crisis resolution skills; 8. Inventiveness, imagination, creativity; 9. Negotiation skills; 10. Results orientation Professional practical competencies - related to areas: scope management, time management, cost management, etc.	Delphi analysis
7	Revisiting the relationship between sustainable project management and project success: The moderating role of stakeholder engagement and team-building [25]	Encouraging interaction with stakeholders, team building skills, and decision-making from the point of view of sustainability	A structured survey method and 323 responses were received from project management specialists in Pakistan
8	Sustainability in Project Management Competencies: Analyzing the Competence Gap of Project Managers [26]	System thinking, a holistic view and interpersonal, normative, proactive and strategic competencies. It is necessary to understand the types of life cycles of the project (product) results, to involve stakeholders in active participation. System thinking competencies, strategic competencies, normative competencies, proactive competencies and interpersonal competencies	Literature-based analysis
9	Exploring Patterns of Sustainability Stimuli of Project Managers [16]	The inclusion of sustainability is a personal trait internally motivated project managers, task-oriented project managers, and pragmatic project managers stimulated by practical knowledge, tools and results	Quantitative pair comparison A quantitative approach using a survey-based strategy

(*continued*)

Table 2. (*continued*)

N	Name of the article	Key competencies (knowledge, skills and personal abilities)	The methodology used in the source
10	The Impact of Project Manager Soft Competences on Project Sustainability [30]	Promoting solutions and actions in the field of sustainable development. The project manager must have a long-term and comprehensive view of the project, and be fully responsible for the consequences of the project, including sustainability-related outcomes. Innovation is directly related to the sustainable development of the project. The results demonstrate the positive impact of the soft competencies of the project manager on the sustainability of the project	A deductive approach was adopted The collected data were then analyzed by doing the structural equation modeling in PLS-SEM in order to examine the relationships
11	Benefits Formulation in Construction Projects: An Exploratory Study through a Social Sustainability Perspective [7]	Identifying at the early stages of the project the intended beneficiaries and the corresponding benefits, as well as the timing and responsibility for the implementation of benefits	SS-centric analysis of project benefits management (PBM) plans for 80 construction projects using content analysis has been conducted
12	The Paradoxical Profession: Project Management and the Contradictory Nature of Sustainable Project Objectives [21]	To simultaneously solve many desirable, but contradictory economic, environmental and social tasks at the level of the firm and society, which operate in different time frames and follow a different logic To make compromise decisions in order to align various sustainable development goals with economic business goals and conflicting requirements of multiple stakeholders	A qualitative methodology based on semi-structured interviews with 14 respondents holding key positions within six different professional project management associations and communities of practice
13	Achieving Sustainability in Railway Projects: Major Stakeholder Concerns [31]	To involve stakeholders, to create stakeholder groups who can share their diverse views on understanding the sustainability of the project, as well as how to achieve sustainability	A triangulated methodology was adopted, including a literature review, a questionnaire survey, and interviews to obtain data from project stakeholders
14	Core capabilities for achieving sustainable construction project management [6]	Sustainability principles in planning, design, construction, and management of projects, social responsibility. Execution-oriented and product-oriented capabilities	A systematic literature review
15	The concept of sustainable construction project management in international practice [29]	Understanding of potential benefits, cooperation between practitioners, research institutes and environmental organizations; a systematic approach to achieving sustainable development goals	Delphi method

(*continued*)

Table 2. (*continued*)

N	Name of the article	Key competencies (knowledge, skills and personal abilities)	The methodology used in the source
16	Can project sustainability management impact project success? An empirical study applying a contingent approach [5]	Project management (PM) process and knowledge areas focus on sustainability, green procurement and partnership, design for the environment, environmental technologies, and social responsibility in a project context	Survey-based research
17	Governing public-private partnerships for sustainability an analysis of procurement and governance practices of PPP infrastructure projects [10]	Knowledge of sustainability considerations, differences between the projects, output specifications and award criteria, sustainability criteria and the influence sustainability criteria have in the final bid evaluation	Survey-based research
18	Interestingly it's innovation: Reviewing sustainability performance management at the base of the pyramid (BoP) [3]	The findings show that the BoP literature discusses only the subcategories of the PM tools and instruments. Moreover, the presence of relational performance measures, such as social capital and trust, indicates their importance and suggests that they should be incorporated into the PM instruments	Structured literature review
19	Key factors of sustainability in project management context: A survey exploring the project managers' perspective [17]	Knowledge of sustainable innovation business management, stakeholders management, Economics and competitive advantages, environmental policies and resources saving	Systematic literature review and the survey-based research
20	Project benefits co-creation: Shaping sustainable development benefits [13]	A long-term view of change, innovation and the creation of benefits. Knowledge of how projects are structured and the composition of project teams	The research is based on a literature review
21	Projects to create the future: Managing projects meets sustainable development [9]	Balancing of economic, social and environmental orientation in the understanding of sustainability. The proactive orientation on stakeholder interests is a defining element of sustainable project management. Emphasizing the values as the basis for considering sustainability in project management, discussing stakeholder co-creation in benefits realization management	The research is based on Q-methodology
22	Project sustainability strategies: A systematic literature review [2]	Knowledge of how sustainability challenges are managed in projects. A more holistic and thorough understanding of the versatility of sustainability	The research is based on a systematic literature review

(*continued*)

Table 2. (*continued*)

N	Name of the article	Key competencies (knowledge, skills and personal abilities)	The methodology used in the source
23	Revolutionize value management: A mode towards sustainability [32]	The knowledge of integration of sustainability into value management. The integration of sustainability required participation from all players of value management to ensure effective integration	The research is based on a literature review
24	Sustainable project management through project control in infrastructure projects [14]	Evaluation of environmentally and socially sensitive context, the influence on various stakeholders. A more holistic control package in sustainable project management, different control mechanisms for the different dimensions of sustainability, sustainability control needs to be integrated as part of general project management, and internal project control to be complemented with effective project sustainability governance	The research is based on a case study methodology
25	The integration of ideation and project portfolio management—A key factor for sustainable success [8]	Not only the execution stages of new products and service innovations but also the early stages, where new opportunities are discovered and new options are developed, require a professional portfolio management system	The research is based on a sound review of the literature

In a large number of studies, the authors draw attention to the fact that considering sustainability requires a change in the thinking of project managers. According to Luca Sabini and Neil Alderman, there is a misconception that sustainability is expensive. The benefits of sustainable project management that it can bring in the medium and long term are ignored.

In their study, Gilbert Silvius and Ron Schipper argue that the behavior of the project manager in relation to sustainability is, first of all, a personal trait. They believe that a project manager should be internally motivated to work on a sustainable project and achieve sustainable results.

Also, many authors emphasize that it is the project manager who makes the greatest contribution to the implementation of sustainable development goals in projects and is personally responsible for the sustainable development of the project.

Scientific sources say that in order to achieve sustainability, the project manager needs to develop: system thinking competencies, proactive competencies, normative competencies, strategic competencies, and interpersonal competencies.

Maria Pilar de la Cruz Lopez and other coauthors identified such sustainable competencies as personal honesty; the ability to interpersonal relationships, participate in project activities and encourage the participation of others; ingenuity, imagination, creativity; focus on results as the most important of personal competencies.

Luca Sabini, Daniel Muzio and Neil Alderman highlighted that the implementation of the sustainable development goals requires certain competencies and skills that are not traditionally taught in the process of training project managers, so they must be achieved at the local level (with experience) or in the learning process (for example, at the university level or with certifications).

Many scientific publications emphasize that the inclusion of sustainable development issues in projects increases the burden on project managers, and excellent communication skills are needed to communicate with a significant number of stakeholders. Thus, communication management becomes vital for the success of the project.

Competencies that are put forward to project managers under the influence of sustainable development are presented in the last row of Table 3 called "Additional requirements from project management standards".

At the second stage of the study, in order to identify the actual competencies of the project manager, three standards in the field of project management were reviewed and analyzed: The International Standard ICB 4.0 IPMA, the American standard PMBOK and the Japanese standard P2M PMAJ. Based on the analysis of the standards, a table of competencies of the project manager was compiled (Table 3). Competencies were divided into three groups: knowledge; skills and abilities; personal qualities.

The analysis of project management standards showed that the standards do not take into account the role of project managers in the implementation of sustainable development projects and programs.

Also, the competencies that the authors of research articles put forward to project managers under the influence of sustainability from the previously compiled Table No. 2 were added to Table No. 3, in a line called "*Additional requirements from sources of literature*". These competencies were also divided into three groups: knowledge; skills and abilities; personal qualities.

After compelling Table 3, a comparative analysis was conducted between the project managers' competencies prescribed in three standards and the competencies found in scientific sources. The competencies that were put forward to project managers under the influence of sustainability were added to the group "Additional requirements from literature sources".

A comparative analysis was conducted to establish the coincidences and identify existing differences in the competencies in knowledge, skills and abilities, and personal qualities of the project manager.

4 Results

Here are some of the main results of the survey:

1. The respondents rated 3.66 points answering the question: "It is noted that the introduction of sustainability principles into project management is accompanied by an increase in the number of stakeholders. Evaluate on a 5-point scale how much more communication (interaction with stakeholders) has become in project management in the context of sustainability, where 1-has remained at the same level, 5-has increased markedly".

Table 3. Competencies of the project manager from different sources.

1	ICB 4.0 IPMA [12]
Knowledge	Strategic PM. Fundamentals of PM. Standards and norms of PM. Sources of interest and power. Green ("sustainable") PM. Theories of culture. Techniques for managing stress, time, and setting goals. The Code of Ethics. Principles of ensuring sustainable development. Knowledge about various techniques and means of communication. Theory of motivation. Leadership models. Coaching. Models of roles in the team and the housing and communal services. Techniques for settling conflicts and crises. Negotiation theories and techniques. Theories of efficiency, effectiveness and productivity. Knowledge about the content, and "boundaries" of the project. Methods of calendar planning. Document management systems. Quality management tools. Techniques for identifying and assessing risks and opportunities, implementing and monitoring risk management plans. Interests and influence of the project stakeholders
Skills and abilities	Implementation of the organization's mission and its sustainable development. Adaptation of standards for a specific organization. Management of power and interests to meet the needs of project stakeholders and obtain mutually beneficial results. The values and culture of the organization and the external environment. Control the behavior. Create an atmosphere of trust. Excellent interpersonal communication, various means of communication. To inspire people to achieve common goals. The ability to overcome resistance. Staff selection and recruitment skills. Facilitation skills. Diplomatic skills, negotiation, discussion and compromise skills, analytical skills. Creative approach. Delegation. The ability to determine the approach of project implementation. Project content management. Determine the sequence, optimize, monitor and control the duration and timing of the project work. Working with the project software. Implementation and adaptation of quality standards, use of quality management tools. Justify and provide sufficient resources for the implementation of the project, monitor the status and prepare the necessary reports. Identify and develop the necessary risk response measures. Presentation skills. The ability to change and transform organizations to realize the intended benefits and goals
Personal qualities	Strategic thinking. Critical thinking. System thinking. Leadership. Sociability. Focus on ensuring sustainable development. Respect for different cultures. Self-awareness. Self-organization. Personal integrity. Reliability. Empathy. Openness. Sincerity. Enthusiasm. Intuition. Sense of humor. Emotional stability. Organization of teamwork. Persuasiveness. Stress resistance. Non-standard, creative thinking. Resourcefulness. Ingenuity. Persistence. Enterprise. Self-confidence, patience. Result orientation. Introspection
2	P2M PMAJ [20]
Knowledge	General principles of management, business management, strategic planning. Project management life cycle. Competently use project resources: vital financial, information and intellectual
Skills and abilities	Competent negotiation. The ability to focus the team's efforts on innovation and breakthroughs. Task structuring and performance monitoring. Planning ability. Execution ability. Strategic thinking (strategic perception, the ability to perceive the strategic elements of the program/project and arrange them according to priorities for proper application). Integral thinking (continuous striving to achieve the results of the project/program, the ability to prevent, evaluate and work with changes in the project environment to achieve its results). Holistic thinking (focus on a holistic mission, ability to identify problems, and their sources and develop solutions to overcome them)
Personal qualities	Innovator. Leadership (leadership to increase added value and introduce innovations). Result-oriented [Focus on achieving results]. Self-realization [Self-discipline]. Harmonization of diverse actions [Ability to balance between different working groups and stakeholders]. Getting rid of tension and competition. High standards of ethics. Foresight. Responsibility. Continuous capacity building, and development. Initiative supports enthusiasm to achieve results through teamwork

(*continued*)

Table 3. (*continued*)

3	PMBoK [1]
Knowledge	Project integration management. Project scope management. Project schedule management. Project cost management. Project quality management. Project resource management. Communication management in the project. Project risk management. Project procurement management. Project stakeholders management
Skills and abilities	Project team management skills. Project lifecycle management skills. Project plan creation skills. Project progress management skills. Skills of dealing with uncertainty. Skills of delivering results and creating value. Skills for evaluating the effectiveness of project implementation. The skill to strive to understand the interests and expectations of stakeholders and involve them in the process of creating value. The skill of being focused on maximizing value for the customer and the organization. The skill to perceive a project as an element of a larger system. Take into account the relationships in the middle of the system when implementing the project. The skill of embedding quality in the process of creating results. The skill to minimize threats and maximize opportunities. The skill of being able to act on a situation while maintaining a focus on creating value
Personal qualities	Sense of humor. Optimism. Respectfulness. Politeness. Friendliness. Kindness. Honesty. Ethics. Initiative. Leadership. Confidence. Persuasiveness. Fluency of speech. Responsibility. Emotional intelligence
4	Additional requirements from sources of literature
Knowledge	Education in the field of sustainability management that meets 5 requirements for competencies: systems thinking, interpersonal, normative, proactive and strategic competencies
Skills and abilities	Take into account the needs of stakeholders throughout the entire housing project, and not only at the beginning of the project. The ability to find a compromise between the goals of sustainability and the triangle of constraints. The competencies and skills required to achieve sustainable goals. Transparency in communication. To determine the intended beneficiaries and the corresponding benefits, the timing and responsibility for the implementation of these benefits. To involve the general public in the early stages of the project assessment
Personal qualities	The behavior of the project manager is primarily a personal trait. Internal motivation is needed to switch on sustainability in the project. Personal honesty. Promotion of solutions and actions in the field of sustainable development. Individual responsibility. Soft competencies have a positive effect on the sustainability of the project

2. More than 80% of respondents agreed that "the requirements for the project manager have changed over the past 5 years (requires the development of new hard and soft skills, and abilities)" and that "project management in the context of sustainability requires a project manager to develop new competencies, knowledge, skills".
3. The respondents evaluated on 6 points the importance of adhering to the principles of sustainable development in project management (on a scale from 1 to 10, where 1 is unimportant, reduces the quality of the project, 10 is very important, improves the quality of projects).
4. Among the key knowledge that a project manager should have, the respondents selected those that were most in demand among project managers 5 years ago: knowledge about project cost management (87%), knowledge about project scope management (83% of respondents), knowledge about calendar planning methods (67%), knowledge about the project management lifecycle (66%).

In comparison, among the key knowledge that a project manager should have, the respondents selected those that are most in demand among project managers today: knowledge about various techniques and means of communication (84%), knowledge about project cost management (67%), knowledge about managing the interests and influence of project stakeholders (61%), knowledge about the project management lifecycle (52%), knowledge about conflict and crisis management techniques (51%), knowledge about time management, goal setting (50%).

5. Among the key skills and abilities that a project manager should have, the respondents selected those that were most in demand among project managers 5 years ago: skills of determining the sequence of work (network planning) (69%), ability to evaluate the effectiveness of the project (67%), implementation skills of achieving results and goals (67%), ability to optimize, track and control the duration and timing of project work (55%), ability to systematically manage projects as a set of interrelated processes (54%), the skill of achieving the expectations of stakeholders and involving them in the process of creating value (54%), negotiation skills (52%).

In comparison, among the key skills and abilities that a project manager should have, the respondents selected those that are most in demand among project managers today: interpersonal communication skills, the use of various means of communication (83%), the skill of achieving the expectations of stakeholders and involving them in the process of creating value (65%), negotiation skills (63%), the ability to act on the situation, while maintaining a focus on creating value (59%), the skills of achieving results and goals (58%), the ability to create and evaluate the value of a product/project (56%), the ability to systematically manage projects as a set of inter-related processes (53%).

6. Among the key personal qualities that a project manager should have, the respondents selected those that were most in demand among project managers 5 years ago: responsibility (100%), leadership (88%), confidence (64%), system thinking (55%), organization (54%), sociability (51%).

In comparison, among the key personal qualities that a project manager should have, the respondents selected those that are most in demand among project managers today: communication skills (83%), leadership (65%), focus on results (64%), emotional stability (58%), confidence (57%), creativity (53%).

5 Conclusion

After analyzing 32 sources of scientific literature on the subject of the study the competencies that are put forward by the authors of scientific sources to project managers under the influence of sustainability were identified.

A comparative analysis of sources of scientific literature and the standards in the field of project management was carried out in order to identify the wide range of competencies of project managers. It showed that the standards do not take into account the role of project managers in the implementation of sustainable development projects and programs. The competencies that were put forward to project managers under the influence of sustainability were added to the group "Additional requirements from literature sources". Many scientific publications emphasize that the inclusion of sustainable development issues in projects increases the burden on project managers, and excellent communication skills are needed to communicate with a significant number of stakeholders. Thus, communication management becomes vital for the success of the project.

At the second stage of the study a survey was conducted among Russian project managers of various economic spheres of activity in order to determine the most important competencies needed by project managers for successful sustainable project management. The sample size of the study was 91 respondents.

The results of the study showed that over the past 5 years, the concept of sustainable development has had a significant impact on the competencies of project managers in Russia and, accordingly, there is an urgent need for project managers to acquire and develop knowledge, abilities and skills, and personal qualities related to sustainable project management. A list of actual professional competencies of project managers in the context of sustainability was determined. In particular, the results showed that among the key personal qualities that a project manager should have, the respondents selected those that are most in demand today: communication skills, leadership, focus on results, emotional stability, confidence, creativity.

References

1. A Guide to the Project Management Body of Knowledge (PMBOK® Guide. 6th edn. Project Management Institute (2017)
2. Aarseth, W., Ahola, T., Aaltonen, K., Økland, A., Andersen, B.: Project sustainability strategies: a systematic literature review. Int. J. Project Manage. **35**(6), 1071–1083 (2016)
3. Aman, S., Seuring, S.: Interestingly it's innovation: reviewing sustainability performance management in the base of the pyramid (BoP). Technovation 102394 (2021)
4. Baba, S., Mohammad, S., Young, C.: Managing project sustainability in the extractive industries: towards a reciprocity framework for community engagement. Int. J. Project Manage. **39**(8), 887–901 (2021)
5. Carvalho, M., Rabechini, R.: Can project sustainability management impact project success? An empirical study applying a contingent approach. Int. J. Project Manage. **35**(6), 1120–1132 (2017)
6. Ershadi, M., Goodarzi, F.: Core capabilities for achieving sustainable construction project management. Sustain. Prod. Consum. **28** (2021)
7. Goel, A., Ganesh, L., Kaur, A.: Benefits formulation in construction projects: an exploratory study through a social sustainability perspective. IIM Kozhikode Soc. Manage. Rev. **9** (2020)
8. Heising, W.: The integration of ideation and project portfolio management – a key factor for sustainable success. Int. J. Project Manage. **30**, 582–595 (2012)
9. Huemann, M., Silvius, A.: Projects to create the future: managing projects meets sustainable development. Int. J. Project Manage. **35**(6), 1066–1070 (2017)
10. Hueskes, M., Verhoest, K., Block, T.: Governing public–private partnerships for sustainability. Int. J. Project Manage. **35**(6), 1184–1195 (2017)
11. Hwang, B.-G., Ng, W.: Project management knowledge and skills for green construction: overcoming challenges. Int. J. Project Manage. **31**(2), 272–284 (2013)
12. Individual Competence Baseline for Project Management. International Project Management Association, The Netherlands (2015)
13. Keeys, L., Huemann, M.: Project benefits co-creation: shaping sustainable development benefits. Int. J. Project Manage. **35**(6), 1196–1212 (2017)
14. Kivilä, J., Martinsuo, M., Vuorinen, L.: Sustainable project management through project control in infrastructure projects. Int. J. Project Manage. **35**(6), 1167–1183 (2017)

15. López, M., Cartelle Barros, J.J., del Caño Gochi, A., Lara, M.: New approach for managing sustainability in projects. Sustainability **13**, 7037 (2021)
16. Marnewick, C., Silvius, A.J.G., Schipper, R.P.J.: Exploring patterns of sustainability stimuli of project managers. Sustainability **11**, 5016 (2019)
17. Martens, M., Carvalho, M.: Key factors of sustainability in project management context: a survey exploring the project managers' perspective. Int. J. Project Manage. **35**(6), 1084–1102 (2016)
18. Nussbaum, M.: Women's capabilities and social justice. J. Hum. Dev. **1**(2), 219–245 (2000)
19. Nussbaum, M.: Creating Capabilities. Harvard University Press (2011)
20. P2M. A Guidebook of Program & Project Management for Enterprise Innovation. Project Management Association of Japan (2017)
21. Sabini, L., Alderman, N.: The paradoxical profession: project management and the contradictory nature of sustainable project objectives. Project Manage. J. **52** (2021)
22. Sabini, L., Muzio, D., Alderman, N.: 25 years of 'sustainable projects'. What we know and what the literature says. Int. J. Project Manage. **37**(6), 820–838 (2019)
23. Sen, A.: Equality of what? In: McMurrin, S.M. (ed.) Tanner Lectures on Human Values, vol. 1, pp. 197–220. Cambridge University Press, Cambridge (1980)
24. Sen, A.: The Idea of Justice. Harvard University Press (2009)
25. Shaukat, M., Latif, F., Sajjad, A., Eweje, G.: Revisiting the relationship between sustainable project management and project success: the moderating role of stakeholder engagement and team building. Sustain. Dev. (2021)
26. Silvius, A.J.G., Schipper, R.P.J.: Sustainability in project management competencies: analyzing the competence gap of project managers. J. Hum. Resour. Sustain. Stud. **2**, 40–58 (2014)
27. Silvius, A., Schipper, R.: Exploring variety in factors that stimulate project managers to address sustainability issues. Int. J. Project Manage. **38**(6), 353–367 (2020)
28. Toljaga-Nikolic, D., Todorovic, M., Dobrota, M., Obradovic, T., Obradović, V.: Project management and sustainability: playing trick or treat with the planet. Sustainability **12**, 8619 (2020)
29. Wang, W.: The concept of sustainable construction project management in international practice. Environ. Dev. Sustain. **23**(11), 16358–16380 (2021). https://doi.org/10.1007/s10668-021-01333-z
30. Xue, J., Rasool, Z., Gillani, A., Khan, A.: The impact of project manager soft competences on project sustainability. Sustainability **12**, 6537 (2020)
31. Yuan, H.: Achieving sustainability in railway projects: major stakeholder concerns. Proj. Manag. J. **48**, 115–132 (2017)
32. Zainul Abidin, N., Pasquire, C.: Revolutionize value management: a mode towards sustainability. Int. J. Project Manage. **25**(3), 275–282 (2007)

Drivers in the Management of Large Projects and Megaprojects

The Challenges of 2021: Uncertainties, Digitalization and Decarbonization

Ekaterina Pujanova(✉), Svetlana Dubovenko, Natalia Olderogge, Nataliia Overchenko, Valeria Ivochkina, and Sergey Sadovnikov

PMSOFT JSC, Saint-Petersburg, Russia
epujanova@pmsoft.ru

Abstract. Multidirectional regional trends will determine the development of primary oil refining for the period up to 2025. Exceeding the cost of the project more than 20% destroys the business case. In addition up to 65% of large industrial projects are completed in the area of 30–40% of the capacity on time. In the coming years, the industry should implement 14 megaprojects with a total investment of about 5 trillion rubles. Such projects are particularly fragile taking into consideration various challenges and risks and a large number of influencing factors.

Since one megaproject can destroy a company Investors and Customers are looking for new ways to manage and control the scheduling and cost, there is a lack of professional competencies in the preparation and planning of such projects.

This paper focuses on why most megaprojects continue to fail to meet stakeholders' expectations despite the experience around the world accumulated by various international associations? What are the unique success factors for those who do manage to avoid significant schedule and cost deviations?

The result evidence from Russian companies showed that there are several significant factors: context factors that require a comprehensive assessment, internal factors associated with high complexity and aggressiveness of schedules and high cost related to sophisticated models for investment and financing. In response to this large number of projects are ready to be implemented according the ecosystem principle. The implementation of projects in the format of an ecosystem and partnership provides an opportunity to manage uncertainty, apply stakeholder management tools, increase transparency and speed of reaction to changes. In addition, the presence of digital solutions is increasing for those companies that demonstrate a high level of maturity of the megaproject management system in order to improve manageability, control over the implementation and decision making process. Without any doubt nowadays one of the most important conditions is adherence to the requirements of ESG and GPM principles. Moreover application of the stage-gate methodology in the implementation of projects, Value Improving Practices, project appraisal process and choice of the optimal ecosystem model for project implementation allows you to reduce risks and increase the project's resilience to external factors.

A model of a project management system that meets these challenges and factors affecting the implementation of projects in the format of an ecosystem and

S. Bushuyev et al. (Eds.): IPMA 2021, LNNS 704, pp. 256–263, 2023.
https://doi.org/10.1007/978-3-031-34629-3_21

partnerships with complex financing conditions makes it possible to control and predict project implementation, reduce risks and uncertainties, take advantage of digital solutions and control decision making process.

Keywords: Project management · Digitalization · Decision-making Process · Megaprojects · VIPs · Schedule and Cost

1 Introduction

In a competitive and chaotic global economy, companies turn to project management to achieve sustainable performance. Project management is no longer a need-driven process. More and more companies are realizing the effectiveness of investing money, time and resources in increasing project management competence. Through the development of project management in the company, it is possible to achieve cost savings, increase efficiency, improve interaction with customers and stakeholders of the company, as well as gain more competitive advantages. And the economic downturn only contributes to an increase in these indicators. Strategic alignment of project, program and portfolio management spans the entire organization, controls project implementation at every level and seeks to achieve benefits at every stage of project implementation.

An Economist Intelligence study shows that 80% of business leaders around the world see project management as a key competence to help them stay competitive during an economic recession. And even when an economic downturn gives way to an upswing there is no doubt that advanced project management applied to all company operations will contribute to better results and long-term business benefits. A McKinsey & Co study found that approximately 60% of top executives call the strengthening of project discipline one of the most important priorities for the development of the company.

Good project management starts at the portfolio level where strategic vision leads to investment and are defined metrics to measure performance.

2 Theoretical Aspects

One of the main reasons why the need for project management is growing from year to year is the shortening of the life cycle of modern products. If we take goods related to the high-tech industry their life lasts from one to three years. Several years ago, the life cycle of such products reached 15 years. A six-month project delay has consistently resulted in a 33% loss in profits. Speed has become the backbone of success, these is why large corporations rely on multidisciplinary project teams. This allows us to speed up the process of developing and launching new types of products on the market. Moreover the modern world is associated with constant technical progress. Advances in science and technology are reflected in projects making them more complex. If 30 years ago the construction of the road was quite a simple process now based on the used innovative technologies, equipment, documentation and standards it is impossible to say so. Therefore, due to the complexity of new projects their management is an important component of their effective implementation.

Climate change and the risk of global warming have also left their mark on project management activities. Now company cannot operate for profit if it damages the environment. Work on the development of methods to reduce the negative impact on the environment should be carried out throughout all stages of the project. Furthermore, modern companies are characterized by a reduction of the staff. This process is critical to the survival of many firms. Project teams are now replacing some of the employees taking over their responsibilities. Downsizing is causing firms to rethink their attitudes toward projects. Besides, constantly, growing competition has led to the dependence on customer satisfaction.

This is a catalyst for the rapprochement between the customer and the contractor. Nowadays it is necessary to take into account all the interests of the stakeholders. In this case, project management is extremely important. It helps to produce products that meet the needs of the customer and are aimed at strengthening relationships. In addition, the speed of change, which characterizes the modern world, affects on the activities of any company, it gives rise to the need for the simultaneous management of several projects. For the head of the project department the most difficult task is the distribution of the available resources. Many companies miscalculate the amount of required resources. As a result, control over such projects is lost and this results in significant financial losses. Obviously, project management is the ideal solution for activities that require speed, flexibility, improvement, accountability and professionalism. However, the management of megaprojects is characterized by particular complexity and high risks, let us consider these factors.

Up to 65% of large industrial projects are completed with an operating capacity of 30–40% after completion of the project. On average 80% of them will reach the declared power indicators within 2–3 years. In addition, exceeding the cost by 20% or more destroys the business case of the project irrevocably. Therefore according data from the Strategic online session (was held 16.12.2020 by PMSOFT JSC) there are special challenges for the large projects and megaprojects:

- megaprojects are counted not by price but by value;
- one project can destroy a company, Investors and Customers are looking for new ways to manage and control the scheduling and cost;
- an acute shortage of professional competencies in the preparation and planning of such projects;
- weak culture of managing contracts, interfaces, contractors, resources and performance;
- weak collaboration competencies (business, government, banks, etc.).

As well as participant survey data are presented in Fig. 1.

Hence, we can highlight the main characteristics of megaprojects making them fragile and very risky. First of all it is a complex context. Since megaprojects are very large-scale it is necessary to take into account a huge number of influencing factors. In addition there is a high internal complexity. To plan and implement such projects the enterprise must have a certain maturity of the team and processes. Furthermore, it is necessary to take into consideration multitude of participants. And here we are talking not only about the project participants, but also about considering all the interests of stakeholders.

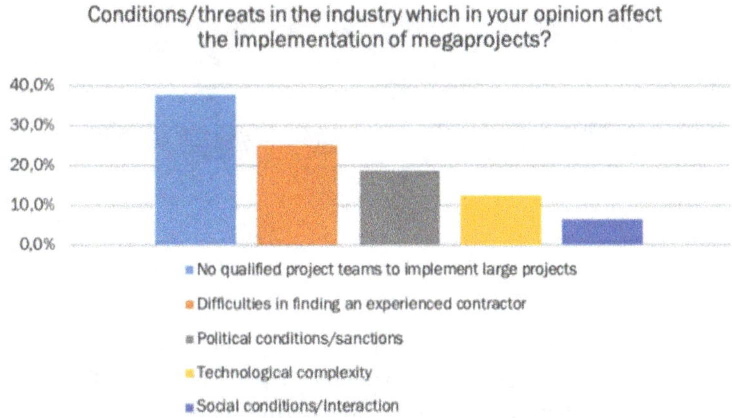

Fig. 1. Conditions/threats in the industry which affect the implementation of megaprojects (Strategic online session (was held 16.12.2020 by PMSOFT JSC)

We must not forget about the aggressiveness of the schedules, the high cost, which entails a complex financing and investment models that requires detailed pre-preparation. All of these lead to a high level of uncertainty in decision-making. The basis of the uncertainty is the incompleteness or inaccuracy of information about the conditions for implementing the decision, as well as the lack of information about the possible results of such decisions and the associated costs, which certainly affects the path of the project. Thus below we be considered a methodology to mitigate the effect of these complicating factors.

3 Project Management Value Improvement (PMVI) Methodology

Nowadays it is difficult to imagine the successful implementation of any project without focusing on the value that is created in the process of project implementation. At the same time, in order to achieve the highest project efficiency and preserve the business case, it is necessary to focus on creating value for the project at each phase of its life cycle. The complex Total Cost Management (TCM) Framework - An Integrated Approach to Project, Program, and Portfolio Management by AACE [1] - allows you to create a system to control and manage for both the value of strategic assets and the implementation of large projects. In this case value is embedded in the fundamental principle of defining a project: if there is no value, there is no project.

From the point of view of project management the value is the combination of the best characteristics of the future project result. Thus speaking about the implementation of a megaproject we can define three stages of the life cycle of a value (see Fig. 2). Moreover, since an unaccounted factor of megaprojects can bring a destructive effect, it is necessary to use any opportunity to improve performance, assess and prevent additional risks, optimize, generate new opportunities and reduce costs and time.

Examples of such values are indicators such as NPV, IRR, ISG, CAPEX, TCO. Additionally how an indicator can be used decrease in time, cost, impact of risks and

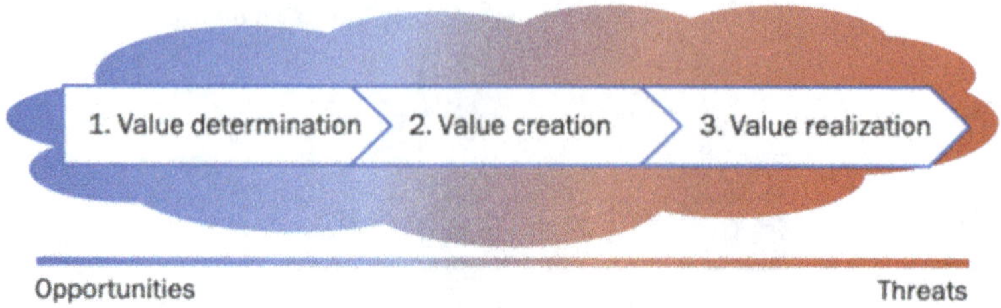

Fig. 2. Three stages of the life cycle of a value

using technological innovation. In order to improve these indicators the development and implementation of large projects makes increasing use of a range of formal review and analytical processes that have been used to improve the chances of achieving successful outcomes. These are generally called Value Improving Practices or VIPs. They are most effective when implemented during the early stages of a project. They are listed below:

1. Target Costing and Cost Deployment: Setting cost targets for costing elements (as a requirement or restrictions) for use in the design, planning and budgeting of the project. Specialized software CostOs is used.
2. Value Analysis and Engineering: An analysis of the function of an asset or asset construction services that establish the value of the function and ensure that the required level of performance is achieved at the lowest total cost. A cost model is used. Specialized software CostOS is used.
3. Operational readiness plan/program: This is the planning of construction readiness and the development of recommendations for commissioning and planning of operational readiness, requirements for contractors and for project management, management of interfaces for interaction with EP and S and other contractors.
4. Constructability review/ analysis: Review of the construction management project or analysis of the manufacturability of construction based on the AWP and Critical Chains methodology with an exit to an optimized project schedule, modularization / construction on site, the use of quarries, optimization of the construction management project. Advanced practice with resource planning (based on performance metrics and work packages). Specialized softwares Oracle Primavera, Deltek Acumen, Safran Risk are used.
5. Project cost verification: This is the verification of the cost model, the applied methods and benchmarks, the basis of the cost estimate for classes 4. It includes assessment of the error and quality check of the completed CAPEX project assessment, preparation of recommendations for reserves and improvement the quality of the assessment. The specialized software CostOS is used.
6. Lookback analysis and lessons learned: An analysis of the decisions made on the project and identification of project trends that affect the effectiveness of the project. This is the identification and analysis of lessons. It also includes training for employee, team and company.
7. Change control plan: This is the selection and analysis of changes in the main physical volumes by design stages. This is development of recommendations for

the cost model of the project and access to value-oriented engineering, for setting up a change management process. Specialized software CostOS is used.

8. Contract Strategy Overview: This is analysis of options for contractual strategies for a project, review of standard contracts (main provisions and their applicability for a project/specific contract for a project) and an assessment of the balance between risk and responsibility sharing. It contains preparation of recommendations on the boundaries of project responsibility and impact on the content of the project (scope of work).
9. Schedule quality verification: This is an assessment and verification of the quality of contractors' schedules in accordance with the key stages of the project, analysis of the feasibility of the project, analysis of compliance with target milestones, adherence to work technology, resource productivity, forecasting quality, change in the scope of work, change in terms, etc. Specialized softwares Deltek Acumen, Powerproject, Oracle Primavera, Project Tracker, MSP are used.
10. 4D modeling: This is an analysis of a model and the formation of construction scenarios in accordance with alternative project schedules. Specialized software Powerproject BIM is used.
11. Modularization principals and logistics: This is an analysis of the chain of fabrication and installation of equipment: installation on site or fabrication of modules and blocks in order to reduce the labor intensity of installation work on the site and provide opportunities for quality improvement. Plus it includes analysis of logistics schemes for the transportation of modules and analysis of the application of robotization and mechanization technologies for the installation of modules on the site. Usually the process consists of generation and evaluation of complex options.
12. Ecosystem and partnership: Creating a partnership it is efficient to use an ecosystem solution when there are many stakeholders and there is a complex supply chain. It allows you to analyze the entire chain of interaction and to choose methods of interaction for a complex management model for large projects and megaprojects. It helps to reduce risks and increase the project's resilience to external factors. Moreover, it provides the ongoing continuity and expertise, which is often missing in the individual member projects.

However, getting additional information, using value improving practices are always requires financial and time resources. Therefore, below will be described a two-part method (see Fig. 3) to help determine the need to use one or another method with the greatest benefits and the lowest cost.

After an assessment the task of determining priorities show up on the basis of which resources will be allocated between chosen objects. These processes help manage uncertainty, apply stakeholder management tools, increase transparency and speed of reaction to changes. Accordingly, two groups of key factors can be identified:

- Value Improving Drivers. Digital solutions: provides an opportunity to improve manageability, control over the implementation and decision making process. The data provided makes it possible to conduct comprehensive analyses and research. ESG and GPM principles: helps companies become more resilient, governments more efficient, and societies more robust, all while safeguarding our natural resources. Quality of the consideration of project documentation: develops sufficient project

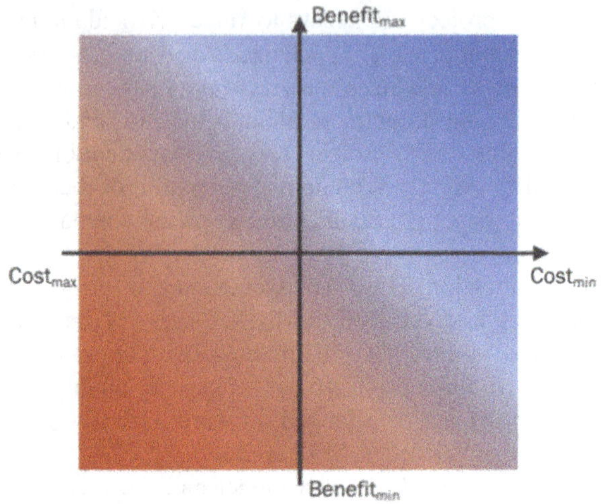

Fig. 3. The two-part method to help determine the need to use one or another method of increasing value in the project

definition. The detailed and tailored data enables the proper organization of process of the implementation of megaprojects. Stage-gate process: it is a project management method in which the project is divided into separate stages or phases, separated by decision points, thus increasing their quality;

- Destructive factors. All in all having analyzed a sufficient number of large projects and megaprojects PMSOFT JSC has developed a multifactor model for benchmarking large projects and megaprojects (PM.benchmarking), which is based among other things on the above factors and best practices for improving the value of projects and project management, as well as on the system of Key Success Factors (KSFs) of project and project management, which measure actual achievement against established Project Management Success Criteria, including criteria such as Effectiveness, Efficiency, and Sustainability. The PM.benchmarking methodology allows you to compare indicators of a specific project with best practices and similar projects and make recommendations for improving the value and success of project management in companies.

4 Conclusions, Limitations and Future Research Lines

It is critically important to build a model of a project management system that meets all challenges and factors affecting the implementation of large projects. Considering the situation nowadays advantageous one is in the format of an ecosystem and partnership since it allows you to use benefits of complex approach. Additionally it is necessary to integrate modern digital solutions in order to improve manageability and decision making process. Also studies have shown the importance of quality of the consideration of project documentation in the early stages of the project and integration of stage-gate process. Moreover it is important not to forget about the ESG and GPM principles since in the modern world no company can conduct its activities without paying attention to these aspects. And of course development of complex financing and investment models that's makes it possible to implement projects.

To sum up, in large projects and megaprojects it is crucial to use Value Improving Practices and PM. Benchmarking methodology in order to measure project systems against the industry average and world's leading industrial owners. Using these comparisons, companies are able to identify areas of weakness that should be reinforced. In addition, these provides an opportunity to learn about the Best Practices that drive superior performance in order to control and predict project implementation, reduce risks and uncertainties and build plans for implementing the learnings with the goal of improving capital effectiveness through better execution of these capital projects.

There are several limitations associated with this study; these limitations, in turn, create suggestions for further research. First limitation is the general nature of the research. In this study were analyzed data of industrial large projects. Due to different particularities of market conditions in different countries (developing and emerging), and due to different particularities of different industries, a more specific approach would be beneficial for other industries. Future research could conduct the similar analysis across various markets and industries and then compare the results; this would allow to make industry and market specific conclusions and implications.

Another limitation of this study is a rather limited amount of performance, risk and control variables. Future research could test more variables not included in this study. Including more variables would enable to track impact on different parts of corporate performance and risk and give more comprehensive results, that in turn, would allow to improve and supplement the results of the empirical analysis.

In a broad sense, as a process, enterprises can conduct Project and Project Management Value Improvement using different strategies and activities for it. In this regard, a very interesting topic for future research would be research how companies could quantify the sufficient conditions for the future Value Improvement process and compare different approaches. These quantified assessments could be in a form of more complex models and would be tremendously useful in helping predict results. The results would be beneficial for managers responsible for implementation of the megaprojects and for decision-making process, as it would allow them to better understanding potential costs and benefits associated with. Finally, if the company decides to implement large project it brings itself a tremendous amount of uncertainty and any possibilities to reduce it would be extremely helpful.

Reference

1. Stephenson, H.L. (ed.): Total Cost Management Framework: An Integrated Approach to Portfolio, Program, and Project Management, 2nd edn. AACE International, USA (2015)

Micro and Nano Projects Based on B.I.Gs

Business Innovation Games

Yuri Kogan(✉)

12757 Seabreeze Farm Dr. #9, San Diego, CA 92130, USA
yurikogan1940@gmail.com

Abstract. In the project movement we think about and work with megaprojects, large, medium and small projects. Besides that, we can work with micro (weeks duration) or/and nano (days duration) projects as well [1]. B.I.G. unites three human activities, labor, study and game, in a productive and attractive activity. B.I.G. responds to starting points of projects, such as: problem solving, needs satisfaction and taking advantage of opportunities. Another problem that occurs is getting the products of the projects, but not always the desired results. Often the task, the process or the post B.I.G. project is not part of the project scope. Initiation of any B.I.G. requires the definition of T.i.P.: Topic, instrument, Product. Instrument always is the same, for any micro or nano project. It's a B.I.G. Topics and Products always are different, specific for any B.I.G. B.I.G. is a hybrid work model which connects waterfall and scrum-agile approaches which makes it possible to achieve flexibility within a rigid framework. B.I.G. is really an innovative process that encourages participants to act intensively and creatively to develop a new product. The article includes introduction, B.I.G. description (questions, answers and examples), B.I.G. management process, post B.I.G activity for achieving results, conclusion and references.

Keywords: Innovation · Micro & nano projects · Game · Topic · Instrument · Product

1 Introduction

Today's business trend is to do everything fast, (ASAP!) B.I.G. responds to the sense of urgency. It is a short nano (days) or micro (weeks) project. B.I.G. is a chain of decision making in small groups and in whole group of the organization's officials, during the execution of the B.I.G. Project, for developing deliverables in each of the steps (phases, stages, sprints [2]) of the B.I.G. Project. B.I.G. Project is a unique process, which has a unique product. B.I.G. is not a traditional consulting. B.I.G. is an involvement of participants, it is self-consulting, it is generating ideas and realizing decision-making by participants (officials of the organizations) in the process. B.I.G. participants become the protagonists of the project, authors, implementers and promoters of the B.I.G.

Project products. B.I.G. is a real innovation, an implemented idea that brings value to the organization, to the society, to the country, to the world. B.I.G. Project requires a

S. Bushuyev et al. (Eds.): IPMA 2021, LNNS 704, pp. 264–274, 2023.
https://doi.org/10.1007/978-3-031-34629-3_22

low inversion, it's a low-cost project. The main B.I.G. Project stakeholders' inversion is in time and effort. B.I.G. Project is a collective wisdom, a co-creation of products and solutions. B.I.G. Project is a real teamwork. B.I.G. Project is an achievement of leadership and group competences [3]. B.I.G. is a serious game for adults. It is an attractive process, is enjoyable. B.I.G. is natural, not artificial, intelligence, which minimize uncertainty. B.I.G. is an instrument of great value: intensive, powerful, valuable, stimulating, effective, attractive, practical.

One of the most important components of the B.I.G. is T.i.P, the main information to initiate B.I.G. T.i.P. includes Topic (the substance), development of such a product: an approach, a system, an offer, a model, a plan, a program, a portfolio, a problem solution… Instrument is a B.I.G. Product, is a final deliverable, it is a consequence of the creation of the project intermediate deliverables. Topic and Product are specific for each B.I.G. B.I.G. as an instrument is the same.

2 B.I.G. Description

Famous English writer Rudyard Kipling said: "I keep six honest servants who taught me what I know: what? when? why? how? where? and who?" [4].

We keep nine honest servants for the B.I.G. description (see Fig. 1).

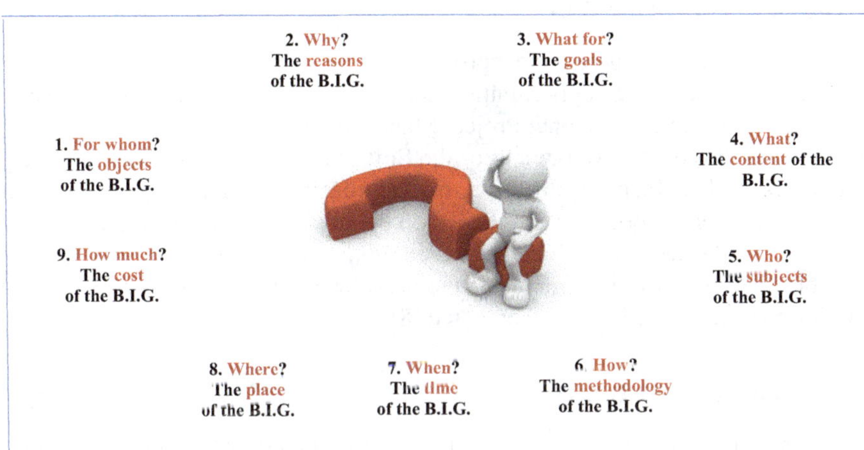

Fig. 1. B.I.G. description. Questions and Components.

2.1 The Objects of the B.I.G.

Micro, small, medium and large companies, public organizations, non-governmental organizations, etc. and their customers.

Case B.I.G. Project "Development of the Innovation System for the Organization". B.I.G. objects are organizational officials who participate in innovation processes and, indirectly, their clients.

Case B.I.G. Project "Latin American Championship over Project Management". B.I.G. objects are youth-representatives (Young Crew) of the Countries of Latin America.

Case B.I.G. Project "New Products Launch". B.I.G. objects are company employees and potential customers who are involved in the production and introduction of new products.

2.2 The Reasons of the B.I.G.

In project movement we think about and work with megaprojects, large, medium and small projects... Besides that, we can work with micro (weeks duration) or/and nano (days duration) projects as well. We need to solve problems and to create intellectual products in a very short time. The real reason we need "to build and seriously play with prototypes is to get a better understanding of ourselves and our priorities" [5].

Case B.I.G. Project "Development of a Model of Implementation of the Agile-Scrum-Kanban Approach in the Organization". Reason – projects not done in time, on budget, without meeting requirements; with interpersonal conflicts within the project team and with other stakeholders.

Case B.I.G. Project "Corporate University". Reason – need to carry out internal teaching-learning processes in the organization [6].

2.3 The Objectives of the B.I.G.

To achieve desire results where project products are the means to achieve them. To solve problems, take advantage of opportunities, satisfy needs, and at the same time acquire some competencies of Professional Project Managers [3].

Case B.I.G. Project "Development of the Soft Skills System for Project Managers and Members of Their Teams". Objective: to act as true professionals increasing different types of benefits for the organization. Based on the implementation of the B.I.G.

Case B.I.G. Project "Development of an Innovation System for the Organization". Objective: Acquire the benefits of the organization through the operation of the Innovation System created and implemented [8].

2.4 The Content of the B.I.G.

B.I.G. is an instrument for producing and managing intentional changes: in being, in relationships, in organization, in context and in changes [7]. B.I.G. is an information and knowledge transmission and exchange, and, at the same time, a productive process (product development), in serious game form, it is an innovative process that encourages participants to act intensively to create a new product.

Case B.I.G. Project "Development of an Audit System for Projects". Content: main concepts, project audit and audit of projects, processes, problems, product(s), results, standardization, audit as a project. Technical process audit, deliverables audit, final product audit, results audit, management process audit, initiation process audit, planning process audit, execution and control process audit, closing process audit.

Case B.I.G. Project "Development of the Soft Skills System for Project Managers and Members of Their Teams". Content: soft skills model (components), soft skills dictionary, soft skills selection, soft skills management, soft skills measurements, performance by competencies and performance evaluation, soft skills development, talent map and path, soft skills implementation plan.

Case B.I.G. Project "Development of an Innovation System for the Organization". Content: introduction into innovation, strategy, processes, metrics, structure, talent, culture, implementation.

Case B.I.G. Project "Development of the Customer Service System for the Ministry of Agriculture and Livestock". Content: team organization; warm up 1; typology of external and internal customers; warm up 2; customer service needs; warm up 3; list of services provided to external and internal customers; warm up 4; movie "Remember me", quality and quantity of the moments of truth discovered by the groups; warm up 5; service cycles (processes) for external and internal customers with their respective moments of truth; warm up 6; subjects and objects of evaluation for servers and clients; warm up 7; indicators and scales for measuring the quality of service to external and internal customers; warm up 8; surveys to measure the quality of customer service.

2.5 The Subjects of the B.I.G.

The subjects of the B.I.G. are (see Fig. 2):

- *B.I.G. Portfolio Manager* leads constructing, managing and controlling Global B.I.G. System (see Fig. 3).
- *B.I.G. Programme Manager* negotiates B.I.G. Projects with potential clients within the country (state) and is responsible of marketing, promotion, sales, country (state) legal issues, relations with the clients and communication with B.I.G. Project Managers.
- *B.I.G. Project Manager* manages B.I.G. with the Client Product Owner during initiation and preparation of the B.I.G. B.I.G. Project Manager guides B.I.G. during execution, control and closing of the B.I.G. (see Fig. 4).
- *Client B.I.G. Product Owner (Technical Expert)* is the main representative of the organization which accepts the B.I.G. B.I.G. Manager negotiate with Client B.I.G. Product Owner about B.I.G. acceptance, initiating and preparation.
- *B.I.G Jury* makes decisions about points won by groups-participants in each step of the B.I.G.
- *B.I.G. Project Team*, as protagonist in the development process and game (see Fig. 5): officials of the organization where prepare and implement a B.I.G., they make decisions creating deliverables and final product(s) of the B.I.G. and take part in the B.I.G. Closing.

One very special characteristic: the participants of the B.I.Gs, in reality, play two roles: objects of the B.I.G. and subjects of the B.I.G. Subjects develop products and / or solve problems for themselves, objects of the B.I.Gs, and for their clients and collaborators. B.I.G. is a kind of internal consulting, but still, it is a self-consulting.

Case B.I.G. Project "Project Portfolio Development". CITEC (Colegio de Ingenieros Tecnólogos de CFIA – Colegio Federado de Ingenieros y Arquitectos). CITEC President as a Product Owner, CITEC Associations representatives as members of the B.I.G. Team.

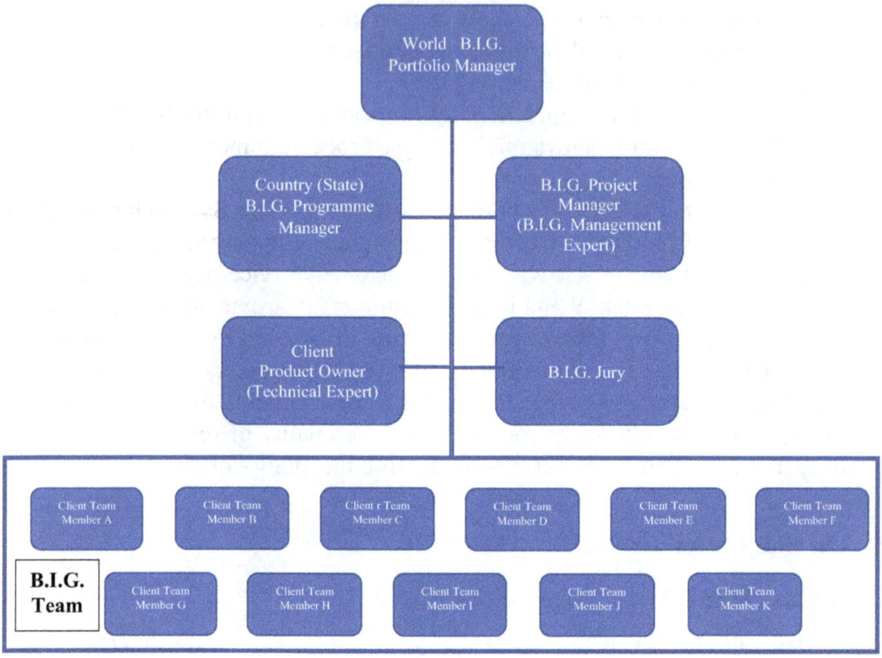

Fig. 2. B.I.G. subjects (principal stakeholders)

Case B.I.G. Project "Project Management Standard Development for ALIARSE" (see Fig. 4). ALIARSE General Manager as a B.I.G. Product Owner, representatives of the different ALIARSE departments as members of the B.I.G. Project Team.

Case B.I.G. Project "Development of a Quality System in Projects and Products". PMO (Project Management Office) Manager as a Product Owner and Project Manager as members of the B.I.G.

2.6 The Methodology of the B.I.G.

It is the action of Pa.Pr.I.C.A. (Participation, Practice. Intensity. Creativity. Action). ("Paprica" is a condiment in Spanish.) It is an action of co-creators of the projects, products and results. By the way, let mi to correct the PMI's definition of project. "A project is temporary endeavor undertaken to create a unique product, service, or result" [9]. Not OR, but AND… It is more still: PROmodel – the conduct of PROtagonists in technical and managerial PROcesses of the B.I.G. PROjects for development of the PROducts or / and solutions of the PROblems, which lead us to the PROgress.

2.7 The Time of the B.I.G.

From three days to three weeks for developing B.I.G. Product. Weeks for B.I.G. Micro Projects. Days for B.I.G. Nano Projects. Duration depends on the topic, on the preparation time, on the process organization, on the number of participants, on the preparation level

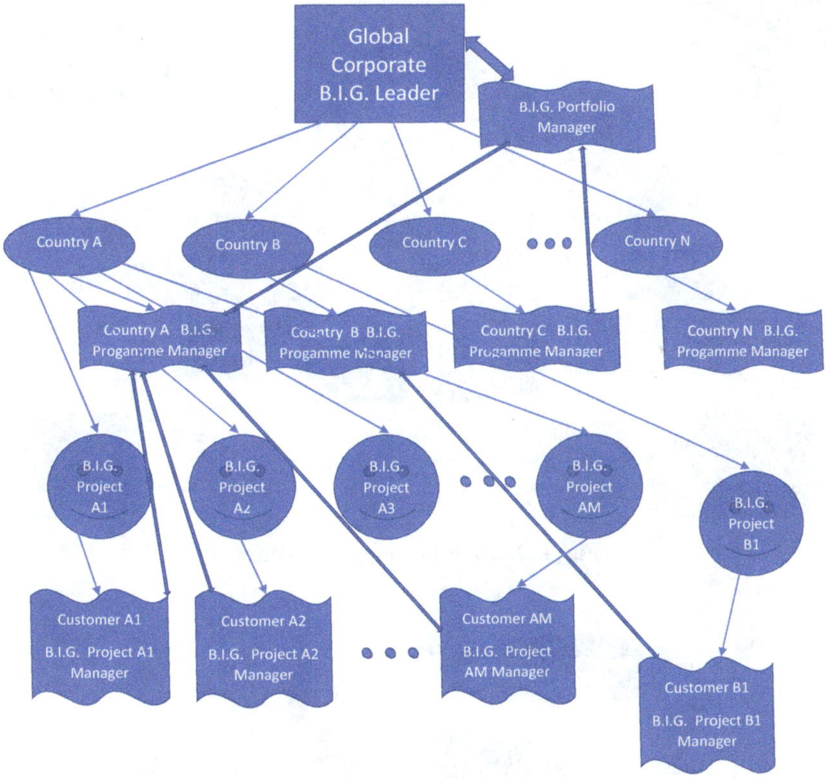

Fig. 3. Global B.I.G. PPP (Portfolio, Programme, Project) Structure

of the participants, on the time at their disposal and so on. Duration of online sessions – steps (phases, stages, sprints) is from 2 to 4 h. Duration of offline sessions is 4 or 8 h.

Case B.I.G. Project "Creation of the Project Management Standard (Protocol) for the Organization". Preparation time in case of the participants know Project Management methodology to be used – 2 days. Execution, control and closing time – 3 days. Preparation time in case of the participants don't know the Project Management methodology to be used – 3 days. Execution, control and closing time – 7 days.

Case B.I.G. Project "Development of a Project Portfolio for CITEC". Duration of the preparation 6 days. Duration of the execution 2 days.

Case B.I.G. Project "Development of a Document System Understandable for the Servers of the Affiliates of a Bank's Pension System". Duration of the preparation 9 days. Duration of the execution 2 days.

2.8 The Place of the B.I.G.

Today is online. Platform must be selected. In case of offline, preferably in out of the organization itself place for greater concentration of participants and with the ban on the use of cell phones.

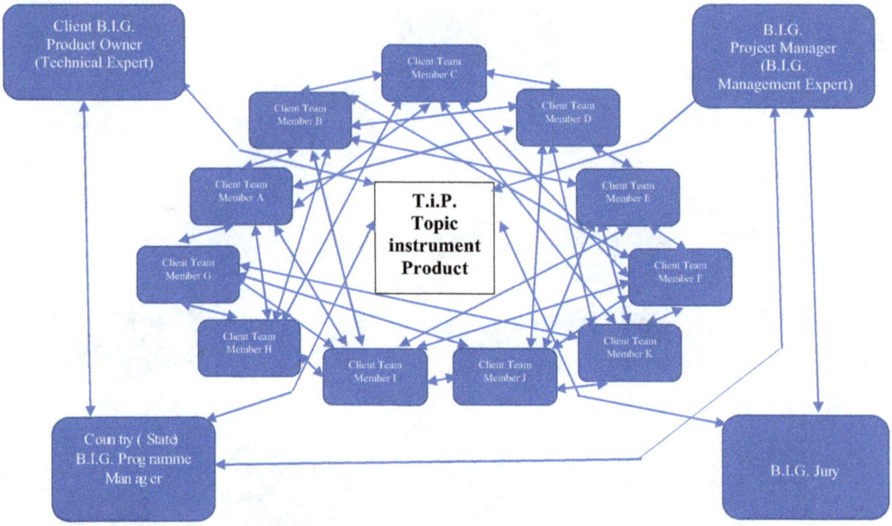

Fig. 4. Local B.I.G. Project Structure

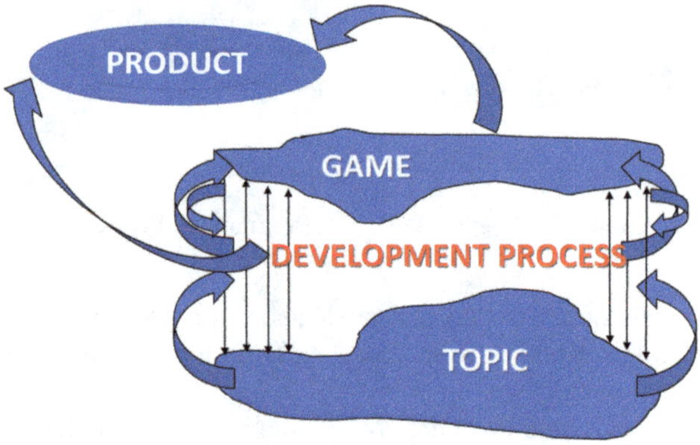

Fig. 5. B.I.G. model *grosso modo*

2.9 The Cost of the B.I.G.

It depends on the planned scope, on the resources used, on the dedicated time, like in any project. And on the cost of the product developed or/and the royalty of its use.

3 B.I.G. Management Process

B.I.G. Management Main Steps [9] are B.I.G. Initiation, B.I.G. Preparation (Planning and Organization), B.I.G. Execution and Control, B.I.G. Closing.

B.I.G. Initiation includes:

- Discussion of the B.I.G. idea for the organization;

- Business case definition: problem solution, satisfaction of a need, seizing an opportunity, fulfilment of a task, product development;
- Determination of an organization goal, which we would like to achieve (desired results);
- Negotiation and decision-making on developing implementation of the BIG.

B.I.G. Preparation (Planning and Organization) includes:

- Determine the gaming idea and objective – the B.I.G. objective as such building the steps (phases, stages, sprints) of the BIG;
- Establish the rules of the game;
- Develop the B.I.G. structure;
- Establish the roles of the participants;
- Design the game process;
- Determine information flow;
- Prepare B.I.G. documentation;
- Prepare instructions for participants;
- Develop participants handbook (optional);
- Develop B.I.G. guide for administrator (optional).

B.I.G. Execution and Control is a process of realizing a chain of creative and well-organized decision-making by B.I.G. Project Team and B.I.G. Project Manager, with acquisition of deliverables at the end of each step (phase, stage, sprint) of the process that leads to a final product of the B.I.G. The processes includes:

- Introduction to the topic or problem, B.I.G. technical characteristic, etc.
- Introduction to the B.I.G.: objective, tasks, decisions, game rules, etc.
- B.I.G. Teams development (natural and well organized): decision-making on team creation tasks. "Before starting engines, it is essential to choose travel companions" [10]. B.I.G. is a real practice of teamwork. "Our behavior towards others determines their behavior towards us" [11].
- Warm ups at the beginning of each of the steps (phases, stages, sprints) in the execution and control of the B.I.G. Verification and refreshing of knowledge about sub topics in each of the steps of the B.I.G. Decision-making by small groups… Objective: check, evaluate and refresh knowledge about sub topics of the B.I.G.
- Work in small groups. Brainstorm and decision-making to develop deliverable offers at the end of each of the steps.
- Plenaries. Final decision-making on group-offered deliverables based on special co-creation behavior, which is a nature for B.I.G. groups.
- Game: a friendly and serious competitiveness between small groups "above" the technical process of the B.I.G. in the decision-making process for the development of the deliverables and the final product of the B.I.G.
- Jury sessions. Decision-making on the points awarded for each group at the end of each step (phase, stage, sprint) of the B.I.G.
- B.I.G. Forms of organization of the Jury: permanent jury (a group of 3–4 internal or external of the organization experts) and changing jury when at each step each team delegates one representative to make jury decisions at this step of the B.I.G., changing their representatives in each step.

- Deliverables: sub products in each step of the B.I.G.
- B.I.G. product: a final deliverable as a consequence of the creation of the B.I.G. Project deliverables.

B.I.G. Closing includes:

- Decision-making about the winner(s) of the B.I.G.;
- Validation of the process and the final product;
- Estimation of the expected results and offers of the ideas of how to carry out project of the achievement of results;
- Making decisions about B.I.G. winner(s);
- Analyzing B.I.G. product;
- Discussing achievements and mistakes made;
- Evaluating the behavior of the players in the course of the B.I.G.;
- Assessing the decisions of the players during the B.I.G.;
- Making the didactic and business conclusions;
- Estimating the expected results and offers of ideas on how to carry out a project, or a process, or a task to achieve results.

4 Post B.I.G.: Results (Impacts and Benefits)

4.1 Possible Results

Improvement of customer service. Market expansion. Increased reputation of the organization. Better brand promotion. Increased profit of the company. Minimization of new products launches times. Effective ROI. Generating ideas and decision-making by participants in the process. Teamwork in reality and human development. Short-time nano and micro projects (from 3 days to 3 weeks), responding the modern tendency of ASAP. Participants are authors, implementers and promoters of B.I.G. project products. B.I.G. Project participants implement their products with a minimum of resistance.

4.2 Post B.I.G.: Implementation

"The real test of individuals and companies is not how much they know or what they know, it is what they do with their knowledge. Making a decision is not the same as implementing it" [5]. And now appears the need of the other project that many times can be neither micro non nano, and it still may not be a project… May be a project, or a process, or simply a task of product implementation. And, if you make this decision and if it's necessary, can be developed a **Bu.G.** (**Bu**siness **G**ame) to teach people how to work with the product of the B.I.G. Project (see Fig. 6).

4.3 Product Implementation Options: Task. Process. Traditional Project. Project as a Bu.G

Once this tool (Bu.G.) is designed, it serves many times for the training of employees. A main Bu.G. idea is to learn making mistakes, making mistakes and making mistakes again, but less, and less, and less. In the "lab", not in real life! Bu.G. is a repetitive

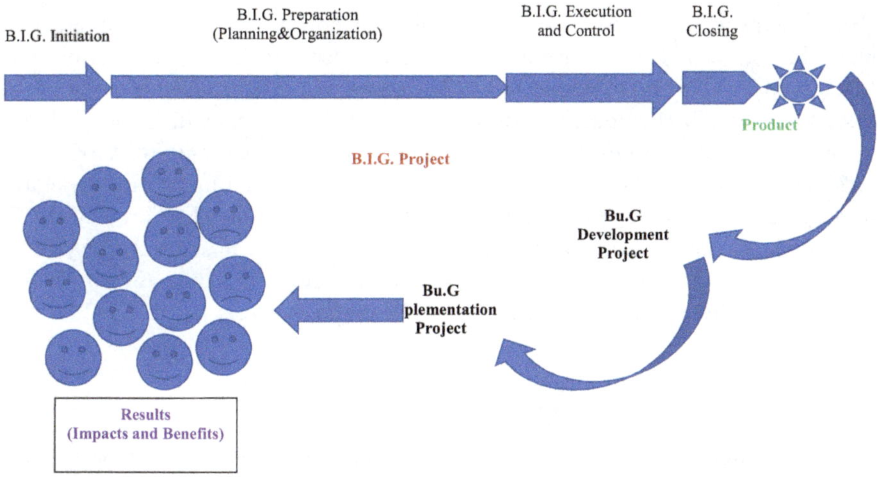

Fig. 6. B.I.G. and Bu.G.

process and for each topic has similar products: competences (see Fig. 7). Bu.G. – Business Game that uses existing or creating information and knowledge, in the form of an already developed model to experiment with this model by doing a simulation for training. "We form our models and then our models form us" [5].

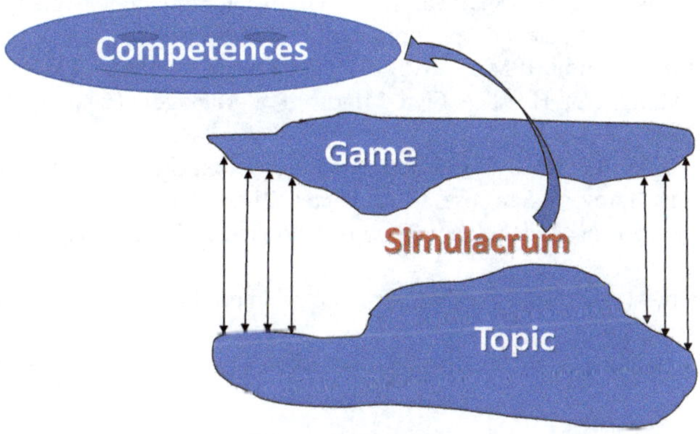

Fig. 7. Bu.G. components *grosso modo*

5 Conclusion

Fifteen projects have been realized using B.I.G. as an instrument for decision-making and developing of the intelectual products, actively involved officials of the organizations in decision-making in each step of the micro or nano project. Some conclusions can be done.

B.I.G. helps us to put into practice the main concept: from an idea to an intellectual product through a micro or nano game project. Successful B.I.G. application allows us to develop, in short time, with collective wisdom of the organization's officials as a self-consulting, an intellectual product creatively, effectively and with pleasure, through decision-making, problem solving, meeting needs and taking advantage of opportunities, for achieving desire results. B.I.G. is not a course, it is not a training, though the learning always exists and knowledge is achieved during the co-creation B.I.G. Project. B.I.G. is a production, is a development of such an intellectual product: an approach, a system, an offer, a model, a portfolio, a plan, a problem solution, etc., involving company officials as internal consultants into the process of the B.I.G. To be Great we need B.I.G.

References

1. Kogan, Y.: BIGs - Business innovation games – as micro and nano projects (how to innovate with BIGs) In: Tsipes, G., Klimenko, O., Abstracts of 32nd World Congress of International Project Management Association, pp. 170–172, SOVNET, Russia (2021)
2. Schwaber, K., Sutherland J.: The Scrum Guide. The Definitive Guide to Scrum. The Rules of the Game. Creative Commons (2017)
3. Individual Competence Baseline for Project, Programme & Portfolio Management. International Project Management Association (IPMA). 4th Version (2015)
4. Kipling R.: The Original 1902 Edition with Illustrations by Rudyard Kipling (2019)
5. Shrage, M.: Serious Play: How the World's Best Companies Simulate to Innovative. Harvard Business School Press, Boston, Massachusett (1999)
6. Tsipes, G., Echkalova, N., Sharova, E., Tovb, A.: Corporate university as a driver of project culture and competence development. In: 29th World Congress International Project Management Association (IPMA) 2015, Elsevier, The Netherlands (2016)
7. Gosling, J., Mintzberg, H.: The Five Minds of a Manager. Harvard Business Review. November (2003)
8. Morales, M., León, A.: Adiós a los mitos de la innovación. Una guía práctica para innovar en América Latina. Innovare. San José, Costa Rica (2013)
9. A Guide to the Project Management Body of Knowledge. 7th edition. Project Management Institute (2021)
10. Perpiñá-Robert, M., Armirall, G.: El nacimiento de un producto. Editorial Planeta. Barcelona, España (2009)
11. Maxwell, J.: Los 5 niveles de liderazgo. Center Street Hachette Book Group. New York, USA (2011)

Phronetic Project Management Leadership

Six Critical Messages to the Project Management Community at the 40th-Year Milestone with IPMA

Hiroshi Tanaka(✉)

Project Management Association of Japan, 1–5-2, Higashi-Azabu, Minato-ku, Tokyo 106–0044, Japan
hirojpmf@gmail.com

Abstract. This paper develops the author's opening keynote presentation titled "Phronetic Project Management Leadership – Five Messages to the Global Project Management Community" delivered at International Project Management Association 32nd World Congress held in St. Petersburg, Russian Federation, on September 21st, 2021. The six messages in this article, with one message having been added in this article, are proposed homework for the global project management community considered critical by the author after having traveled in the global project management arena over the past half century as a practitioner, operator of a national project management association, academic and global trainer of project and program management. The six messages are being proposed as an essay based on the author's qualitative analysis and continuous industry monitoring via the opportunities provided in the above-mentioned capacities and represent discrete homework topics analyzed on their own although the messages except the first topic are related with each other in one common issue of a narrower-than-potential focus of the project management discipline.

Keywords: Project Management Vision into 2020's · Narrow Focus of Project Management Discipline · Reverse Analysis of Project Management Potential · Phronetic Project Management in Fuzzy Days of VUCA or BANI

1 Conservation of Robust EPC Project Management Capability

I mentioned in the keynote paper presentation at PMI® Research Conference 2006 that the EPC (engineering-procurement-construction) industry engineers and builds national and social infrastructure, as well as industrial facilities, which generate notable economic outputs. In doing so, it contributes, in both industrialized countries and developing countries, a gross domestic product (GDP) ranging between seven and ten percent. This industry is also a major employer of a workforce that pioneers project management practices and transnational business operations. As a pioneer in project management practice, the global EPC industry offers robust project management characteristics compared with those of the information services (IS) industry, which employs the largest number of

S. Bushuyev et al. (Eds.): IPMA 2021, LNNS 704, pp. 275–286, 2023.
https://doi.org/10.1007/978-3-031-34629-3_23

project managers globally [Tanaka, 2006]. The EPC industry has grown steadily while the average physical size, investment costs and complexity of EPC projects are increasing, and has been going through rugged performance of projects which has triggered assorted new technology and project execution and management methods.

Looking at a recent brilliant record of an EPC project, the Yamal LNG Project in Russia, with US27 billion investment costs, successfully started LNG production from its first train in late 2017, the second train in August 2018, and full commercial operation in early 2019 [NOVATEK, 2019]. This success of the project has enabled Russia to substantially increase its LNG production from 9.6 million ton/year by Sakhalin LNG Plant to 26.1 million tons/year in January 2019 with Sakhalin and Yamal LNG Plants together, thereby strengthening Russia's LNG market retention power in the world (see Fig. 1).

Fig. 1. Overview of Yamal LNG Complex (courtesy of Novatek).

Yamal LNG project has realized such project management and trade innovations as conquest of harsh conditions of construction and logistics, e.g. the remote site on the Arctic Sea, winter temperature of -50 degree C that the site construction must face, full use of modules fabricated in Far East and Europe and transported to the site by specially built ships qualified for Arctic Sea navigation; LNG shipment to Far East and Europe via the Arctic Sea even during winter; and marking the first benchmark of China's "One belt, one zone" program realized (idem).

In this EPC industry, similarly, a series of huge natural gas projects such as US$25 billion PEARL GTL -- natural gas to liquid -- in Qatar, US$40 billion Ichthys LNG in Australia as well as complete refinery projects such as Nghi Son Refinery Project in Vietnam, all exceeding US$10 billion in investment costs according to economic journals, have been completed successfully [Tanaka, 2018]. We can infer from these cases that investment costs, sizes and complexity of projects do not affect linearly probability of success or otherwise of projects. Mediating variables should include an application area (industry branch) of the project in question.

International Energy Agency' World Energy Outlook [IEA, 2015] estimated that the global oil and gas industry, the most representative EPC project arena, was expected to invest US$ 25 trillion from 2015 to 2040 in development projects which stands at US$ 1.02 trillion average per year. This expected oil and gas development investment volume has been drastically reduced in 2020 and 2021 due to two reasons: the drastically reduced demand for fuel caused by the prolonged COVID-19 Pandemic and accelerated global policy on energy transition forcing a shift of gravity of oil and gas companies' investment from oil and natural gas development to that of renewable energy or its derivatives such as green hydrogen and bio fuels by which international oil and gas companies have frozen a considerable part of oil and gas development spending, as of May 2021. This current situation is drastically reducing the size of EPC market [Tanaka, 2021].

A strong warning is that loosing or weakening EPC project management capability will be a serious harm to industrial project management base as this is the most rigorous project management capability, and it would cause a huge loss to our society. Owner companies, EPC contractors and concerned governments should stage concerted efforts to conserve the EPC operations and management capability in facing the 'existential crisis' of capital program delivery [ECI, 2018]; such efforts include closer owner and contractor alignment initiatives for sound growth of the EPC industry and particularly an industry initiative to renovate the structures of capital program (FEED + EPC) delivery for reducing transaction costs in the EPC chain [CII, 2018] while continuing prudent investment to safeguard supply of fuel.

2 Care of Young Project-Connected Businesspersons Alienated from Project Management

We in the project management community should take appropriate care of young businesspersons engaged in new types of projects who are alienated from the professional project management community.

I argued why average project management performance in the world has not improved, or has even been deteriorated over the past half-century in my article [Tanaka, 2018] and raised an issue that the average project management skill is being diluted by a sharply increased entrance of young businesspersons in soft, creative projects, who are ignorant of project management or have not been prepared to practice it due to sudden placement in project-oriented work. This sharp increase of new entrants to projectified business reflects the fact that the 63.6% (the world), 70.9% (European Union), 80% (USA) or 68.7% (Japan), respectively, of GDP is contributed by tertiary industry economic output [World Factbook, 2021].

Figure 2 which indicates the increase of projectification of business versus three layers of project-related practitioners, has been drafted by the author as a symbolic trend to address this issue based on the author's qualitative analysis. The ratio of projectifiction of business reflects a series of IPMA research reports [IPMA, 2018] and the author's monitoring of economic growth of Asian countries by way of industry news. The density

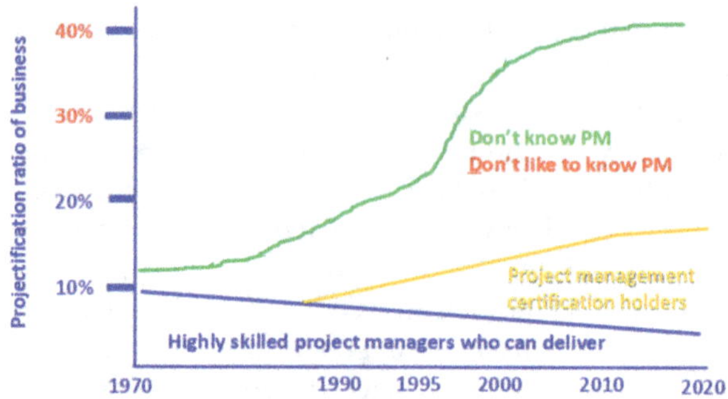

Fig. 2. Projectification of Business and Society vs. Skill Layers of Project Management Population.

of project management population represents the author's feeling of project industry population mix shift over decades.

A projectificaiton ratio of economy and society is increasing constantly, to assumedly around 40% of GDP, e.g. from 33 to 34% in Germany, Norway, Iceland and Italy in mid-2010's which are generally expected to increase to 40% plus in 2020's [Schopper, 2019] to the author assessed close-to-50% in Asian countries - Asia is home to 56% of the world's fastest growing companies [Nikkei Asia, 2019]. Meanwhile, there is no accurate demography of project managers in the world. PMI estimates that there are 16.5 million project managers in the world and of which 8.8 million to 11million are located in North America [Garret, 2021].

This number is too small if we consider China, Japan and the rest of Asia, which all in all are currently a growth engine of the world economy. I estimate the ratio of very experienced project managers who can constantly delivery a project with little influence by changing socio-economic environment is time-relatively decreasing; that of certified project managers is increasing with currently one million + PMP®s [PMI, 2021], one million + PRINCE 2 certified practitioners [PRINCE2, 2021], 250,000 IPMA 4-Level credential holders [IPMA, 2021] and 50,000 other association project management certificate holders (author estimate), totaling 2.3 million, or just 14% of the PMI estimated project manager population - project managers in the bottom layer may or may not be project management credential holders as commonly seen in heavy project industries such as oil and gas industry EPC and societal infrastructure construction.

A fundamental problem to the project management community is the assumedly thickest layer of the project management population at the top who are largely engaged in soft and creative projects, which need more conceptual thinking or design thinking than management. Being the founder and former president of Japan's professional project management association, I have noticed over the past quarter century that extremely few persons in this project category have joined the association.

During the key panel of 30th IPMA Word Congress in Astana, titled "Breakthrough Competences in the Development of the Main Sectors of the Economy", there were voices from the hall that the era of command and control is over and that we need facilitation type project management with the advent of young project team members as a new trend of project management. We should note that this was a hall dialogue among traditional project management professionals. My interviews with young businesspersons and graduate students over the past ten years since 2000 indicate that young players in soft (not heavy industrial or ICT) projects like more creativity and design thinking than project management as everything labeled as management connotes bureaucracy.

I consider that there are two reasons that project management community cannot embrace these fresh players. One lies in the economics of project management associations or any other project management promotion platforms; they should target mid-range project players, if not high end, to secure revenue with standardized project management methods and credentials, and are not financially resourceful to provide opportunities to entice those fresh players to join the platform as there are very few experts to convince them with potentials of project management applicable to their business operations. The second reason is that there are no project management standards' methodologies to address conceptualization and ideation to create an innovative project. The traditional project management communities take it for granted that their members have no question as to what to make in a project and how to create a project, and do not have an incentive to explore a new project management market by way of time-taking and painful investment of effort. This analysis of project management economy applies to the mentality of project management leaders and trainers.

Somebody should volunteer to start a dialogue with this potentially great population of tomorrow's project community. In addressing this issue, the author is carrying a series of easy-to-understand creative project design and project leadership seminars with graduate students not majoring in project management studies, and young businesspersons in Japan and Russia. As shown in Fig. 3 (in Russian), I have started this seminar series in European Division of Russian Federation via Russia Japan Center with support of Alexandr Tovb, President of SOVNET.

Fig. 3. Easy-to-Understand Creative Project Design and Leadership Seminars in Russia.

3 Cross-Fertilization of the Project Management Discipline with Other Disciplines to Survive in the New Normality

Three researchers from HEC Paris argued in 2016 that *"(Project Management) Researchers have long recognized that standard approaches to project management are ill-suited to address changes in the environment or business needs, particularly in innovative contexts characterized by uncertainty and complexity*" [Mahmoud-Jouini, et al. 2016].

In fact, except for casual collaboration cases with other classical management disciplines such as operations research (OR), industrial engineering (IE) and systems management (or theories), there have been no evident cross-fertilization vehicles in the part of the project management discipline. Project management has proven to be more effective in a stable economy in the past where a considerable number of the projects are for the expansion of production capacities for certain good-selling products or services, or to develop civil infrastructure of which definition is not difficult thanks to well-developed common global engineering standards. However, in the contemporary world, service economy is remarkably more than production economy as we have seen in previous Sect. 2 of this article, and the world is characterized by such acronyms as VUCA (volatility, uncertainty, complexity and ambiguity) [U.S. Army Heritage and Education Center, 2018] or BANI (brittle, anxious, non-linear, incomprehensible) [Grabmeier, 2020]. We see dwindling opportunity to deploy hard-core project management approaches on the assumption that what to make is clear to all the stakeholders and the question is how to define the scope, delivery timeline and budget of a project and how to efficiently deliver the plan.

As I argued in Sect. 2, to those fresh players in soft, creative projects, a critical question is what value added and innovative project to create; here the process of conceptualization is more, or at least as important as, planning, and the former is not found in global project management standards. Program management addresses benefit identification or value identification but in a generic, superficial manner.

For instance, there are vast opportunities for programs and projects which provide responses to climate change, corresponding to all sorts of sustainability in society, industry, and human life. United Nations SDGs - Sustainable Development Goals [United Nations, 2015] are gaining momentum influence in the society and industry throughout the world; not only governments and NGOs but corporations are bound by the market, financial institutions, and community concerns to honor the SGGs as a condition for corporations to maintain sustainable business.

As shown in Fig. 4, we need here a multi-disciplinary approach in terms of target application areas of sustainability; sense-making from policy and enterprise strategy and availability of finance, be government funding, green funds, or private company funds; enabling means such as eco and smart technology, digital transformation (DX), circular economy and sharing economy as if each of them was an application software while we propose to use Program Management as the Operating System. Among the enabling means, digital transformation is a powerful accelerator [Tanaka, 2022].

Figure 5 shows a constellation of traditionally peripheral and emergingly peripheral disciplines supplemental to the project management studies that the author has considered as efficient if cross-fertilization of the disciplines came true; this is my concept

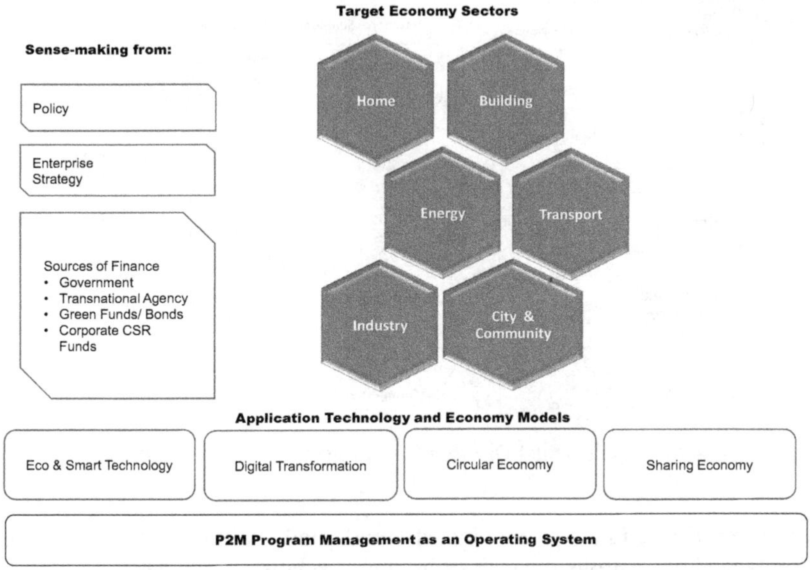

Fig. 4. Multi-disciplinary Approach to Sustainability Projects with Program Management as Core

acquired through teaching at graduate schools in Russia, Ukraine and Japan in which researchers and practitioners of different specialty came with unique ideas [Tanaka, 2022].

Currently, study of theories of knowledge (TOK) is a focus subject at universities and International Baccalaureate® diploma program. Theory of knowledge (TOK) plays a special role by providing an opportunity for students to reflect on the nature of knowledge, and on how we know what we claim to know [International Baccalaureate, 2021].

In summary, to expect healthy growth of the project management discipline, both researchers and expert practitioners should come out of silos, or comfort zones, and cross-fertilize them with other disciplines.

4 A Proposal for a Mezzanine of Project Management Research

Project management studies have established themselves as pure social science thanks to the development of Tier 1 journals such as International Journal of Project Management and Project Management Journal. Rigorous methodologies and robust and transparent data support the scientific quality of papers accepted for publication. Is it, however, a prejudice to mention that this rigorousness of research causes a time lag of content from the state-of the art of project management practice in the industry or eliminates the contributions from such branches of industry where data sharing is not encouraged such as the oil and gas industry?

Meanwhile practitioner papers are interesting to practitioners but most those papers are case studies and are questionable regarding general applicability of the case. The author proposes a mezzanine of project management research, which can accommodate conceptual papers or position papers contributed by authors having more than 20 years

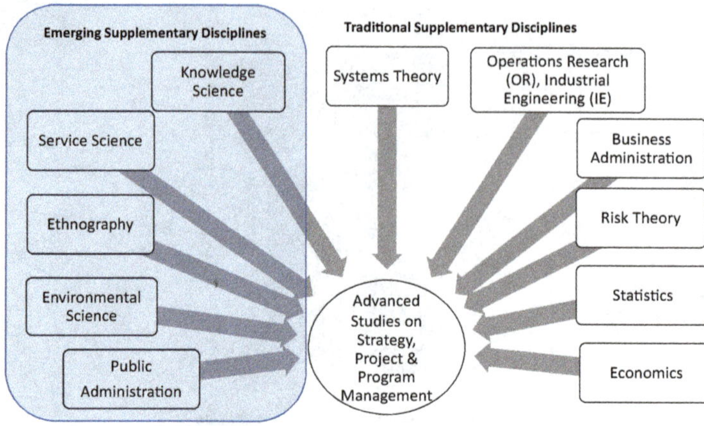

Fig. 5. Supplemental Disciplines to Project Management Studies.

of first-tier experience in the project industry and good qualitative analytical capability of project behaviors or project management trends keeping up to date with the industry. If any research is to improve real world performance of industry or human beings, we need papers, which give a sharp and hot insight into essence of management in project context.

5 Insufficient Effort Being Spent for Project Management, Especially Front-end Planning

In my article [Tanaka, 2018] argued that "while project management standards in the world, either knowledge standards or competency standards, have been developed by a group of experts representing broad application areas of project management and are hence expected to help uniformly extract good project management performance out of the practice founded on the standards, average project management performance in one application area to another differs considerably. It is necessary to analyze gross causes in the respective application areas concerned that result in such variations in performance. Here it should be worthwhile arguing that top-down analysis based on differences of industry systems and culture could render a more balanced view regarding differences in average performance of projects and project management by branch of industry than detailed performance analysis of actual individual projects, or that an apple to pair comparison does not make sense". A simple question representing my objective of the research is why a US$20 + billion base-load LNG plant project such as Yamal LNG project in Russia, being cited in Sect. 1 of this article, can be completed successfully with respect to plant performance, delivery time and meeting the sanctioned budget, and why only one third of software projects are success [The Standish Group, 2018]. Details of my argument are found in (Tanaka, 2018).

One additional observation as to why project management so frequently fails is that projects in most industry branches spend so less time in project management than is required for successful delivery of a project. Dr. Morten Fangel, one of the grandfathers of

IPMA, exactly mentioned the same comment in presenting his paper "Plan and Evaluate your PM Effort as well" at 32nd IPMA World Congress [Fangel, 2021]. Many practitioners of project operations mix time being spent for project execution work such as front-end technical planning, basic design engineering, detailed or production engineering, procurement of materials and construction or installation with that for project management ranging from project direction or project integration management, front-end project planning, systems configuration management (scope management), time-resources-cost management, risk management and organization and communication management. On a project exceeding a certain investment cost amount or required person-hours, separation of project management function from project execution function is lesson one of project success.

In the same article and others by the author, knowing from industry data that 80% of project lifecycle performance is determined during the initial 20% progress of an industrial project, namely, prior to EPC, or project implementation, stage [Davis, 2003], I have consistently voiced the importance of rigorously practicing front-end planning --- front-end engineering in industrial plant projects --- which echoes with consistent same views by Dr. Morten Fangel who advocates project start-up planning from 1970's, and Professor Knut Samset (2020) discussing paradoxes of front-end management of projects arguing that while public projects represent major complex investment and there has been much written about how to develop, manage and deliver such projects, practice still does not match up with expectations and that the researchers from the Concept Research Program brought together some of this mismatch in "Ten Paradoxes" paper.

6 Phronetic Project Management Leadership

Nonaka published his original model of a phronetic leader as a type of leader required to prudently guide societal and business management activities in the contemporary environment characterized by the acronym VUCA and amid the Japanese corporate suites where immoral management activities were often found [Nonaka, 2007], and expanded his model in a Harvard Business Review article in 2011[Nonaka, Takeuchi, 2011]. Phronesis is one of the three teachings by Aristoteles and means practical wisdom which wisely connect Epsiteme, scientific knowledge and Techne, technical knowhow. It is a type of wisdom or intelligence relevant to practical actions; it implies both good judgment and excellence of character and habits, and it was a common topic of discussion in ancient Greek philosophy [Engberg-Pedersen, 1983].

Nonaka and his colleagues argue phronesis in the contemporary context as a virtuous habit of making decisions and taking actions that serve the common good and as a capability to find a "right answer" in particular context, for instance trying to find a right answer for business decision making and prudent reasoning, and posits that phronesis can be acquired only through high quality direct experiences of a leader [Nonaka, Takeuchi, 2011]. They argue that wise leaders empowered by phronesis:

- can judge goodness,
- can grasp the essence,
- creates shared contexts; this comes from the theory of "Ba," meaning a shared platform for collaborative creation of knowledge and value according to Nonaka's definition,

- communicates the essence,
- exercises political power - this does not mean that wise leader should wear a political clout but he or she knows how to properly use authority to get though goodness, and
- fosters practical wisdom in others.

Also, a relational study of wise (phronetic) leadership, knowledge management capability, and innovation performance has been conducted by Weixu Ding et al. [Ding et al., 2019]. This research is based on Nonaka's explanation of wise (phronetic) leadership and uses empirical methods to verify the correlation of wise leadership and innovation performance. 144 valid samples have been applied for analyzing. The research results illustrate that wise leadership has positive effects on organizational innovation performance. Moreover, knowledge management (KM) capability has been evidenced in this research having part mediating effect between wise leadership and innovation performance.

The world is highly turbulent as of year 2021 as it is damaged by the COVID-19 Pandemic and certain business models and societal norms are affected seriously. I believe that where organizations and their management wander around criteria of decision-making, the phronetic leadership model bridging scientific knowledge and technological or more broadly technical knowhow in a critical way is an efficient solution. Japanese pioneering modern philosopher Nishida, K. explored the virtue of truth, goodness and beauty in the dawn of modern Japanese context [Odin, 1987]. In project management context, phronetic leadership would prompt organizations to create projects which meet the virtues of truth, goodness and beauty. For instance, United Nations' 17 Sustainable Development Goals (SDGs) provide good ground for creating projects for truth, goodness and beauty. Also, in summer 2021, was announced that Russia will build a large cluster of green hydrogen production in Russian Far East by involving Japanese and Chinese companies (PROSPERO, 2021). This is another eloquent example for a project of this category.

7 Conclusion

The project management discipline is not on the rise as of 2021 and the project management community's focus is narrower than to-be considering the pace of changes in the political – economic – social – technological – legal – ecological (PESTLE) scenarios and shift of majority players in project business.

The author lists six agendas of homework for the global project management community. Also, the author emphasizes phronetic project management leadership in these days of VUCA or BANI.

References

CII: "OPERATING SYSTEM 2.0" unlocking project value. Construction Industry Institute (2018). https://slideplayer.com/slide/14254402/. Accessed 28 July 2020

Davis, M.: Pre-project planning. In: Research presentation at CII Conference 2003, Construction Industry Institute (2003). https://www.slideserve.com/halima/pre-project-planning. Accessed 28 July 2020

Ding, W., Choi, E., Aoyama, A.: Relational study of wise (phronetic) leadership, knowledge management capability, and innovation performance. Asia Pac. Manage. Rev. **4**(24), 310–317 (2019)

Engberg-Pedersen, T.: Aristotle's Theory of Moral Insight. Oxford University Press, UK (1983)

ECI: Existential Crisis – Rethinking how capital programmes are delivered. European Construction Institute (2018). http://www.eci-online.org/news/rethinking-how-capital-programmes-are-delivered-event-report/. Accessed 28 July 2020

Fangel, M.: Plan and evaluate your PM effort as well. In. Tsipes G., Klimenko, O., Abstracts of 32nd World Congress of International Project Management Association, pp. 129–130. SOVNET, Russia (2021)

Garret, D: How many Project Managers are in the US and Canada? Project Management 2.0. Blog by Dave Garrett (2021). https://www.projectmanagement.com/blog-post/4226/How-many-Project-Managers-are-in-the-US-and-Canada. Accessed 28 July 2020

Grabmeier, S: BANI versus VUCA: a new acronym to describe the world (2020). https://stephangrabmeier.de/bani-versus-vuca/. Accessed 28 July 2020

International Baccalaureate: Theory of knowledge (2020). https://www.ibo.org/programmes/diploma-programme/curriculum/theory-of-knowledge/what-is-tok/. Accessed 10 Oct 2021

IEA: Energy and Climate Change. World energy outlook Special Report. International Energy Agency, France (2015)

IPMA: Project Management and its Impact on Societies. 6th IPMA Research Conference. IPMA. The Netherlands (2018)

IPMA: IPMA® Certification Yearbook 2020 (2021). https://www.vzpm.ch/fileadmin/dokumente/downloads/Deutsch/IPMA_Yearbook_2020.pdf. Accessed 31 Jan 2023

Mahmoud-Jouini, S.B., Midler, C., Silberzahn, F.: Contributions of design thinking to project management in an innovation context. Proj. Manage. J. **47**, 144–156 (2016)

Nikkei Asia: Asia is home to 50% of world's fastest growing companies (2019). https://asia.nikkei.com/Business/Companies/Asia-is-home-to-50-of-world-s-fastest-growing-companies2. Accessed 10 Oct 2021

Nonaka, I.: Phronetic leadership (in Japanese). https://scienceportal.jst.go.jp/explore/highlight/20070507_01-2/. Accessed 01 Nov 2021

Nonaka, I., Takeuchi, H.: The big idea: the wise leader. Harv. Bus. Rev. **89**(5), 58–67 (2011)

Odin, S.: An explanation of beauty – Nishida Kitaro's Bi No Setsumei. Monumenta Nipponica. Sophia University, Tokyo, **2**(42), 211–217 (1987)

NOVATEK: Project Yamal LNG (2019). https://www.novatek.ru/en/business/yamal-lng/. Accessed 10 Oct 2020

PRINCE2: What is PRINCE2 (2021)? https://www.prince2.com/uk/what-is-prince2. Accessed 15 May 2021

PMI: PMP Certification (2021). https://www.pmi.org/certifications/project-management-pmp. Accessed 10 Sept 2021

PROSPERO: Russia launched new hydrogen strategy, focuses on development infra and export. https://www.prosperoevents.com/russia-launches-new-hydrogen-strategy/. Accessed 10 Sept 2021

Schoper, Y.: Light and shadows of Projectification for the Societies. In:. Schoper, Y., Project Management and its Impact on Societies. 6th IPMA Research Conference, pp.115–128. IPMA, The Netherlands (2018)

Tanaka, H.: The engineering & construction industry: a mature model for the future of project management success (2006). https://www.pmi.org/learning/library/engineering-construction-model-project-management-success-8083. Accessed 21 July 2020

Tanaka, H.: Insight into project performance: sector-to-sector review and a Japanese case study on a multi project management model (Part I). Project and Program Management, Moscow. Grebennikon, **04**(56), 246–263 (2018)

Tanaka, H., Bushuyev, S.: Project management in the global oil and gas industry – by co-creating strategic value for the industry. Sci. J. Astana IT Univ. **6**, 59–75 (2021)

Tanaka, H.: Impact of the new normality on strategic management and project & program management. Bull. Nat. Tech. Univ. "KhPI". Strateg. Manage. Portfolio, Program Proj. Manage. **1**(5), 3–22 (2022)

The Standish Group: Chaos report (2018). https://www.standishgroup.com/benchmark. Accessed 20 Mar 2019

United Nations: The 2030 Agenda for Sustainable Development (2015). https://sdgs.un.org/goals. Accessed 02 Feb 2023

U.S. Army Heritage and Education Center: Who first originated the term VUCA (Volatility, Uncertainty, Complexity and Ambiguity)? USAHEC Ask Us a Question (2018). https://www.armywarcollege.edu. Accessed 10 July 2018

World Factbook: GDP- composition, by sector of origin. https://www.cia.gov/the-world-factbook/field/gdp-composition-by-sector-of-origin/. Accessed 07 Dec 2021

Cooperation Between Organisations and Projects

Hans Knoepfel(✉)

Rosenthaler + Partner AG, Zurich, Baumackerstrasse 24, 8050 Zurich, Switzerland
kn@rpag.ch

Abstract. The project governance is a competence element group of the IPMA OCB® and a key aspect of a competence element of the IPMA ICB®. It is the subject of the ISO Standard 21505 Project, Programme and Portfolio Management "Guidance on Governance". The discussion about the cooperation between organisations and projects and the project governance models is reflected in hundreds of pictures concerning the structure of permanent organisations and project organisations and a large volume of rules and templates in project management handbooks, contracts and regulations and guidelines for organisations. Governance is also the topic in professional books.

The difference of the recommendations from theory and practice and the unsatisfactory coordination, clarity and concretisation of the governance models and realisations in organisations, portfolios, programmes and projects are an incentive for further research in this matter.

Keywords: Project Governance · Project Context · Organisational Competence · Competence Framework

1 Introduction

No project is an island (Engwall, 2003); A project is no island (Knoepfel, 2016). Several organisations are cooperating for defining and delivering a project. Project management must take the project context into account.

There are recommendations for the cooperation between organisations and projects, see (Mueller, 2009), (PM4, 2019), (Butler, 2019), (Bea et al., 2020) and (Wikipedia, 2021). There is experience from many projects in the practice.

From this literature research and practical experience, it seems to be worth to enhance the performance of the project, programme and portfolio management (PP&PM) by further research on relevant topics with specific hypotheses.

The first part of the paper deals with 3 examples of governance topics. Other topics, for example the definition of the project objectives, the management of opportunities and risks and the achievement of the sustainability with the project impacts, would be relevant as well.

The second part of the paper deals with a competence framework for housing the many topics.

S. Bushuyev et al. (Eds.): IPMA 2021, LNNS 704, pp. 287–294, 2023.
https://doi.org/10.1007/978-3-031-34629-3_24

2 Governance Topics

2.1 Relations for the Cooperation

The situation is that there is always a cooperation between a project and the organisations which contribute to it. Theoretically, a project can depend on an organisation

- not at all (0% relations), this means that the project is an island indeed;
- completely (100% relations), this means that the project is a fully internal project of one organisation.

The task is to investigate what relations between a project and an organisation should exist. Theoretically, the cooperation can take place through

- one channel (bi-polar model, see Fig. 1), also called 'single point of contact';
- pre-defined relations between responsible persons of the organization and roles in the project organization (matrix model);
- situational relations between responsible persons of the organization and roles in the project organization (e.g. for delivering a work package).

A possible hypothesis could be that the bi-polar model for the whole project duration leads to the best performing governance. The measure could be the % of the achievement of the (final, i.e. including changes agreed) project objectives.

A challenge could be that the project manager on the purchaser side and the project manager on the supplier side could be overloaded with too many different responsibilities.

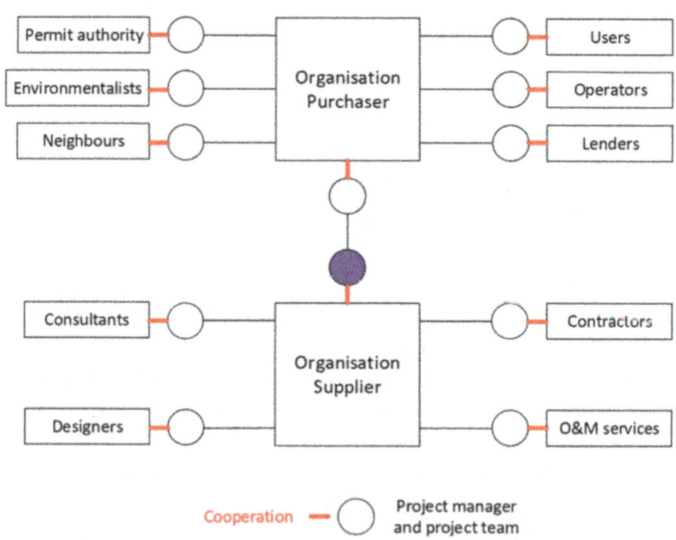

Fig. 1. Bi-polar model

2.2 Definition of Responsibilities

The situation is that the work for the project has to be done and the decisions have to be made. Theoretically, the works can be assigned, and the decisions can be made

- completely within the project (100% decisions), this is the 'pure project organisation';
- wholly outside the project (0% decisions), this would mean that the project needs to get the decision for all details from outside.

The task is to investigate, for what decisions and to what degree (e.g. the approval for proposals is mandatory) the project depends on which organisation. Theoretically, the task assignment and decision-making can be

- done by one institution, e.g. the Steering Committee (see Fig. 2), or the project team;
- defined for each type of work and decision with the responsibility matrix;
- a situational organisation for each work package or work task.

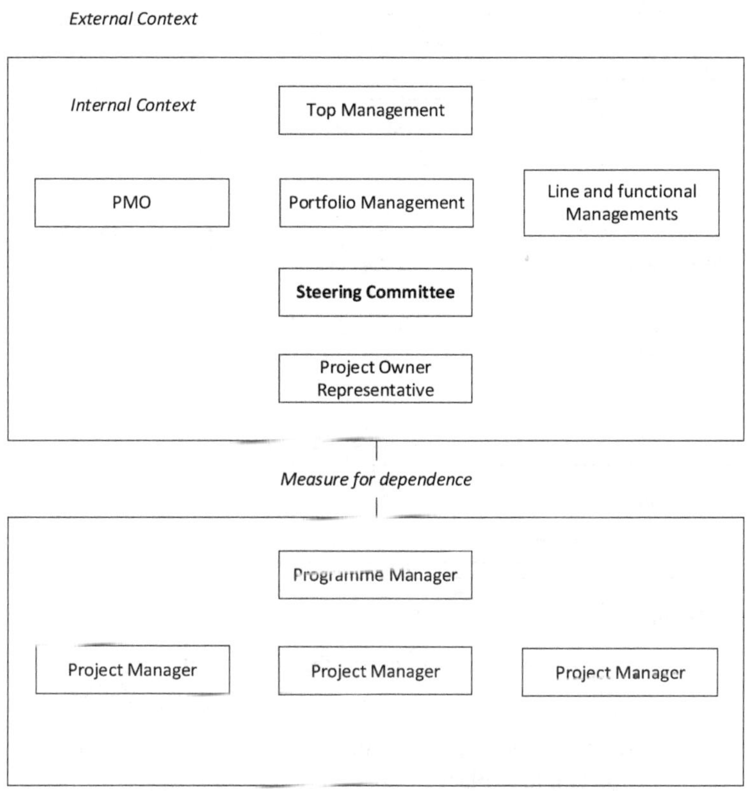

Fig. 2. Authority with the Steering Committee

A possible hypothesis could be that the task assignment and decision making, after the decision for the project definition (project charter), in the project team, leads to the best performing governance. The measure could be the % of the stakeholder satisfaction at the project end.

A challenge could be that the governance model is not applicable to the most relevant initial phase of the project and to the changes of the stakeholders' objectives during the project delivery.

2.3 Feasibility of Changes

The situation is that in almost every project, stakeholders want changes during the project life cycle. Theoretically.

- all change proposals can be rejected (0% consideration);
- change proposals are accepted for analysis and decision anytime in the project life cycle (100% consideration), see Fig. 3.

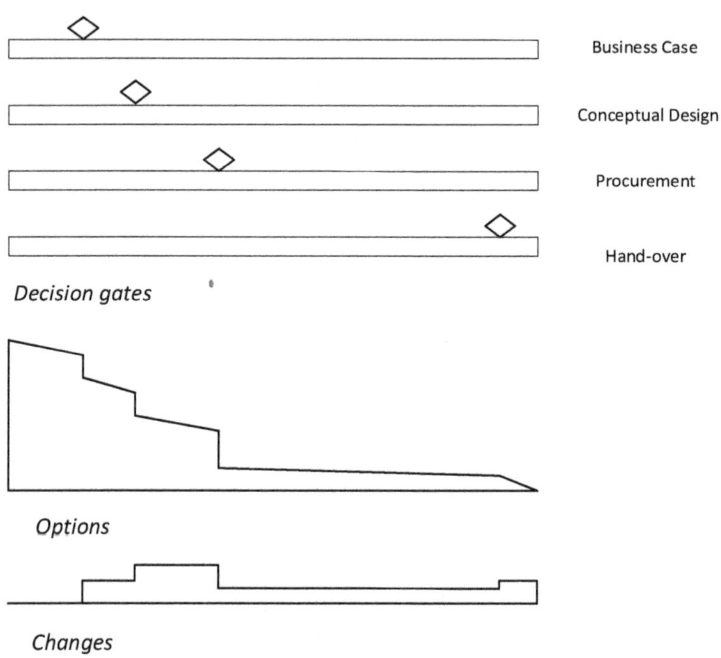

Fig. 3. Decisions during the project cycle

The task is to investigate what changes in projects should be considered. Theoretically, the consideration can take place.

- in an independent Change Management Board for the whole project;
- by those people who had made the preceding decision which needs to be changed;

- by winning a sufficient majority of people for making the change decision for each case.

A possible hypothesis could be that the decisions by the people who had made the preceding decision leads to the best performing governance. The measure could be the total net benefit from changes.

A challenge could be that the people who made the preceding decision have the tendency to reject useful changes and maybe they are no longer available for change decisions.

3 Competence Framework

The background for the cooperation between organisations and projects consists of.

- The competence of the organisations for running their business;
- The competence of the individuals and project teams for managing the projects.

The cooperation itself is the main subject of the IPMA Delta® assessment process. This is done with the following main steps.

1. **Inventory.** The content and structure of the PP&PM in an organisation is understood with the detailed lists of the projects and people involved in their management.
2. **Self-assessments.** The individuals and projects respond to questions about their interaction with the organisation (such as governance, enabling), concerning their PP&PM. The structure of their PP&PM is the IPMA Individual Competence Baseline and the IPMA Project Excellence Baseline.
3. **Interviews.** The assessors discuss the content of the OCB Competence Elements with the interviewees, starting with key questions and continuing with specific applications, methods, tools and experience.
4. **Scoring.** All responses are measured with the Competence Class Scale of IPMA Delta®. Averages are calculated and depicted automatically for describing the competence.
5. **Report.** Organisational competence strengths and potentials for the further development are collected from the responses and scores.

The recent experience is a multi-case evidence from assessments in Italy, Spain, Russia, Sweden, Switzerland, Portugal, Finland, Iran and Brasil. The assessment partners for each case were the representatives of the customer organisation (such as the supplier of power in the Energy sector) and the manager and assessors of the National Certification Body. The author of this paper was involved as assessor in 4 cases.

4 Conclusions

The following theses are suggested for discussion and further development:

- Situational PP&PM competence needs research and experience in many government topics (such as responsibility).

- The cooperation between organisations and projects is the main subject of the organisational competence.
- Die IPMA Baselines OCB, PEB and ICB can be an appropriate framework for the assessment and certification of organisations in PP&PM.
- The IDR2 is likely to be effective and efficient due to its rich and direct data collection and its evaluation focus on the OCB competence elements and scoring.
- The growing influence of Project Management Offices (Egeland, 2021) promotes the PP&PM competence of organisations.
- A summary of key questions concerning project governance, taken from the (OCB, 2016) is shown in the Appendix A of this paper.
- Examples for the expectations of organisations and project teams for the cooperation are listed in the Appendix B of this paper.

Appendix A: Summary of Key Questions Concerning Project Governance [OCB, 2016]

1. Projects, Programmes and Portfolios Mission, Vision, Strategy
   - Does top management effectively communicate the PP&P mission, vision and strategy to all stakeholders and provide the necessary resources?
   - Is the PP&P mission, vision and strategy evaluated and updated on a regular basis?
   - Does top management involve relevant stakeholders (e.g. PP&P managers and staff) while developing or updating the PP&P mission, vision and strategy?
2. Projects, Programmes and Portfolios Management Development
   - Does the organisation have objectives for PP&P management development?
   - Does the organisation have a process for developing the PP&P management, collecting, analysing, evaluating and using all relevant information, including a lessons-learned process?
3. Leadership
   - Do top managers and senior executives define and communicate goals, and clarify their expectations for PP&P management?
   - Does the organisation enable the development of leadership competences in PP&P?
   - Does the organisation foster a robust decision-making process linking strategic and operational levels?
4. Performance
   - Do top managers and senior executives define performance targets (e.g. using KPIs) and clarify their expectations for performance?
   - Does the organisation monitor and control performance on a regular basis?
   - Do PP&P managers and staff provide feedback on performance?

Appendix B: Expectations of Organisations and Project Teams

What do the Organisations Expect from the Project Teams?

1. Content

   - Respect mission, vision, strategy and project objectives.
   - Achieve objectives concerning PP&PM.
   - Realistic estimates for and information about functional, deliverable, time, cost, risk, sustainability etc. parameters.
   - Feedbacks and, if needed, requests for support.

2. Resources

   - Adequate usage of management rules, processes and tools and reasonable feedbacks and requests for changes and additions.
   - Proposals for competence and terms of engagement for the project management staff.
   - Inquiries for information for the project.
   - Realistic estimates of the resources from the project planning.

3. Performance

   Timely and complete change requests.
   Reliable status reports and forecasts and, if necessary, requests for escalation.
   Delivery of information from the project and appreciation for advice.
   Realistic estimates for the project status and the forecast.

What do the Project Teams Expect from the Organisations?

1. Content

   - Strategy that is reliable, beneficial and feasible and project objectives that can be achieved.
   - PP&PM objectives and conditions that are agreed on and realistic.
   - Work tasks that they can manage.
   - Sufficient and clear leeway and authority.

2. Resources

   - Human resources that are competent and available for their work tasks and decisions.
   - Management rules, processes and tool that are appropriate for the PP&PM.
   - Correct information from the project context and to the project context.
   - Reliable financing.

3. Performance

   - Response to status reports and requests for escalation.
   - Response to change requests.

– Response to information requests.
– Recognition of achievements and indication of dissatisfaction.

References

Bea, F.X., et al.: Projektmanagement, 3. Auflage, UVK Verlag München (2020)
Butler, S.: Portfolio Governance. In: Lock, D., Wagner, R. (eds.) The Handbook of Project Portfolio Management, pp. 155–166. Routledge, Milton Park (2019)
Egeland, B.: 7 Key considerations for building a perfect PMO (2021). https://www.projectmanagementupdate.com/2021/pmo/?open-article-id=20277397&article-title=7-key-considerations-for-building-the-perfect-pmo&blog-domain=liquidplanner.com&blog-title=liquidplanner. Accessed 15 Dec 2021
Engwall, M.: No project is an island. Res. Policy **3**(5), 789–808 (2003)
ICB: Individual Competence Baseline for Project Management. International Project Management Association, The Netherlands (2015)
ISO 21505: Project, Programme and Portfolio Management – Guidance on Governance. International Organization for Standardization (2017)
Knoepfel, H.: A project is no island. In: Future Trends in Project, Programme and Portfolio Management, IPMA Expert Seminar. IPMA, pp.119–127. Zurich (2016)
Mueller, R.: Project Governance. Routledge, UK (2009)
OCB: Organisational Competence Baseline. International Project Management Association, The Netherlands (2016)
IDR2 Public, IPMA International Delta Certification Regulations, Version 1.3, April 2021
PM4: Kompetenzbasiertes Projektmanagement. Deutsche Gesellschaft für Projektmanagement e.V., Nürnberg (2019)
Wikipedia: Project Governance. https://en.wikipedia.org/wiki/Project_governance. Accessed 15 Sept 2021

Translating Sustainability Strategy into a Digital Transformation Project in Public Sector

Oxana Klimenko(✉)

Project Alliance, Novaya Street, 17B-367, 140054 Kotelniki, Russia
o.klimenko@pmalliance.ru

Abstract. Public sector organisations undertake projects, programmes and portfolios with the sustainability principles in their strategies as the main focus due to their big impact on the economy and society. To speed up with the results, large number of public sector organisations realize projects, programmes and portfolios for digitalization of their processes and activities. However, there is a lack of tools and methods for translating sustainability strategy into digital transformation type of projects, programmes and portfolios in public sector.

The purpose of the research is to study the public sector organisation's practice in breaking down the sustainability strategy into the digital transformation projects, identifying the success practice and the challenges and to suggest the recommendations for further development of the specific tools and methods to be applied. The three cases of public sector organisations have been analyzed and the results presented in the paper.

The study demonstrates the lack of tools and methods applied for translating the organizational sustainability strategy into digital transformation project portfolio in public sector. The proposed model for defining the step by step process of translating the strategy into a digitalization project has been suggested by the author. The contribution to the development of the sustainability theory and project management application in specific types of projects has been made. The author's proposals for public sector could be of interest for the large audience including the project management professionals as well as linear managers of the organisations in transferring the sustainability strategy into digital transformation in practice.

Keywords: Sustainability Strategy · Sustainable Project Management · Digitalization Projects · Portfolio Management

1 Introduction

Sustainable development concept that was developed and published as early as in 1987 in the report by the World Commission on Environment and Development (WCED), had been spreading globally and locally in different countries and economy sectors. The main meaning of the sustainability concept is that the activities of the organisations must meet the needs of the current generations without stopping the future generations from meeting their own needs in return. The speed of development depends on the projects, programmes and portfolios realisation which makes the large impact on the Gross Domestic Product growth [1].

S. Bushuyev et al. (Eds.): IPMA 2021, LNNS 704, pp. 295–305, 2023.
https://doi.org/10.1007/978-3-031-34629-3_25

Projectification of economy and society is another concept that has been spreading recently that leads to the penetration of the project, programme and portfolio management tools and methods into different sectors of economy and other social – oriented activities [2]. Projectification indexes have been measured by a number of authors in the countries with different levels of development located in different places over the globe [3].

In this context to perform the projects, programmes and portfolios in public sector the digital transformation is being used at all the levels of government starting from the federal through regional and going to municipal and corporate ones. Thus, organisations develop and accept the sustainability strategy that should be translated into the digitalization projects to achieve better results. The certain set of sustainability competences is needed such as strategic, system thinking, normative, anticipatory and interpersonal [4].

The purpose of the research is to study the public sector organisation's practice in breaking down the sustainability strategy into the digital transformation project, identifying the success practice and the challenges and to suggest the recommendations for further development of the specific tools and methods to be applied.

The main methods applied for the research purpose are content analysis, comparative analysis and case study. The three cases of public sector organisations have been analyzed and the results presented in the paper. The author describes the step by step model of transferring the sustainability strategy into projects based on the three case studies of the digitalization projects in public sector.

2 Methodology Applied

The descriptive analysis was applied to study the cases of the three digitalization projects in public sector in Russian Federation, open sources and semi-structured interview were used for making the comparative analysis and developing the author's proposal.

The sample of the projects includes three projects performed on the different levels - federal, regional and corporate ones and shows the similarities in the processes, methods and tools of translating the sustainability strategies of the organisations into the digitalization projects.

For describing the process of translating the sustainability strategy into the project the model proposed by Van den Brink [5] with scope/context and time horizon dimensions was adapted. The "onion model" was developed and suggested by the author, it includes the global, sector, organizational, project portfolio and project levels (Fig. 1).

2.1 Digitalisation Projects in Public Sector

The three projects were chosen in the public sector to represent the differences and similarities in their approach to breakdown the strategical sustainability directions into the project's level.

One of the projects is being conducting on the federal level and is called "National Project" that in fact is a programme or a mega-project. As part of the implementation of Presidential Decrees and the national development objectives of the Russian Federation for the period until 2030, including the goal of ensuring accelerated implementation of digital technologies in the economy and social sphere, the national programme

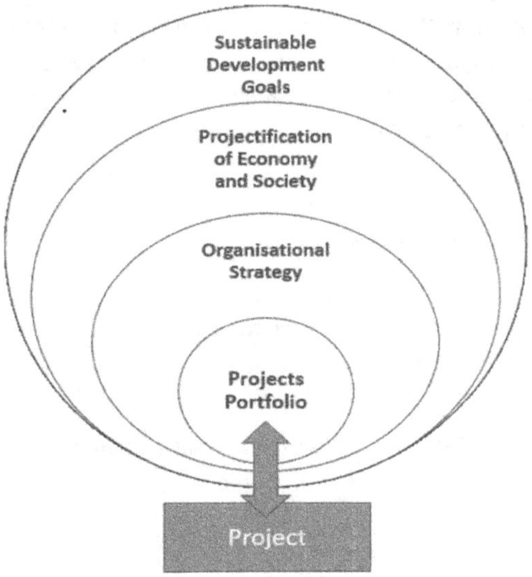

Fig. 1. Sustainability contexts levels.

"Digital Economy of the Russian Federation" has been launched. The strategic goal is defined as Digital Transformation. The Digital Economy national programme includes the following federal projects:

- "Regulatory regulation of the digital environment";
- "Personnel for the Digital Economy";
- "Information Infrastructure";
- "Information Security";
- "Digital technologies";
- "Digital Government";
- "Artificial Intelligence";
- "Internet access through satellite communications development";
- "Developing the human resources capacity of the IT industry".

The author suggests to consider this programme as the federal level mega-project.

The second case project is begin conducted on the regional scale, relates to Smart City topic. The regional project period is defined from 09.01.2019 till 31.12.2024.

This is the regional project for the digitalisation of urban economy in the Krasnodar region.

The third case project has been conducted at the organisation level in the centralized accountant department at Yamalo – Nenets Autonomous okrug. The project's goal is to develop and make the digital processes in the organisation to provide the reliable and sustainable self – developing organisation with book keeping and accounting reporting system.

All the three projects data in the open source have been collected and analysed, the semi-structure interview with the project manager of the digitalization project at Yamalo – Nenets Autonomous okrug has been conducted with the data being clarified.

Due to the ethical matters the research carried out on the basis of the open sources and the permissions of the organisations for the analysis not needed. The study conducted has guaranteed anonymity to all the three chosen organisations, to ensure this the coding system has been used further: project number 1, project number 2 and project number 3.

2.2 Research Questions Structure

The comparative analysis was conducted according to the sustainability contexts levels with the main research questions structure for each of the sustainability context. The structured questions as described in the Table 1 below.

Table 1. Research questions structure.

№	Sustainability context level	Research question
1	Sustainable Development Goals (SDGs)	Does the project relate to the SDGs?
2	Projectification of Economy and Society	Does the project contribute to the projectification of economy and society?
3	Organisational strategy	Does the organizational strategy contain the sustainability aspects?
4	Projects Portfolio	What tools and methods are used to translate the sustainability strategy into the project portfolio?
5	Project	What tools and methods are used to translate the sustainability strategy into the digitalization project?
6	Good practices	What good practices does the organisation use to breakdown the sustainability strategy into a digitalization project?
7	Challenges	What main challenges does the organisation encounter while transferring the sustainability strategy into a digitalization project?

3 Analysis and Results

When analyzing the three digitalization projects the good practices and challenges have been defined and studied. The research questions for each context were answered by conducting the content analysis and comparative analysis.

3.1 Sustainable Development Goals

The studying of the sample of the three digitalization projects shows they all have the defined objectives in relation to the Sustainable Development Goals (SDGs).

Project number 1 shows the relation to the following SDGs:

- 9 Industry, Innovation and Infrastructure;
- 11 Sustainable Cities and Communities;

Project number 2 relates to the following SDGs:

- 9 Industry, Innovation and Infrastructure;
- 11 Sustainable Cities and Communities;
- 12 Responsible Consumption and Production;

Project number 3 relates to the following SDGs:

- 9 Industry, Innovation and Infrastructure;
- 11 Sustainable Cities and Communities;
- 12 Responsible Consumption and Production.

The conclusion is that though there exist the difference of the main project objectives, such as digital transformation of economy, smart city and digitalization of the accountant department, the SDGs from the global context are taken seriously into consideration by the organisations in public sector and show the similarities of all the three projects. The government departments responsible for the launch of the digitalization projects pay attention to the SDGs as this contributes to the international community and not only to the national ones. Thus, the similarity is defined in all the three sample projects.

3.2 Projectification of the Economy and Society

Sustainability strategy contribute to the projectification of the economy and society, conceptualizing sustainability in the project management contexts [6, 7].

In terms of the projectification context the mega project on the federal level demonstrates the big impact on the projectification of economy and society as it has a long-term life cycle and has been develop and launched for the period from 01.10.2018 till 31.12.2024. The mega project has clearly defined objectives broken down to the KPIs to be achieved and spread per year. The project has the well-defined organizational structure with the project manager, project sponsor, steering committee, team members, experts' communities and project management office. While executing the projects that are part of the mega project, the reporting system has been introduced on the yearly and quarterly basis. In this case each of the project within the mega project has its own project manager that increases the projectification level of the economy and society for the period of the programme realisation.

Two other projects in the sample have less impact on this context as they require one project manager and comparing to the first one, they both contribute less to the projectification of economy and society.

3.3 Organisational Strategy

The organizational strategies in public sector are usually developed and introduced on the country and regional levels.

The strategic goals set for the period till 2030 on the country level are taken seriously when developing the regional strategies. The strategic goals of the country are defined in a number of documents and at the same time there is no one solid document totally dedicated to the sustainability topic. When moving from the country level to the regional, each region has to develop the strategy that should include the sustainability aspects. This approach enables to balance the country needs and demands and the regional ones.

In the regional project the sustainable aspects are part of the social and economic strategy of Krasnodar region development approved till 2030. The main goals have been defined and structured as follows:

- G1. Markets (products and economic complexes);
- G2. Institutions (state, private and public);
- G3. Human capital;
- G4. Innovation and information;
- G5. Natural resources and sustainable development;
- G6. Space and Real Capital;
- G7. Investment and financial capital.

The digitalization project Smart City has been initiated based on the strategic goal G4 Innovation and information.

As for the third project on the organizational level, the parent organisation of which has been initiated on the basis of the strategy of the regional government that contains the sustainability aspects including smart cities and smart government.

Thus, on the organisational context the sustainability aspects are reflected in the strategy that enables the public sector organisations to create the programmes and projects for its realisation and achieving the sustainability goals.

3.4 Project Portfolio

In the context of project portfolio, the main interest of the study has been focused on tools and methods being used to translate the sustainability strategy into the project portfolio.

When studying the tooling of project portfolio, the analysis of the project management documentation has been made as well as the semi-structured interview with the project manager conducted. In the analysis of documentation, the following tools have been used to define the project portfolio:

- Lessons learned review;
- Big data analysis;
- Finance analysis and finance modelling;
- Meetings and experts round tables;
- Mechanisms of feedback and soundings with the local communities;
- Open sounding rounds (e.g. "Active citizen" platform);
- Brainstorming;
- SWOT – analysis;
- Verifying the projects goals against the strategic goals in social, environmental and economic areas (social aspect is seen as the key one);

- Prioritization of the projects.

There is a lack of information though of the internal processes and procedures as well as decision making tools while forming the project portfolio and putting the sustainability goals into it. Strategic sessions have been used in two cases to identify the digitalization projects to be launched. These mechanisms and tools should be further explored and analysed to complete the list of the tools for translating the sustainability strategy into the whole project portfolio.

3.5 Project

In the context of a project itself, the main interest of the study has been focused on tools and methods used to translate the sustainability strategy into the digitalization project. While analyzing the documentation from the open source and the semi-structured interview the following tools and methods have been defined:

- Strategic session;
- Big data analytics;
- Prioritization of the SDGs and defining the relation between them and the sector;
- Estimating the resources within the portfolio;
- Stakeholder engagement: meetings with the local communities, feedback, open discussions;
- Justification of the project;
- Setting the goals and objectives of the project;
- Roadmap of the project;
- Milestone schedule;
- Project scheduling;
- Piloting of the solutions.

In the chapters of the regional project the environmental and social objectives are clearly defined. In the organizational project social, environmental and economical aspects are well defined. As for the mega-project of the federal level the social aspect is defined as the main one, environmental impact and economical impact are also stated but with less priorities than the social one. The conclusion could be that all the three projects have similarities in breaking down the sustainability goals into the projects.

3.6 Good Practices

Studying the sample project of digitalization the good practices of organisation could be seen in the documentation analysed. On the level of organisation the sustainability principles are used, the strategy of organisation sets the goals, objectives and KPIs per each year. The projects to fulfil the strategy are set according to the category and priority principles. Top-down approach has been used as good practice in the projects in public sector as shown at the Fig. 2 below.

Though there is a lack of information about the tools that lead to the breakdown of the sustainability strategy into a digitalization project form the two projects.

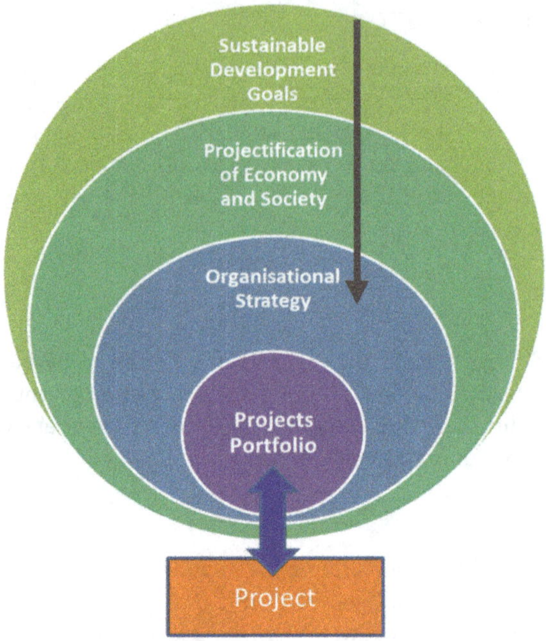

Fig. 2. Top-down approach in translating sustainability into digitalization project.

The organizational level project has shown such practices as:

- Defining competence level of the project team at the beginning, including project management and sustainability topics;
- Planning the competence development of the team members, including the sustainability in project management course;
- Conducting the competence development plan such as project management training and certification both in project management and sustainability management;
- Strong support and engagement of the project sponsor;
- Developing and application of the specific project management system (new processes have been introduced as well as corporate PM methodology adapted);
- Weekly reporting on the project progress, including the sustainability KPIs;
- Weekly project team meetings and project review sessions.

At the same time, there has no evidence found during the research about the monitoring and control of the project within the project portfolio as the project has been fulfilled in a specific environment and not within the project portfolio.

3.7 Challenges

During the translation of the sustainability strategy into a project and planning the deliverables may lead to the necessity of developing the new practices and create new organizational model as well as frameworks and working environment as well as the brining new challenges. One of the challenges is that managers discuss the strategy and at the same time may forget to translate strategy into a plan and put it into action [8]. While implementing the sustainability strategy the leaders have to have the certain competences and capabilities.

During the study of the three cases the following list of challenges have been defined:

- High level of uncertainty of the digitalization projects at the initiation stage.
- Competence level of the project team members in project management and sustainability needs to be leveled up.
- The project portfolio is not managed according to sustainability principles. There us a gap between the sustainability strategy and project portfolio.
- The project has been included into the PMO's project portfolio (it was initiated by the project, not by the PMO). That means on the organizational level the digitalization projects could be initiated by using bottom-up approach.
- The issue with assigning the resources have been arisen and needed to be solved as it is stated in the semi-structured interview with the project manager.
- The top-down approach and the bottom-up approach are not fully aligned and there is a gap at the project portfolio level. Thus, the breakdown of the sustainable strategy into a project does not show the sustainability objectives of the whole portfolio. The KPIs on sustainability form the project level contribute directly to the achievement of the strategic sustainability goals and not to the project portfolio first (Fig. 3).

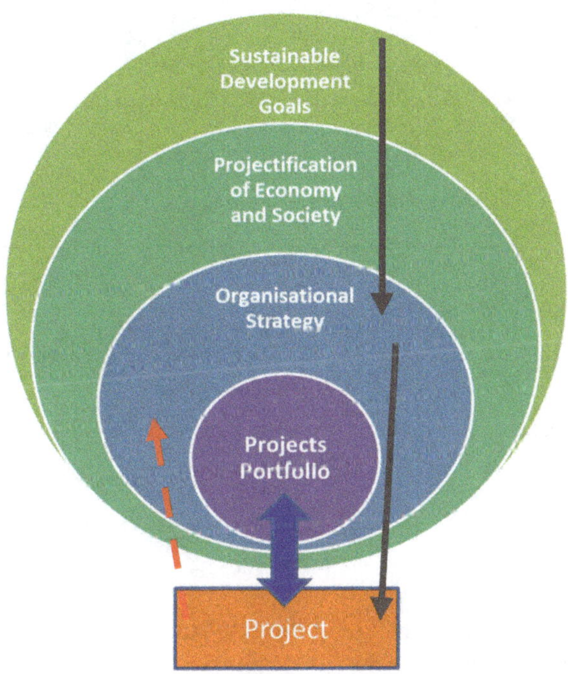

Fig. 3. Translating the sustainability strategy into the digitalization project.

The challenges been described of the sample projects need to be addressed by the organisations in public sector while launching the digitalization transformations.

4 Proposed Model for Translation the Sustainability Strategy into the Project

To identify the steps that could be used for the translating the sustainability strategy into the digitalization project the similarities of the three cases have been combined. The model is described in 5 steps as shown at the Fig. 4 below.

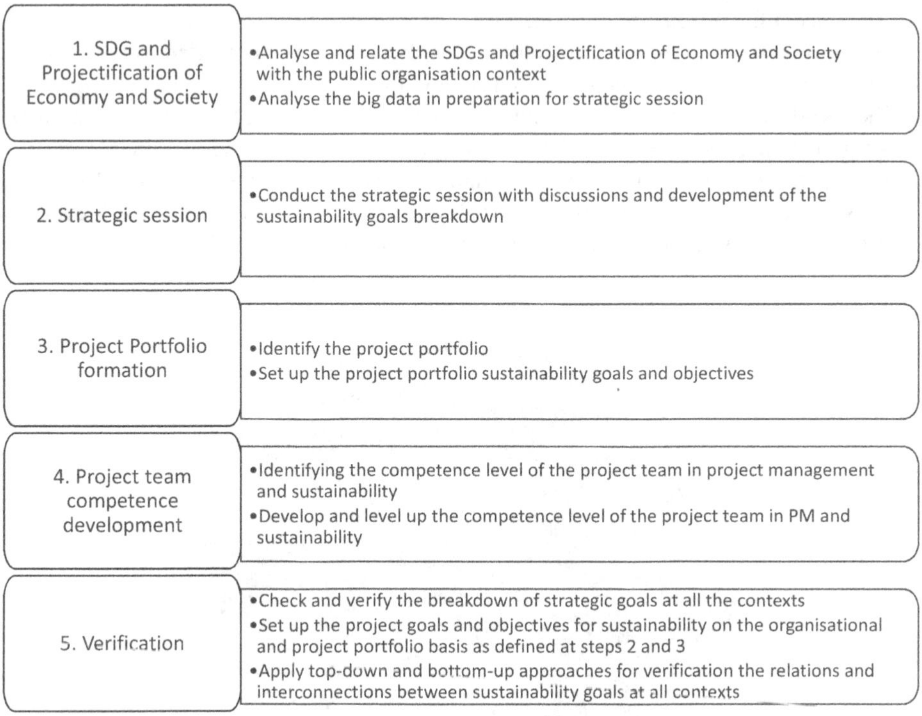

Fig. 4. Translating the sustainability strategy into the digitalization project model.

5 Conclusion

The results and key findings achieved at the case study research of the three cases of digitalization projects in public sector:

- there is the relationship between the sustainable development strategy in the organisation and the digital transformation project;

- the implementation of the sustainability principles into the strategy of organisation helps organisation to form the programmes and define the projects with the three main aspects taken into consideration (social, environmental and economic);
- the practice of translating the sustainable strategy into the project portfolio of the organisation is not clearly defined and there is a lack of transparent processes, mechanisms and criteria in the organisations;
- the contribution of the project to the realization of the sustainable development strategy of the organisation has been well fulfilled, though there are areas for improvement in the corporate project management system, namely at project portfolio area.

The good practices and challenges in the process of breaking down the strategy of sustainability into project portfolio and projects have been defined, both differences and similarities of the projects of three different levels have been identified.

The proposed model of translating the sustainability strategy into digitalization projects has been proposed based on the case studies. This could be a contribution to the sustainability theory as well as to the practices for project management professionals in public sector.

References

1. Parsi, N.: Global jobs report 2015. PM Netw. **29**(1), 28–39 (2015). https://www.pmi.org/learning/library/global-jobs-report-2015-9441. Accessed 02 Mar 2023
2. Schoper, Y.G., Ingason, H.T.: Projectification and the impact on societies. Int. J. Manag. Proj. Bus. **12**(3), 517–521 (2019)
3. Schoper, Y.G., Wald, A., Ingason, H.T., Fridgeirsson, T.V.: Projectification in western economies: a comparative study of Germany, Norway and Iceland. Int. J. Project Manage. **36**(1), 71–82 (2018)
4. Silvius, G.: Sustainability as a competence of project managers. PM World J. **5**(9) (2016). https://pmworldlibrary.net/wp-content/uploads/2016/09/pmwj50-Sep2016-Silvius-sustainability-as-project-manager-competence-featured-paper.pdf. Assessed 09 Mar 2023
5. Van den Brink, J.C.: How positive psychology can support sustainable project management. In: Silvius, A.J.G., Tharp, J. (eds.) Sustainability Integration for Effective Project Management, pp. 245–260. IGI Global Publishing, Hershey (2013)
6. Sabini, L.: Project management and sustainability (2016). https://www.researchgate.net/publication/313726106. Assessed 09 Mar 2023
7. Sabini, L.: Project of sustainability: role of project management in developing a more sustainable economy and society (2016). https://www.researchgate.net/publication/308337951. Assessed 09 Mar 2023
8. Verweire, K.: The challenges of implementing strategy. Adv. Proj. Manage. PM World J. **7**(5), 1–12 (2018). https://pmworldlibrary.net/wp-content/uploads/2018/05/pmwj70-May2018-Verweire-challenges-of-implementing-strategy-series-article.pdf. Assessed 09 Mar 2023

Author Index

S. Bushuyev et al. (Eds.): IPMA 2021, LNNS 704, p. 307, 2023.
https://doi.org/10.1007/978-3-031-34629-3

CPSIA information can be obtained
at www.ICGtesting.com
Printed in the USA
LVHW051459190723
752608LV00016B/29

9 783031 346286